CLARENDON LECTURES IN GEOGRAPHY AND ENVIRONMENTAL STUDIES

*General editors*

Gordon Clark   Colin Clarke   Andrew Goudie

CLARENDON LECTURES IN GEOGRAPHY AND ENVIRONMENTAL STUDIES

*The Geographical Structure of Epidemics*
Peter Haggett (2000)

*Integrated Land-Change Science and Tropical Deforestation in the Southern Yucatán*
Edited by B. L. Turner II *et al.* (2004)

*The New Imperialism*
David Harvey (2003)

# Integrated Land-Change Science and Tropical Deforestation in the Southern Yucatán

Final Frontiers

*Edited by*

B. L. TURNER II
JACQUELINE GEOGHEGAN
and DAVID R. FOSTER

OXFORD
UNIVERSITY PRESS

OXFORD
UNIVERSITY PRESS

Great Clarendon Street, Oxford OX2 6DP

Oxford University Press is a department of the University of Oxford.
It furthers the University's objective of excellence in research, scholarship,
and education by publishing worldwide in

Oxford New York

Auckland Bangkok Buenos Aires Cape Town Chennai
Dar es Salaam Delhi Hong Kong Istanbul Karachi Kolkata
Kuala Lumpur Madrid Melbourne Mexico City Mumbai Nairobi
São Paulo Shanghai Taipei Tokyo Toronto



Published in the United States
by Oxford University Press Inc., New York



First published 2004



British Library Cataloguing in Publication Data
Data available

Library of Congress Cataloging in Publication Data
Data available

ISBN 0–19–924530–4

1 3 5 7 9 10 8 6 4 2

Typeset by Hope Services (Abingdon) Ltd.
Printed in Great Britain
on acid-free paper by
Biddles Ltd,
King's Lynn, Norfolk

# PREFACE AND ACKNOWLEDGEMENTS

This book reports on the first phase of a project entitled Land-Cover and Land-Use Change in the Southern Yucatán Peninsular Region (LCLUC-SYPR), a large, integrated, multi-institutional, team-based study designed to understand and project land changes in a 'development frontier'—one that pits the rapidly growing needs of *campesinos* or smallholder farmers to cut forests for cultivation against federally sponsored initiatives committed to various international programs of forest preservation and complementary economic programs. The specifics of the project are detailed in Ch. 1, including the regional background, research design and products, host and participating institutions, and the primary and ancillary sponsors of the project. It is sufficient to note here that the SYPR project is a response to interdisciplinary defined research themes deemed critical to global environmental change and complementary international research agendas (e.g. environment and development, ecosystem assessment, biotic diversity). Pivotal among these agendas are those posed by the Land-Use/Cover Change (LUCC) effort of the International Geosphere-Biosphere Programme and the International Human Dimensions of Global Environmental Change Programme as it is linked through such US sponsors as the National Aeronautics and Space Administration (NASA) and the National Science Foundation (NSF). The themes (i.e. questions and subjects) posed by these programs and organization are 'integrated' or 'synthesis' in kind—meaning that they rest within the intersection of formal disciplines and are intended to fit into a larger, systems framework about human–environment relationships and the structure and function of the biosphere, or what the Russians of yesteryear called the noösphere. The kinds of problems addressed about land change and the systemic linkages sought in regard to global environmental change, interpreted broadly, gives rise to 'integrated land science'. This science requires large research teams, coupled into a tightly knit, problem-solving unit—a 'big science' approach common in the biophysical sciences but atypical, even viewed with suspicion, throughout much of the human sciences.

The editors of this volume, as most of its contributors, come from the disciplines of geography, ecology, and economics. The first two fields of study have long sought 'synthesis' understanding through 'integrated' problem framing and approaches. For at least two centuries, if not more, geography has positioned itself among the disciplines as a 'synthesis' science, ultimately seeking to understand the 'human–environment condition', often through a set of problems and approaches variously referred to as *landschaft* or landscape, and doing so by bridging various ways of knowing from the natural sciences to the humanities. No matter what the geographic specialties of the project members (e.g. cultural-political ecology, remote sensing-GIS, historical landscape studies), addressing the character of land-change dynamics, principally deforestation and agricultural expansion, by way of integrated historical, socio-economic, spatial, and environment analysis, is simply business as usual. Indeed, the lead editor, the geographer B. L. Turner II, has spent most of his career in pursuit of understanding different aspects of tropical deforestation and agriculture, be it among the ancient Maya civilization, among contemporary smallholders throughout the tropics, or in the development of the international agendas on land-use and land-cover change.

Contemporary ecology too has become increasingly expansive in the integrated problems it tackles and the approaches used. If once focused on the structure and function of 'natural' systems, ecologists increasingly recognize that few such systems have no human impress, and many are the product of 'co-evolution' between this impress and natural processes. Moreover, ecologists worldwide have taken leadership roles in making the case for the need to examine the biosphere as a whole, linking its many parts and processes to understand the systemic changes wrought by natural flux at various timescales and human action. Also, they have expanded into questions of the economic value of nature and ecosystem services. The lead ecology editor of this volume, David R. Foster, is no stranger to these contemporary dynamics in the discipline. For 10 years he has headed the one Long-Term Ecological Research Site in the United States, Harvard Forest, dedicated to study of forest ecology in a landscape that was once denuded by New England farmers, and he has spent the majority of his career examining forest ecology here and elsewhere from a historical perspective, that incorporating the critical role of human action.

Economics may be seen as insular by other social sciences, but this view is surely inappropriate for the subdiscipline dealing with resources and the environment, which has a long history of integrated and interdisciplinary scholarship. Similar to the disciplines noted above, many environmental and resource economists have worked hand-in-hand with other researchers to develop the international agendas dealing with global environmental change, and are even more deeply involved in environment and development studies worldwide. Indeed, some of the most long-lasting, if contested, central concepts about these relationships have emanated from this subfield, including the role of the open-access resources and induced innovation on landscapes. The economics editor of this volume, Jacqueline Geoghegan, the recent recipient of a NASA New Investigator Fellowship, represents a new generation of environmental and resource economists who apply econometric analysis specifically to the question of land change, thus engaging geographical information systems (GIS) and remote sensing to assist in their analyses.

The practitioners of these (sub)disciplines are not only interested in the outcomes exposed through 'integrated' research efforts, they increasingly recognize that much of the environmental synthesis they seek is predicated on the spatial patterning and resulting dynamics of the landscape. In this case, understanding 'where' land is changing is not only significant to understanding 'why' it is doing so, but is essential to understanding the systemic, human, and environmental consequences. The SYPR project at large, therefore, seeks to examine deforestation and agricultural change at various spatial scales of analysis, moving from the region (*c.* 22,000 $km^2$) to pixel in the TM Landsat imagery (*c.* 900 $m^2$). To move to this micro-spatial scale requires the project as a whole to link its findings to satellite imagery through GIS, again an analytical mode used throughout geography, ecology, and spatial econometrics, and a strong integrative component of the project.

A major aim of the global environmental change research programs worldwide is to unite the various disciplines and research fields towards cooperative studies of the major themes in question. These themes and this 'union' typically involve 'big science' in the sense that the various research projects buy into an overall goal, even standards of data collection and analysis, and that the individual projects themselves require large teams of researchers. In phase one of the SYPR project, for example, three research institutions and about 20 individual researchers, including econometricians, community and ecosystem ecologists, and geographers representing such subfields as historical geography, cultural-political ecology, and remote sensing-GIS, joined together in a three-year field and laboratory study with direct expenditures in excess of $1.5 million. Compared to many physical science, medical, and engineering projects, the number of institutions and researchers engaged, and surely the level of funding for phase one constitute a modest 'big science' endeavor. For the human sciences and especially for integrated land science, however, the SYPR project and its allied projects elsewhere represent a scale of research for a specific problem that remains uncommon, outside a few long-term data and applied development programs.

The amount of data generated, the range of specific analyses, and the overall level of 'synthesis' obtained by the project to date are remarkable, observations repeated to the lead editor in various presentations on the project made to different human science programs. He has followed such observations with the question: 'Are you interested in putting together such a project and competing for the next round of funding?' The almost universal answer from social scientists, especially geographers, is an emphatic 'NO'. The start-up costs and administrative headaches are viewed as simply too large, the explanation goes. This explanation obscures a deeper issue, however: the long-standing tradition in much of the human sciences to value individual over team-based research and to judge scholarship in terms of individual contributions over that of a team. This 'antisocial' behavior is ironic for practitioners labeling themselves social or human scientists and is illustrative of a humanities more than a science tradition. If, as the editors argue in Ch. 1, the value of integrated and synthesis science has been rediscovered, taking a shared cornerstone position in global change and sustainability studies with reductionist science, then surely the human sciences must undergo a modest change in attitude and reach out to the integrated science community and its teams of researchers.

As to the selection of the case study and the research units, both involved calculated and serendipitous elements. Anticipating the challenges posed by integrated land science and the various funding opportunities complementing it, Turner encouraged one the project's initial researchers, Yelena Ogneva-Himmelberger, to undertake dissertation research in the southern Yucatán peninsular region, funded by US Man and the Biosphere Programs Tropical Directorate (TEDFY94-003) and aimed at imagery analysis and land change. The area was selected because Turner had worked there in 1971–2 and 1974, just after the initial paved highway was completed through the region. That research dealt with ancient land-use in the region, providing a backdrop that has proved important for the current study (Chs. 2 and 9). Ogneva-Himmelberger developed strong institutional ties to a research unit that subsequently became a research institute known as ECOSUR-UC (El Colegio de la Frontera Sur-Unidad Chetumal or Southern Frontier College-Chemutal Unit). Moreover, these developments took place as regional tensions mounted between the increasing number of rural immigrants and federal decisions to make much of the region a UN-registered biosphere reserve. The frontier settlers were knocking down the forest at a clip that placed the region on virtually every list of hot spots of tropical deforestation.

Geoghegan was recruited to the Department of Economics, Clark University, in a post designed to foster linkages between economics and Clark's large community of environmental researchers. Her expertise in econometric approaches to land-change modeling fit perfectly with developments in the Graduate School of Geography and the George Perkins Marsh Institute in regard to integrated land science. Joining Turner and Geoghegan at Clark was Basil Savitsky, a researcher of GIS and remote sensing. This Clark group turned to David Foster, Director of Harvard Forest, for expertise on forest ecology. Turner and Foster had been involved in several activities together, and both sought a way to bring together the different areas of expertise that their units held. This linkage proved exceptionally fortuitous, as Harvard Forest was able to attract outstanding postdoctoral researchers to the project (Chs. 4 and 5).

With only ephemeral ties to ECOSUR, Clark University and Harvard University procured base funding for the project, and worked with ECOSUR to develop the basis for a long-term research effort. In this regard, the former head of ECOSUR, Dr Pablo Farias, was enormously helpful in promoting the linkages among the institutions, and Pedro Marcario Mendoza took it upon himself to facilitate the project within ECOSUR-Unidad Chetumal, despite the fact that his principal research lay outside the study region. He was joined in much of his work by Sergio Cortina Villar, which led the way to a larger ECOSUR research input in phase two of the project, currently underway and including Co-Principal Investigator (Co-PI) and forest ecologist, Hans Vester, and lead social scientist researcher, anthropologist Francisco Gurri (ECOSUR-Unidad Campeche).

Phase two of the project also involves two other Co-PIs joining Turner, Geoghegan, and Vester. J. Ron Eastman (Clark Labs., George Perkins Marsh Institute, and Graduate School of Geography) takes over the remote sensing component of the project, and Deborah Lawrence, (Department of Environmental Sciences, University of Virginia), formerly working through Harvard Forest, leads research on forest nutrient dynamics and carbon flux. As well, Diego Pérez-Salicrup, also a former Harvard Forest researcher, adds links to Departamento de Ecología de los Recursos Naturales, Universidad Nacional Autónoma de México.

A project of this size requires considerable financial support, and the number of different programs sponsoring various parts of phase one of the SYPR project are too many to list in detail here, but are reviewed in Ch. 1. It is important to recognize that base funding was provided by NASA's LCLUC program under grant NAG 54406 and by the Center for Integrated Studies of the Human Dimensions of Global Environmental Change, Carnegie Mellon University, under grant NSF-SBR95-21914. We are indebted to the support of these and other programs listed in Ch. 1, and welcome the receptivity of the federal agencies in the United States to multiple-sponsor activities of SYPR sort.

Most of the graduate students and program researchers who worked on the project are authors of various papers, and their credits are found in that vein. Deserving special recognition because of their critical roles in phase-one activities but having moved on to other programs and responsibilities are Basil Savitsky (to Florida State University), Yelena Ogneva-Himmelberger, Colin Vance (to US Environmental Protection Agency), Peter Klepeis (to Colgate University), and Pedro Macario Mendoza (Ph.D. candidate, La Universidad Autónoma de Yucatán). Assisting in the final preparation of the volume were Steven McCauley, Luis (Ciro) Marcano, Suchandra Basu, and Anne Gibson.

Finally, the lead editor expresses his deep appreciation to the School of Geography, Oxford University, for selecting him to deliver the Clarendon Lectures on Geography and Environment, Oxford University (Fall 2000), a set of lectures based on the SYPR project and from which this volume issued. Throughout the development of the lectures and the volume, Anne Ashby, Oxford University Press, was instrumental in providing support and guidance.

B.L.T.II
J.G.
D.R.F.

*Worcester and Petersham, Mass.*
*2002*

# CONTENTS

## PART III. AGENTS AND INSTITUTIONS OF LAND CHANGE: HOUSEHOLD ECONOMY AND CULTIVATION



## PART IV. SPATIAL MODELING OF LAND CHANGE: EMPIRICAL APPROACHES IN DATA-SPARSE ENVIRONMENTS

# LIST OF COLOR PLATES

# LIST OF FIGURES

# LIST OF TABLES

# THE EDITORS

**B. L. Turner II** (bturner@clarku.edu) is the Higgins Professor of Environment and Society, in the Graduate School of Geography and George Perkins Marsh Institute, Clark University. He is author of numerous books, articles, and reports on the ancient Maya, tropical agriculture and smallholder behavior, and land-use and land-cover change.

**Jacqueline Geoghegan** (jgeoghegan@clarku.edu) is Associate Professor, Department of Economics and George Perkins Marsh Institute, Clark University. She specializes in developing spatially explicit econometric models of land-use change in the contexts of both tropical deforestation and suburban sprawl in the United States.

**David R. Foster** (drfoster@fas.harvard.edu) is Director, Harvard Forest, Harvard University. He is the author of books and numerous articles on forest ecology, forest history, and the role of human disturbance on temperate, tropical, and boreal forests.

# THE CONTRIBUTORS

**Audrey Barker-Plotkin** (aabarker@fas.harvard.edu) is a research assistant in forest ecology at the Harvard Forest where she serves as co-ordinator of the Long Term Ecological Research program.

**Rinku Roy Chowdhury** (rroychowdhury@miami.edu) is Assistant Professor, Department of Geography, University of Miami. She works on problems at the intersection of ecology, social institutions, and remote sensing-GIS.

**Sergio Cortina Villar** (hcortina@sclc.ecosur.mx) is a Senior Researcher in El Colegio de la Frontera Sur, Unidad San Cristobál. He is a specialist in remote sensing analysis.

**Eric Keys** (eric.keys@asu.edu) is an Assistant Professor, Department of Geography, Arizona State University. He studies agricultural decision-making and environmental change in Latin America, focusing on Middle America. He has conducted research and published on house gardens, sustainable development, and ethnobotany.

**Peter Klepeis** (pklepeis@mail.colgate.edu) is Assistant Professor, Department of Geography, Colgate University. He examines the history of human–environment relationships in Latin America, especially in the Maya and Patagonia regions.

**Deborah Lawrence** (dl3c@virginia.edu) is an Assistant Professor of Environmental Science at the University of Virginia. She has conducted studies on the biogeochemical and ecological effects of human disturbance in the tropical forests of Cameroon, Costa Rica, Indonesia, and Mexico.

**Steven M. Manson** (manson@umn.edu) is an Assistant Professor, Department of Geography at the University of Minnesota. He is author of several articles and reports on integrated assessment, land-use and land-cover change, geographic information systems, modeling, and complexity theory.

**Pedro Macario Mendoza** (pmacario@ecosur-qroo.mx) is a Senior Researcher at El Colegio de la Frontera Sur, Unidad Quintana Roo. He is a specialist on forest ecology and on swidden cropping systems in the region.

**Yelena Ogneva-Himmelberger** (Yogneva@clarku.edu) is a Research Associate in the George Perkins Marsh Institute, Clark University. She specializes in remote sensing and GIS analysis and undertook a trial study of the SYPR before the formal project began.

**Diego Pérez-Salicrup** (diego@oikos.unam.mx) is Assistant Professor at the Institute of Ecology, at the National Autonomous University of Mexico. He has worked on the effects of human activities on forest structure in the tropical forests of Bolivia, Brazil, and his native Mexico.

**Birgit Schmook** (Birgit_Schmook@yahoo.com) is a researcher at El Colegio de la Frontera Sur, Unidad Quintana Roo and a doctoral graduate student in the Graduate School of Geography, Clark University. She examines the role of women and household demography.

**Laura Schneider** (lschneider@clarku.edu) is a doctoral candidate, Graduate School of Geography, Clark University. She employs remote sensing and GIS methods to solve land-change problems and is working on invasive species in the region.

**Colin Vance** (colin.vance@dlr.de) is a Reasearch Scientist with the Institute of Transportation Research, German Aerospace Center, Berlin, Germany. He is a spatial and resource econometrician.

# 1

# Three Frontiers of the Southern Yucatán Peninsular Region and SYPR Project

*B. L. Turner II, D. R. Foster, J. Geoghegan*

Frontiers advance and retreat, both figuratively and literally. At this moment they are advancing in three ways relevant to the subject of this book and the ongoing project on which it is based. First, after more than a century of reductionist hegemony, various science communities worldwide increasingly recognize the need to improve complementary, *synthesis* understanding—a way of putting the reductionist pieces of the problem back together again in order to understand how the 'whole' system works and to identify the emergent properties that follow from the complex interactions of the pieces. Synthesis understanding is not, of course, new. In the late eighteenth century, Immanuel Kant argued for it as one of the pillars of science in the reorganization of knowledge in the European academy (Turner 2002*a*) and designated *geography* as one of the 'synthesis sciences'. Its contemporary rediscovery, however, rests in the science of global environmental change (Lawton 2001; Steffen *et al.* 2002), especially efforts to model complex systems, such as those in ocean–atmosphere–land interactions, and has been expanded by emerging research agendas seeking to couple human and environment systems, often registered under the label of 'sustainability science' (e.g. Kates *et al.* 2001; NRC 1999). Second, within these developments land-use and land-cover change (or, simply, land change) is singled out because of its centrality to a wide range of environmental concerns, including global climate change, regional–local hydrological impacts, biodiversity, and, of course, human development and ecosystem integrity (e.g. Brookfield 1995; NRC 2000; Watson *et al.* 2001). The need to advance an *integrated land-change science* is also increasingly recognized, one in which human, ecological, and remote sensing and geographical information systems (GIS) sciences are intertwined in problem-solving (Liverman *et al.* 1998; Klepeis and Turner 2001; Turner 2002*b*). And central to this effort is the need to advance geographically (spatially) explicit land-change models that can explain and project coupled human-ecological systems, and thus serve a wide range of research and assessment constituencies, from carbon to biodiversity to human vulnerability (IGBP 1999; Irwin and Geoghegan 2001; Kates *et al.* 2001; Liverman *et al.* 1998; Veldkamp and Lambin 2001). These two developments—synthesis science and integrated land science directed towards geographically explicit land-change models—constitute the broader intellectual and research frontiers to which this work contributes.

The third frontier is substantive and registered in at least two fundamental ways. The *southern Yucatán peninsular region* constitutes the southeastern end of the tropical forest zones in Mexico, one that remained sparsely occupied well into the twentieth century (e.g. Edwards 1957), and awaited modern development of the last thirty years for the contemporary 'agricultural frontier' to reach the region. This frontier is beset by unusual circumstances, however. As it was advancing inland from the Gulf and Caribbean littorals, Mexican interests shifted and the interior of the region became a pivotal part of international programs concerned with forest preservation, carbon sequestration, and the protection of wildlife corridors (De Jong, Tipper, and Montoya-Gómez 2000; Hughes, Kauffman, and Jaramillo 1999, 2001;Turner *et al.* 2001). The Calakmul Biosphere Reserve, biological corridors, and associated archaeo-tourism programs now compete with farming and agroforestry for use of the landscapes within or abutting agricultural settlements (Primack *et al.* 1998). The juxtaposed needs to cultivate and preserve have created a problematic 'land-use frontier': how to create and maintain a regional landscape that simultaneously serves the needs of the growing numbers of farmers and the ecological integrity of the reserve and corridors. This last frontier is precisely the kind of synthesis problem for which integrated land science is intended (Klepeis and Turner 2001; Lambin *et al.* 2001; Turner 2002b), especially the potential spatially explicit, integrated land models that inform both science and practice.

## 1. *The Significance of Tropical Deforestation*

Land change in the southern Yucatán peninsular region is symptomatic of stresses on tropical forests worldwide. The rate and magnitude of deforestation in the tropics are the highest of any land-cover change worldwide and the focal point of international conservation concern for various reasons (Achard *et al.* 1998; Cincotta and Engelman 2000; FAO 1999). Tropical forests are a large source and sink of atmospheric carbon (Houghton *et al.* 2000; Pacala *et al.* 2001; Skole and Tucker 1993; Watson *et al.* 2001) and support the majority of the world's terrestrial biotic diversity and net primary productivity (NPP) (DeFries *et al.* 1999; Vitousek *et al.* 1997). Their loss worldwide, therefore, contributes to global climate warming and diminished biodiversity, and the land uses replacing them markedly increase the portion of net primary productivity captured globally by human activity. Tropical deforestation at any spatial scale commonly involves land degradation (Barrow 1991) and threats to critical ecosystem services (Daily *et al.* 2000; Kremen *et al.* 2000), and may induce pronounced changes in local-to-regional rainfall and feedbacks to the earth system processes (IGBP 1999; Lawton *et al.* 2001).

These consequences of deforestation and land conversion increase the vulnerability of various human–environment systems to the perturbations and stresses of climate and other kinds of changes (e.g. Angelsen and Kaimowitz 1999; Downing 1996; Geist and Lambin 2001; Rosenzweig and Parry 1994; Kasperson, Kasperson, and Turner 1995). Compounding the complex effects of tropical deforestation is the fact that much, if not most, of it takes place under the auspices of economic development (Brookfield, Potter, and Byron 1995; Jepsen *et al.* 2001; Utting 1993). Many of the tropical forests of the world have been sparsely occupied and accrue little economic benefit to the controlling polity, not accounting for the ecosystem goods and services they provide (Daily *et al.* 2000), save for their extractive resources, such as precious woods and minerals. It is not surprising, therefore, that governments, especially within poorly developed or developing economies, envision tropical forest lands as potential panaceas for socio-economic relief and long-term gain, and ignore the externalities associated with the expansion of the development frontier into them. While some development programs aim for sustainable off-take of timber and tree products in these lands (see Anderson 1990), more often than not the dynamics of projects, infrastructure development, and 'inexpensive' land lead to forest conversion (e.g. Angelsen and Kaimowtiz 1999; Kummer 1991), complete with the environmental consequences noted as well as additional impacts, such as invasion by exotic species (e.g. Mooney and Hobbs 2000; Uhl *et al.* 1982).

## 1.1. *Tropical Deforestation in Mexico and the Region of Study*

Tropical deforestation in Amazonia (e.g. Andersen 1996; Hecht and Cockburn 1989; Nepstad *et al.* 1999; Pfaff 1999; Skole and Tucker 1993; Smith *et al.* 1995) and Indonesia (e.g. Brookfield, Potter, and Byron 1995; Jepson *et al.* 2001) dominates global attention, but few if any tropical frontiers have escaped development pressures. These pressures have been long-standing in Mexico, beginning in the Gulf Coast lowlands and subsequently spreading eastward through the Lacandon forests of Chiapas to the Yucatán Peninsula (Cairns, Dirzo, and Zadroga 1995; Dirzo and García 1991; Hughes, Kauffman, and Jaramillo 1999; Masera, Ordóñez, and Dirzo 1997; Merino 1996; O'Brien 1998). Indeed, Cairns and colleagues (2000) estimate the annual rates of deforestation in Mexico between 1977 and 1992 to have been 1.9 per cent, most of this occurring in tropical forests. Pressures on regional tropical forests are not limited to Mexico, however, but include the southern and central Petén of Guatemala where highland Mayan immigration has been significant (Sader *et al.* 1994). These vectors of human pressures on forest lands, including the 'development successes' of Cancún and the northeastern Yucatán coastline, have now advanced to southern Campeche and Quintana Roo—the southeastern limit of tropical forest lands in Mexico and, with the adjacent northern Petén, Guatemala, the last remaining portion of the ancient Maya lowlands to feel the impress of contemporary deforestation.

The southern Yucatán peninsular region dominates the interior uplands of this once 'empty quarter' (Edwards 1957) of Mexico (Fig. 1.1). The subject of various resource extraction schemes during the first half of the twentieth century (Ch. 3; Klepeis 2000), occupation was restricted to several sawmill-based *ejidos* (communal lands) and chicle camps located, respectively, on the eastern and western flanks and the interior of the region.[1] The near region-wide exhaustion of mahogany and Spanish cedar from selective but extensive logging coincided with the devastation of Hurricane Janet in 1955, which exposed the isolation and underdevelopment of the region, especially the inability of Mexican assistance to reach southeastern Quintana Roo in an effective manner. One result was the construction of Highway 186, fully paved in 1972, cutting straight across the heart of the region (Fig. 1.1).

[1] Chicle refers to both the tree (*Manilkara zapota* L. Van Royen, formerly, *Achras sapota* L.) and the resin taken from it that was the original substance from which chewing gum was made (see Ch. 3). Locally the species is also called *sapodilla* or *chicozapote*.

The highway opened the region to extensive occupation and development, and was immediately followed by the establishment of agricultural *ejidos* along the east–west swath that the road cut through the region and of several large-scale development projects (rice cultivation) on the eastern and western edges of the region. These actions triggered a boom in land clearing and elevated the annual rate of deforestation to 2 per cent along its 270km route through the region (Cortina Villar, Macario Mendoza, and Ogneva-Himmelberger 1999: 46).[2] The subsequent Mexican debt crisis dampened forest cutting in the early 1980s, and the annual rate of deforestation fell to 0.2 per cent during that time. After various adjustments, local farmers began to move into commercial chili (pepper) production and lay the groundwork for future livestock production by creating widespread pasture. Both these activities increased deforestation rates as well as the conversion of lands from swidden or slash-and-burn cultivation to permanent or near-permanent 'open' land during the 1990s (Chs. 6, 8, 9, 10). Consequently, the southern Yucatán peninsular region has been identified as a hot spot of tropical deforestation, forest fragmentation, and biodiversity loss in numerous recent studies (e.g. Achard *et al.*, 1998; Cincotta and Engelman 2000).

Compounding the significance of these land-use and land-cover changes is the new national vision for the region. Mexico's largest tropical biosphere reserve, Calakmul, was established in the heart of the region in 1989 (Ch. 7; Ericson, Freudenberger, and Boege 1999; Primack *et al.* 1998; Snook and Zapata 1998). Linked to other reserves and conservation areas to north and south, Calakmul is a critical segment of the major north–south biological corridor across the Maya lowlands (Kaiser 2001; Miller, Chang, and Johnson 2001).[3] Development of the reserve and corridor, in turn, are consistent with national and international plans to make the region part of El Mundo Maya (the Maya world), an international archaeo-ecotourism scheme relying on the diversity of forest biota and the multitude of Maya ruins within them (Garrett 1989; Primack *et al.* 1998).[4]

[2] Mas Caussel (1996) calculated a staggering annual rate of deforestation of 4.5% for the state of Campeche between 1978/80 and 1992. This calculation includes that portion of the study region currently designated as the Municipio of Calakmul.

[3] The move to make much of the region a model of 'green' or sustainable development has led to research on biotic diversity in the region (e.g. Escamilla *et al.* 2000; Pozo 1998).

[4] During the late stages of the development of this volume, American Airlines ran a feature story on the southern Yucatán, complete with a picture of the first up-scale hotel in the center of the region (Stanton 2002). El Mundo Maya appears to be becoming a reality.

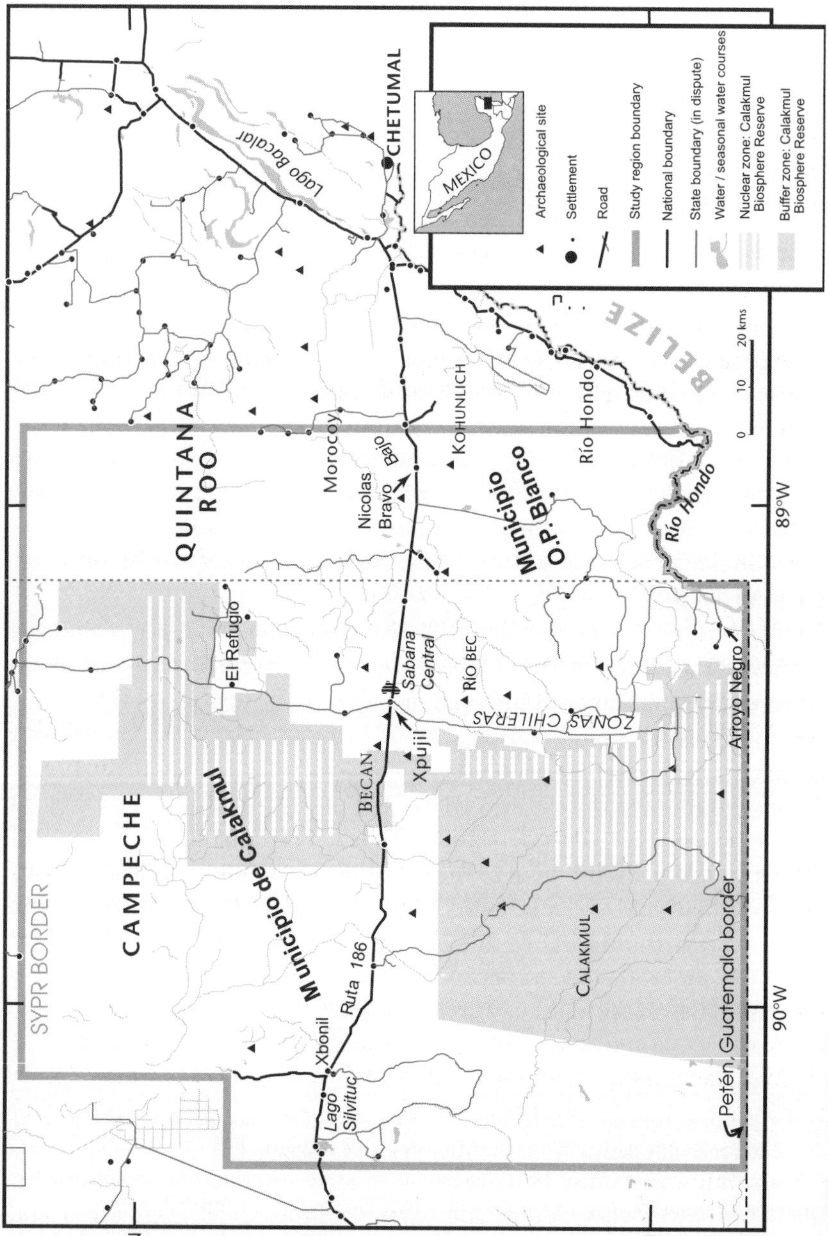


**Fig. 1.1.** The southern Yucatán peninsular region.

## 2. *The Study Problem and Approach*

The current setting, therefore, juxtaposes the pressures to increase forest preservation and economic development, and provides the substantive problem confronting the Southern Yucatán Peninsular Region (SYPR) project, established in 1997. These conditions also inform the major questions guiding the SYPR project:

- What are the trajectories of change in land-use and land-cover across the region?
- What natural and social processes and forces contribute to these changes, and what are their impacts on forest maintenance and recovery, and on household economic well-being?
- What suites of models best explain and project land-use and land-cover change in the region in terms of type, magnitude, and geographical location?

These questions and the approaches used to address them follow from experiences in global change and sustainability science (e.g. Kates *et al.* 2001; NRC 1999) indicating that there is much to be learned of value for research and application by addressing the totality of the integrated or coupled systems in question. The SYPR project has followed the lead of the Land-Use/Cover Change (LUCC) effort of International Geosphere-Biosphere Programme (IGBP) and International Human Dimensions Programme (IHDP) (SSC-LUCC 1999; Turner *et al.* 1995), the Population, Land, and Environmental Change (PLEC) effort sponsored by the United Nations University (Brookfield 1995), and various initiatives sponsored by federal agencies in the United States (e.g. Liverman *et al.* 1998). Specifically, the project has been undertaken in the belief that a critical yet underdeveloped element of research is that on 'integrated' land-use and land-cover change, especially the need to explain and project the mosaic of use and cover changes across areas, landscapes, and regions. These mosaics of land surface changes cannot be adequately understood and modeled by way of approaches emphasizing an economic sector, households, and/or ecosystems alone, but must be integrated or coupled (Fig. 1.2; Bockstael 1996; Chomitz and Gray 1996; Geoghegan, Wainger and Bockstael 1997; Hall 2000; Irwin and Geoghegan 2001; Lambin 1994; Veldkamp and Lambin 2001), giving rise to 'integrated land science' (Klepeis and Turner 2001; Turner *et al.* 2001).

## 2.1. *Specific Research*

The SYPR project posits that the dual regional goals of forest conservation/preservation, and economic development require foundational studies of the basic land dynamics at play now and in the past, a calibrated means of detecting, observing, and monitoring land change, and improved means of explaining and projecting land change in geographically (spatially) explicit ways (e.g. Calhoun 1998). The project's method of integration joins the research and findings of the social, economic, ecological, remote sensing, and GIS (geographical information systems) sciences (Fig. 1.2). They are joined through a set of specific research problems (subquestions) that, when fit together, permit advances in models of land change at different scales of analysis.

*Land cover—ecological questions.* What is the current regional variation in forest structure, composition, and function, and how are these shaped by historical processes, modern environment and ongoing land-use activities? How do critical forest ecosystem processes and environmental conditions change through the cycle of the traditional and widespread land-use practice of swidden agriculture?

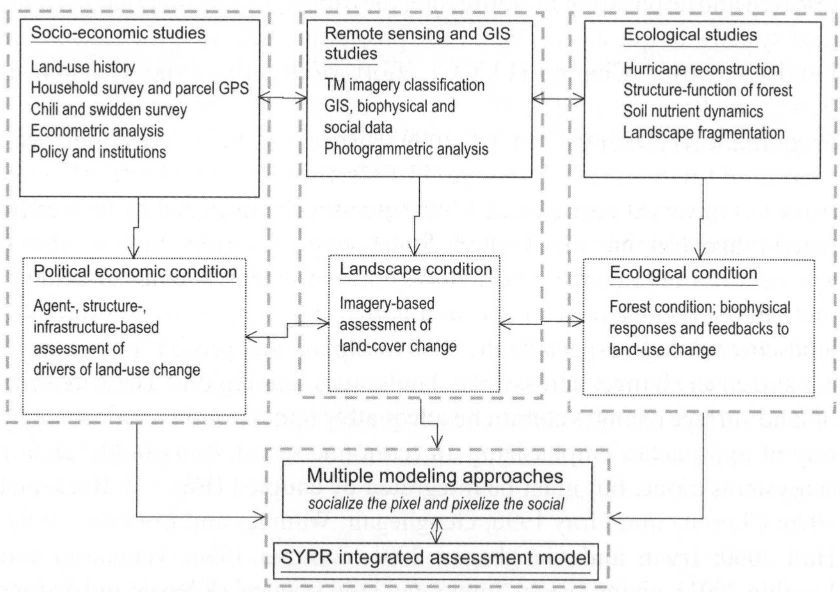


**Fig. 1.2.** The four-part SYPR project study.

*Land use—socio-economic questions.* What macro-structures and household characteristics affect land manager decisions and in what way? What principal strategies of cultivation are employed and what are their implications for forest and household well-being?

*Land-use and land-cover distribution and change.* What are the regional distribution and variation in land-cover changes across the region during the period of intensive land-use activity since the 1970s? What land classifications can be accurately assessed through Landsat TM imagery, and are these classes sufficiently fine-tuned to serve research needs that include farmers' decision-making? How well does Landsat TM land classification and time-series analysis of images capture the land-use and land-cover dynamics documented from historical, air photo, ground-based, and oral accounts?

*Land-change modeling questions.* Which kind of explanatory models best capture the observed changes in land use and land cover in the region at different spatial and temporal scales? How can these models be linked directly to the pixel in the satellite imagery? Can an integrated assessment model of land change be constructed that provides robust spatially explicit outcomes?

## 2.2. *Project Parts and Integration*

A project as ambitious as that of the SYPR requires a coordinated team of researchers and research units (Geoghegan *et al.* 2001; Turner *et al.* 2001). Programmatic, funding, and research realities, however, necessitate that the research questions posed above be undertaken by specific units or groups of researchers (Fig. 1.2). Consequently, the project may be viewed as containing four basic study parts: forest ecology, imagery analysis, socio-economic and historical studies, and land-use modeling. While individual and group research contains and produces stand-alone elements, each is designed to take output from and provide input to other parts of the project. The information gained and our understanding of key processes are used to inform the various modeling efforts that link to pixels or pixel clusters in Landsat TM imagery, a linkage labeled elsewhere as 'socializing the pixel' and 'pixelizing the social' (Geoghegan *et al.* 1998).

## 3. *The Region*

The southern Yucatán peninsular region is fabricated for this study. Its boundaries (Fig. 1.1), however, are not arbitrary but positioned to encompass (1) the rolling uplands or hilly lands (≥ 100m amsl) of southern Campeche and Quintana Roo, that form the spine of the peninsula and define the eastern and western margins of the study zone (Ch. 2), and (2) the Calakmul Biosphere Reserve, whose limits define the north–south boundaries of the region, the southern border of both coinciding with the international border of Mexico and Petén, Guatemala. In effect, this delimitation captures the old growth forests that were largely untouched by significant human activity from ancient Maya times until the opening of Highway 186 beginning in the late 1960s (above). These old forests form the matrix from which the reserve was carved, and in which land pressures in *ejidos* located within and around the reserve continue to mount.

The borders of the region were also established to exclude those lower-lying and hilly areas that experienced extensive settlement and intensive development prior to the completion of Highway 186. Laguna Silvituc in the west and Bajo Morocoy in the east are convenient markers for this exclusion and permit the study to examine the large-scale projects undertaken at each of these locales that helped to spur post-1960 deforestation in the region. And, the borders of the northwest corner of the region were reconfigured slightly to omit a large development project that began there earlier and that could not be reached directly from the remainder of the region until later in the development of southern Campeche.

The demarcated region falls primarily within two *municipios* (large administrative districts). The Municipio of Calakmul covers the southeastern portion of the state of Campeche, containing but not solely restricted to the boundaries of the Calakmul Biosphere Reserve (Fig. 1.1). Immediately across the border in southwestern Quintana Roo, the Municipio of O. P. Blanco dominates most of the study region.

## 4. *Programmatic Elements and Sponsorship*

The SYPR project began in 1997 and will continue at least to 2004 under the auspices of two major awards from NASA's LCLUC (National Aeronautics and Science Administration, Land-Use and Land-Cover

Change) program and from sustained support from the Center for Integrated Studies on the Human Dimensions of Global Environmental Change, Carnegie Mellon University (National Science Foundation (NSF) supported). The scope of the project, however, requires additional funds, and these were provided for various activities and events in phase one of the SYPR project by the following entities, awarded to individuals and units in regard to the projects:

El Colgeio de la Frontera Sur, Unidad Chetumal (ECOSUR)
Clark University, George Perkins Marsh Institute
Conservation Research Foundation
Fulbright Fellowship (Garcia Robles)
Harvard Forest, Harvard University
NASA
- New Investigator Fellowship
- Earth System Science Fellowships

NSF
- Decision, Risk and Management Science
- Environmental Biology
- Geography and Regional Science
- Inter-Americas Program
- International Programs

A. W. Mellon Foundation

Three research units took primary responsibility for the research. Clark University, through the George Perkins Marsh Institute, provided economists and geographers who undertook work on imagery analysis, historical studies, socio-economic household and agricultural systems research, and modeling. Harvard Forest (Harvard University) provided expertise on forest ecology and aerial photograph analysis. El Colegio de la Frontera Sur-Unidad Chetumal provided agronomic, forest management, and ecological expertise, especially in regard to regional and Maya conditions. The project's field center remains in Zoh Laguna, in the center of the study region, but ECOSUR-Unidad Chetumal provides much needed communication facilities and administration. As the new phase of the project began in 2001, the participating institutions have been expanded to include the following: Department of Environmental Science, University of Virginia, nutrient cycling studies; the Institute of Ecology, Universidad Nacional Autónoma de México, forest structure research; and ECOSUR-Unidad Campeche, social survey and household expertise. Forest ecologists and social scientists from ECOSUR-Unidad Chetumal remain instrumental to the project and researchers from Unidad Villa Hermosa have been added to the study.

## 5. *State of the Study and Organization of the Book*

These research units and the various researchers on the SYPR project seek to explain, model, and project the land changes underway in and the implications for the forests of the southern Yucatán peninsular region. While the project remains in progress, four years of fieldwork and analysis provide a wealth of information and insights that justify the integrated and detailed presentation offered here. The focus of this volume is on the foundational data and the means by which it has been and will continue to be integrated to understand land-change dynamics in the region. Its substantive goals are to document the nature, pattern, and scale of deforestation in the southern Yucatán peninsular region, provide an understanding of the combined forces triggering this land change and its implications for forest and household well-being, and to reformulate this understanding in a way that permits fine-tuned assessments of land change to the level of the pixel or pixel cluster in Landsat TM imagery. The analytical power and insights gained in achieving this level of integration should be obvious, ranging from the question of forest health to sustained economic development. Most importantly, the goals of the project and book point the way towards critical components by which the tensions between forest preservation and cultivation can be monitored, assessed, and altered based on a deep understanding of the potential impacts of the changes made.

Part I provides the human–environment history of the region, illustrating how it has affected the 'natural' setting, with a focus on the forests. Chapter 2 addresses this history over the long term, stretching back from the end of eighteenth century to the ancient Maya, indicating how land practices made over a millennium ago affect species abundance today. The chapter also provides an introduction to the current environmental conditions in the region. Chapter 3 details the modern human–environment history of the region, starting at the transition between the eighteenth and nineteenth centuries. It demonstrates the existence of three different human–environment episodes, each predicated on different views of how to use the region and each with different forest consequences. The most dramatic of these consequences, selective logging of hardwoods, affects species abundance in the forest today.

Part II treats current forest conditions, land cover, and land-cover changes across the region. In Ch. 4, the structure of the region's forest is addressed. This work demonstrates the difference between two basic forest types, upland (*mediano*) and *bajo* (seasonal wetland), and provides information on the length of recovery to older growth conditions, once disturbed. Chapter 5, in contrast, examines changes in forest ecosystem processes, especially nutrient cycling, with the traditional *milpa* or swidden land-use cycle. It also highlights variations in biomass dynamics across rainfall gradients and the implications of this for forest recovery. Land-cover change is documented in Ch. 6 by way of aerial photo and TM imagery analysis. The image analysis methods are described, and the overall trajectories of different kinds of land-cover change are identified.

Part III provides the details of the region's political economy, demonstrating which 'human' factors are currently driving deforestation or intensifying what was once swidden or slash-and-burn cultivation predicated on successional forest growth. Chapter 7 documents the major components (i.e. policy, institutions, and initiatives) of the socio-economic structures prevalent in the region. It demonstrates that these 'macro-structures' have been and remain important in explaining why certain kinds of land-use changes are undertaken, even when their immediate rewards to households making them may be suspect. The 1997 and 1998 household survey undertaken by the SYPR project is introduced in Ch. 8. It not only establishes the kind of land manager present in the region, but points to the nature of household decision-making. Chapters 9 and 10 address the two major kinds of cultivation undertaken by households, indicating the crop–fallow cycles followed and trajectory of changes underway in each. Chapter 9 addresses maize swidden cultivation, employed by every household minimally for consumption or subsistence production, while Ch. 10 considers commercial chili production which has spread like wildfire among the region's households, despite minimal state or NGO support for the activity. Finally, Ch. 11 provides an econometric test of household decision-making, demonstrating that local farmers follow a production rationale that mixes subsistence and commercial objectives, providing strong clues about the pathways for future modeling efforts. Part IV presents two different spatially explicit land-use modeling approaches under development that use the data created by the project. The first, in Ch. 12, employs an econometric approach to test hypotheses concerning the factors that affect the probability of land-use change at different temporal and spatial scales. Chapter 13 presents a spatially explicit, agent-based integrated assessment (simulation) model of land change in which the 'rules' guiding the agents may change as conditions in the region change. These two chapters illustrate the strengths and weaknesses of different modeling approaches. A major reason for developing spatially explicit land-change models is to undertake 'synthesis' studies; for example, region-wide assessments of changes in carbon stocks and biodiversity—both subjects of future SYPR studies.

The editors offer commentary and linkages between the major research components of the project in four 'book part' introductions, and a retrospective view of phase one of the SYPR project at the end of the volume. Definitions of major Spanish and other foreign terms as well as acronyms are located in the Glossary.

## REFERENCES

Achard, F., H. Eva, A. Glinni, P. Mayaux, T. Richards, and H. J. Stibig (eds.) (1998). *Identification of Deforestation Hot Spot Areas in the Humid Tropics.* Brussels, European Commission.

Andersen, L. E. (1996). 'The Causes of Deforestation in the Brazilian Amazon'. *Journal of Environment and Development* 5: 308–28.

Anderson, A. B. (ed.) (1990). *Alternatives to Deforestation: Steps Toward Sustainable Use of the Amazon Rain Forest.* New York, Columbia University Press.

Angelsen, A., and D. Kaimowitz (1999). 'Rethinking the Causes of Deforestation: Lessons from Economic Models'. *The World Bank Research Observer* 14: 73–98.

Barrow, C. J. (1991). *Land Degradation: Development and Breakdown in Terrestrial Environments.* Cambridge, Cambridge University Press.

Bockstael, N. E. (1996). 'Modeling Economics and Ecology: The Importance of a Spatial Perspective'. *American Journal of Agricultural Economics* 78: 1168–80.

Brookfield, H. (1995). 'PLEC and Global Environmental Change'. *Global Environmental Change Part A* 5: 263–6.

Brookfield, H., L. Potter, and Y. Byron (1995). *In Place of the Forest: Environmental and Socio-economic Transformation in Borneo and the Eastern Malay Peninsula.* Tokyo, United Nations University Press.

Cairns, M. A., R. Dirzo and F. Zadroga (1995). 'Forests in Mexico: A Diminishing Resource'. *Journal of Forestry* 93: 21–4.

Cairns, M. A., P. K. Haggerty, R. Alvarez, B. H. J. De Jong, and I. Olmsted (2000). 'Tropical Mexico's Recent Land-Use Change: A Region's Contribution to the Global Carbon Cycle'. *Ecological Applications* 10: 1426–41.

Calhoun, J. M. (ed.) (1998). *Forest Policy: Ready for Renaissance.* Seattle, College of Forest Resources, University of Washington.

Chomitz, K., and D. Gray (1996). 'Roads, Land Use and Deforestation. A Spatial Model Applied to Belize'. *World Bank Economic Review* 10: 487–512.

Cincotta, R. P., and R. Engelman (2000). *Nature's Place: Human Population and the Future of Biological Diversity*. Washington DC, Population Action International.

Cortina Villar, S., P. Macario Mendoza, and Y. Ogneva-Himmelberger (1999). 'Cambios en el uso del suelo y deforestacion en el sur de los estados de Campeche y Quintana Roo, Mexico'. *Boletin del Instituto de Geografia de la UNAM. Investigaciones Geograficas* 38: 41–56.

Daily, G. C., T. Söderqvist, S. Aniyar, K. Arrow, P. Dasgupta, P. R. Ehrlich, C. Folke, A. Hansson, B.-O. Jansson, N. Kautsky, S. Levin, J. Lubchenco, K.-G. Mäler, D. Simpson, D. Starrett, D. Tilman, and B. Walker (2000). 'The Value of Nature and Nature of Value'. *Science* 289: 395–6.

De Jong, B. H. J., R. Tipper, and G. Montoya-Gomez (2000). 'An Economic Analysis of the Potential for Carbon Sequestration by Forests: Evidence from Southern Mexico'. *Ecological Economics* 33: 313–27.

DeFries, R., C. Field, I. Fung, G. Collatz, and L. Bounoua (1999). 'Combining Satellite Data and Biogeochemical Models to Estimate Global Effects of Human-Induced Land Cover Change on Carbon Emissions and Primary Productivity'. *Global Biogeochemical Cycles* 13: 803–15.

Dirzo, R., and C. García (1991). 'Rates of Deforestation in Los Tuxtlas, a Neotropical Area in Southeast Mexico'. *Conservation Biology* 6: 84–90.

Downing, T. E. (1996). *Climate Change and World Food Security*. Berlin, Springer.

Edwards, C. R. (1957). *Quintana Roo: Mexico's Empty Quarter*. Berkeley, University of California.

Ericson, J., M. S. Freudenberger, and E. Boege (1999). 'Population Dynamics, Migration, and the Future of the Calakmul Biosphere Reserve. *Occasional Paper No. 1, Program on Population and Sustainable Development*. Washington DC, American Association for the Advancement of Science.

Escamilla, A., M. Sanvicente, M. Sosa, and C. Galindo-Lela (2000). 'Habitat Mosaic: Wildlife Availability, and Hunting in the Tropical Forest of Calakmul, Mexico'. *Conservation Biology* 14: 1592–601.

FAO (1999). *State of the World's Forests, 1999*. Rome, Food and Agricultural Organization of the United Nations.

Garrett, W. E. (1989). 'La Ruta Maya'. *National Geographic* 176: 424–79.

Geist, H., and E. Lambin (2001). *What Drives Tropical Deforestation? A Meta-Analysis of Proximate and Underlying Causes of Deforestation Based on Subnational Case Study Evidence*. Louvain-la-Neuve, LUCC International Project Office.

Geoghegan, J., L. Wainger, and N. Bockstael (1997). 'Spatial Landscape Indices in a Hedonic Framework: An Ecological Economics Analysis Using GIS'. *Ecological Economics* 23: 251–64.

Geoghegan, J., L. J. Pritchard, Y. Ogneva-Himmelberger, Roy Chowdhury, S. Sanderson, and B. L. Turner II (1998). '"Socializing the Pixel" and "Pixelizing the Social" in Land-Use Cover Change', in D. Liverman, E. Moran, R. Rindfuss, and P. Stern (eds.), *People and Pixels: Linking Remote Sensing and Social Science*. Washington DC, National Academy of Sciences Press, 51–69.

Geoghegan, J., S. Cortina Villar, P. Klepeis, P. Macario Mendoza, Y. Ogneva-Himmelberger, R. Roy Chowdhury, B. L. Turner II, and C. Vance (2001). 'Modeling Tropical Deforestation in the Southern Yucatán Peninsular Region: Comparing Survey and Satellite Data'. *Agriculture, Ecosystems and Environment* 85: 25–46.

Hall, C. A. S. (2000). *Quantifying Sustainable Development: The Future of Tropical Economies*. New York, Academic Press.

Hecht, S. B., and A. Cockburn (1989). *The Fate of the Forest: Developers, Destroyers, and Defenders of the Amazon*. London, Verso.

Houghton, R. A., D. L. Skole, C. A. Nobre, J. L. Hackler, K. T. Lawrence, and W. H. Chomentowski (2000). 'Annual Fluxes of Carbon from Deforestation and Regrowth in the Brazilian Amazon'. *Nature* 403: 301–4.

Hughes, R. F., J. B. Kauffman, and V. J. Jaramillo (1999). 'Biomass, Carbon, and Nutrient Dynamics of Secondary Forests in a Humid Tropical Region of Mexico'. *Ecology* 80: 1892–907.

——(2001). 'Ecosystem-Scale Impacts on Deforestation and Land Use in a Humid Tropical Region of Mexico'. *Ecological Applications* 10: 515–27.

IGBP (1999). 'Special Issue on BAHC'. *Global Change Newsletter 39 (September)*. International Geosphere-Biosphere Programme.

Irwin, E. G., and J. Geoghegan (2001). 'Theory, Data, Methods: Developing Spatially Explicit Economic Models of Land Use Change'. *Agriculture, Ecosystems and Environment* 85: 7–23.

Jepson, P., J. K. Jarvie, K. MacKinnon, and K. A. Monk (2001). 'The End of Indonesia's Lowland Forest'. *Science* 292: 859–61.

Kaiser, J. (2001). 'Bold Corridor Project Confronts Political Reality'. *Science* 293: 2192–4.

Kasperson, J. X., R. E. Kasperson, and B. L. Turner II (eds.) (1995). *Regions at Risk: Comparisons of Threatened Environments.* Tokyo, United Nations University Press.

Kates, R. W., W. C. Clark, R. Corell, J. M. Hall, C. C. Jaeger, I. Lowe, J. J. McCarthy, H. J. Schellenhuber, B. Bolin, N. M. Dickson, S. Faucheaux, G. C. Gallopin, A. Grübler, B. Huntley, J. Jäger, N. S. Jodha, R. E. Kasperson, A. Mabogunje, P. Matson, H. Mooney, B. Moore III, T. O'Riordan, and U. Svedin (2001). 'Sustainability Science'. *Science* 292: 641–2.

Klepeis, P. (2000). *Deforesting the Once Deforested: Land Transformation in Southeastern México*. Worcester, Mass., Clark University.

Klepeis, P., and B. L. Turner II (2001). 'Integrated Land History and Global Change Science: The Example of the Southern Yucatán Peninsular Region Project'. *Land Use Policy* 18: 27–39.

Kremen, C., J. O. Niles, M. G. Dalton, G. C. Daily, P. R. Ehrlich, J. P. Fay, D. Grewel, and R. P. Guillery (2000). 'Economic Incentives for Rain Forest Conservation across Scales'. *Science* 288: 1828–32.

Kummer, D. M. (1991). *Deforestation in the Postwar Philippines*. Chicago, University of Chicago Press.

Lambin, E. F. (1994). *Modeling Deforestation Processes: A Review*. Luxembourg, European Commission.

Lambin, E. F., B. L. Turner II, H. Geist, S. Agbola, A. Angelsen, J. W. Bruce, O. Coomes, R. Dirzo, G. Fischer, C. Folke, P. S. George, K. Homewood, R. L. J. Imbernon, X. Li, E. F. Moran, M. Mortimore, P. S. Ramakrishnan, J. F. Richards, H. Skånes, W. Steffen, G. D. Stone, U. Svedin, T. Veldkamp, C. Vogel, and J. Xu (2001). 'Our Emerging Understanding of the Causes of Land-Use and Cover Change'. *Global Environmental Change: Human and Policy Dimensions* 11: 5–13.

Lawton, J. (2001). 'Earth System Science'. *Science* 292: 1965.

Lawton, R. O., U. S. Nair, R. A. Pielke, Sr., and R. M. Welch (2001). 'Climatic Impact of Tropical Lowland Deforestation on Nearby Montane Cloud Forests'. *Science* 294: 584–7.

Liverman, D., E. F. Moran, R. R. Rindfuss, and P. C. Stern (eds.) (1998). *People and Pixels: Linking Remote Sensing and Social Science*. Washington DC, National Academy Press.

Mas Caussel, J. (1996). 'Estimación preliminar de las tasas de deforestación en el estado de Campeche'. *Jaina* 7: 5–6.

Masera, O., M. J. Ordóñez, and R. Dirzo (1997). 'Carbon Emissions from Mexican Forests: Current Situation and Long-Term Scenarios'. *Climatic Change* 35: 265–95.

Merino, L. (1996). 'Los bosques de Mexico, una perspectiva general'. *Cuadernos Agrarios* 14: 157–62.

Miller, K., E. Chang, and N. Johnson (2001). *Defining Common Ground for The Mesoamerican Biological Corridor*. Washington DC, World Resources Institute.

Mooney, H. A., and R. J. Hobbs (eds.) (2000). *Invasive Species in a Changing World*. Washington DC, Island Press.

Nepstad, D. C., A. Veríssimo, A. Alencar, C. Norte, E. Lima, P. Lefebvre, P. Schlesinger, C. Potter, P. Moutinho, E. Mendoza, M. Cochrane, and V. Brooks (1999). 'Large-Scale Impoverishment of Amazonian Forests by Logging and Fire'. *Nature* 398: 505–8.

NRC (National Research Council) (1999). *Our Common Journey: A Transition Toward Sustainability*. Washington DC, National Academy of Sciences Press.

——(2000). *Grand Challenges in the Environmental Sciences*. Washington DC, National Academy of Sciences Press.

O'Brien, K. L. (1998). *Sacrificing the Forest: Environmental and Social Struggles in Chiapas*. Boulder, Col., Westview Press.

Pacala, S. W., G. C. Hurtt, D. Baker, P. Peylin, R. A. Houghton, R. A. Birdsey, L. Heath, E. T. Sundquist, R. F. Stallard, P. Ciais, P. Moorcroft, J. P. Caspersen, E. Shevlia Kega, B. Moore, G. Kohlmaier, E. Holland, M. Gloor, M. E. Harmon, S.-M. Fan, J. L. Sarmiento, C. L. Goodale, D. Schimel, and C. B. Field (2001). 'Consistent Land and Atmosphere-Based US Carbon Sink Estimates'. *Science* 292: 2316–20.

Pfaff, A. (1999). 'What Drives Deforestation in the Brazilian Amazon?' *Journal of Environment and Economic Management* 37: 26–43.

Pozo, C. (1998). 'Zona Sur del Estado de Quintana Roo, México', in L. Herrera-MacBryde (ed.), *Maya Forest Biodiversity Workshop: Inventoring and Monitoring*. Washington DC, Smithsonian Institution, 50–52.

Primack, R. B., D. Bray, H. A. Galletti, and I. Ponciano (eds.) (1998). *Timber, Tourists, and Temples: Conservation and Development in the Maya Forests of Belize, Guatemala, and Mexico.* Washington DC, Island Press.

Rosenzweig, C., and M. Parry (1994). 'Potential Impact of Climate Change on World Food Supply'. *Nature* 367: 133–8.

Sader, S. A., T. Sever, J. C. Smoot, and M. Richards (1994). 'Forest Change Estimates for the Northern Petén Region of Guatemala—1986–1990'. *Human Ecology* 22: 317–33.

Skole, D., and C. Tucker (1993). 'Tropical Deforestation and Habitat Fragmentation in the Amazon: Satellite Data from 1978 to 1988'. *Science* 44: 1905–10.

Smith, N. J. H., E. A. S. Serrão, P. T. Alvim, and I. C. Falesi (1995). *Amazonia: Resiliency and Dynamism of the Land and Its People*. Tokyo, United Nations University Press.

Snook, A., and G. Zapata (1998). 'Tree Cultivation in Calakmul, Mexico: Alternatives for Reforestation'. *Agroforestry Today* 10: 15–18.

SSC-LUCC (Scientific Steering Committee-Land-Use/Cover Change) (1999). *Land-Use and Land-Cover Change, Implementation Strategy*. Stockholm and Bonn, International Geosphere-Biosphere Programme and International & Human Dimensions Programme.

Stanton, T. (2002). 'Remains of the Day'. *American Way*, 4 Jan. 2002: 29–32, 35–6.

Steadman-Edwards, P. (1997). 'Socioeconomic Root Causes of Biodiversity Loss: The Case of Calakmul, Mexico'. México DF, World Wildlife Fund. Unpublished.

Steffen, W., J. Jäger, D. Carson, and C. Bradshaw (eds.) (2001). *Challenges of a Changing Earth: Proceedings of the Global Change Open Science Conference, Amsterdam, NL, 10–13 July 2002*. Heidelberg, Springer-Verlag.

Turner, B. L., II (2002*a*). 'Contested Identities: Human–Environment Geography and Disciplinary Implications in a Restructuring Academy'. *Annals of the Association of American Geographers* 92: 52–74.

—— (2002*b*). 'Toward Integrated Land-Change Science: Advances in 1.5 Decades of Sustained International Research on Land-Use and Land-Cover Change', in W. Steffan, J. Jäger, and D. Carson (eds.), *Challenges of a Changing Earth: Proceedings of the Global Change Open Science Conference, Amsterdam, NL, 10–13 July 2002*. Heidelberg, Springer-Verlag, 21–26

Turner, B. L., II, D. Skole, S. Sanderson, F. G. L. Fresco, and R. Leemans (1995). *Land-Use and Land-Cover Change: Science/Research Plan.* Stockholm and Bonn, International Geosphere-Biosphere Programme and International & Human Dimensions Programme.

Turner, B. L., II, S. Cortina Villar, D. Foster, J. Geoghegan, E. Keys, P. Klepeis, D. Lawrence, P. Macario Mendoza, S. Manson, Y. Ogneva-Himmelberger, A. B. Plotkin, D. Pérez Salicrup, R. Roy Chowdhury, B. Savitsky, L. Schneider, B. Schmook, and C. Vance (2001). 'Deforestation in the Southern Yucatán Peninsular Region: An Integrative Approach'. *Forest Ecology and Management* 154: 343–70.

Uhl, C., C. Jordan, K. Clark, H. Clark, and R. Herrera (1982). 'Ecosystem Recovery in Amazonia Caatinga Forest after Cutting, Cutting and Burning, and Bulldozer Clearing Treatments'. *Oikos* 38: 313–20.

Utting, P. (1993). *Trees, People and Power: Social Dimensions of Deforestation and Forest Protection in Central America*. London, Earthscan.

Veldkamp, A., E. Lambin (guest eds.) (2001). 'Special Issue: Predicting Land-Use Change'. *Agriculture, Ecosystems and Environment* 85.

Vitosuek, P. M., H. A. Mooney, J. Lubchenco, and J. M. Melillo (1997). 'Human Domination of Earth's Ecosystems'. *Science* 277: 494–500.

Watson, R. T., I. R. Noble, B. Bolin, N. H. Ravindranath, D. J. Verardo, and D. J. Dokken (eds.) (2001). *Land Use, Land-Use Change and Forestry.* Cambridge, Cambridge University Press.

# PART I

## *Human–Environment History in the Southern Yucatán Peninsular Region*

The southern Yucatán peninsular region was ephemerally occupied for more than a millennium from the collapse of the Classic Lowland Maya civilization until the middle of the twentieth century. From the lay perspective, the 'old growth' forests of the region are seen as relatively free of human impact, in part justifying the large expanse of them set aside as the Calakmul Biosphere Reserve. This caricature simplifies a much more complex and significant human–environment history, perhaps dating back four millennia or more, and one of which the forest still bears the marks.

The SYPR project employs both the older and more recent human–environment histories to help understand the composition of the forests present today as well as identify various processes that assist the project in determining various impacts on forest recovering under modern disturbance in the region as well as on the cropping systems employed. In Part I the forest impacts of the ancient Maya and modern human activities are identified. The original occupants denuded much of the region's landscapes but apparently managed orchard-gardens and forest patches for economically significant species and wood fuel. The extant forests bear the imprint of this ancient use in terms of the abundance of favored Maya species and those species edaphic to the conditions of Maya disturbance, especially the limestone remains of their abundant relics. Subsequently, logging throughout the region reduced, in some cases almost eradicated, mahogany and Spanish cedar, both of which were identified as exceptionally abundant in these forests at the beginning of the last century. Thus the forest that is under protection and used today is not pristine and may not be that which would be present had its human–environment history been different.

*The Editors*

# 2

# The Long View: Human–Environment Relationships in the Region, 1000 BC–AD 1900

*David R. Foster and B. L. Turner II*

## 1. *The Lay of the Land*

The Yucatán Peninsula is a thick, low-lying, limestone shelf that rose fully above the Caribbean Sea and Gulf of Mexico during the Pleistocene (Folan 1983; Wilson 1980).[1] Given its calcite and dolomite base, the entire peninsula northward from the Libertad Arch, which extends east–west across the central Petén, is a karstic landscape of coastal plains and interior uplands or hills dominated by solution features and subsurface drainage (Finch 1965; Jennings 1985; Weidie and Ward 1976; Wilhemy 1981).[2] Permanent rivers and streams appear only in the southern, high rainfall, portion of the peninsula, largely along the lower-lying coastal plains and adjacent edges of the interior uplands. For the most part, the southern Yucatán peninsular region (Fig. 1.1) constitutes an undulating upland (about 80% of the landscape) composed of ridges and, to the south, cone karst (Figs. 2.1 and 2.2: upland-*bajo* distribution). Interspersed among the uplands are large, surface solution sinks or poljes (Weidie and Ward 1976), locally known as

[1] Virtually all the study region is composed of Eocene (50 million years ago) limestone and dolomites. Further to the south, in central Petén and in the Maya Mountains of Belize, respectively, Cretaceous and Paleozoic rock occurs, respectively, in the Libertad Arch and Maya Mountains (Lopez Ramos 1973).

[2] Karst dominates most of the extreme southern part of the peninsula as well, although igneous and metamorphic rocks occur in the Maya Mountains, and various quartz sand and gravel outwashes from these mountains exist.

**Fig. 2.1.** The rolling uplands.

*bajos* (sometimes, *akalches*). These features infill with sediments, predominately montmorillonite clays, which impede subsurface drainage and retain surface water built up during the rainy season, creating seasonal wetlands. Otherwise, water percolates through the upland karst to subterranean aquifers, which generally lie deep below the surface. In the uplands proper, these aquifers may be more than 200m below the surface (Turner, 1983).

The region occupies a transitional position between tropical monsoon (Am, Köppen classification) and tropical wet-dry (Aw) climates to the south and north respectively. The critical distinction between the two is the length and severity of the dry season, which increases northward. Overall, the climate is characterized by hot, humid summers and warm, dry winters, in which *nortes* (cold northern airstreams) penetrate the region for brief periods. Average annual precipitation ranges from about 900mm to in excess of 1,400mm, increasing to the south, the large majority of which falls during the wet season (mid-May through October) (Garcia 1970). Water deficits exist throughout the region during the mid-to-late dry season, stressing vegetation, wildlife, agriculture, livestock, and human settlement.

During periods of peak precipitation and especially during hurricanes, some *bajos* overflow and link with one another by way of intermittent channels, permitting water to flow towards the coast. *Bajos* may hold water well into the dry season, although almost all fully desiccate every year. However,

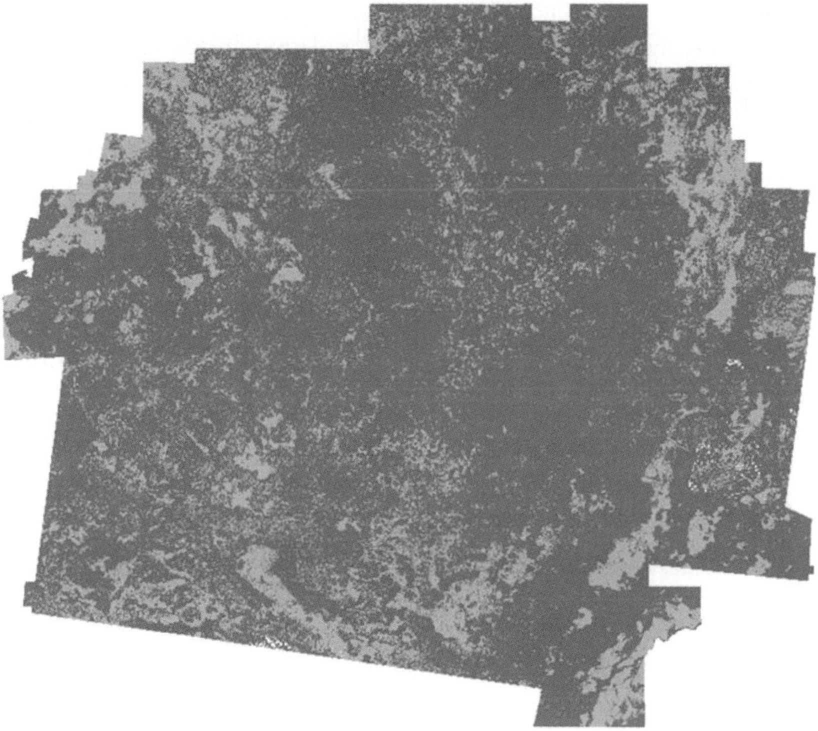

**Fig. 2.2.** The distribution of uplands and *bajos*. Created from Landsat TM imagery and various soil maps, the image covers only those administrative units that fall fully within the borders of the region indicated in Fig. 1.1.

some of the smaller and deeper solution features, locally known as *lagunas*, retain water in their deeper parts throughout the year.[3] The largest such feature, the *laguna* Lago Silvituc, is located on the western edge of the study region (Fig. 1.1). In addition, the ancient Maya constructed large numbers of reservoirs (or *aguadas*) throughout the region, and while frequently infilled with vegetation, these depressions may also retain surface water well into the dry season (Scarborough and Gallopin 1991).

The weathering of the limestones gives rise to rendzinas or mollisols (specifically, troporendolls) throughout most of the well-drained uplands

[3] Small surface cavities that can hold water throughout the dry season are common throughout the region. Most of these are Maya-made reservoirs from a millennium past, usually called *aguadas*. Occasionally, small *lagunas* may be called *aguadas* as well.

(Turner 1983; Wilson 1980). These soils are suitable for cultivation but are shallow, ranging from 5 to 50cm in depth, and rest immediately upon limestone bedrock that is either hardened or partially eroded, in which case it is locally known as *sascab*. The rendzinas are alkaline (pH 7.4–8.0) silt and clay loams of colluvial origins (Turner 1983). In scattered locations, limestone punctures the surface, interspersed by shallow litosols. Soils tend to be shallow on the crowns of ridges and hills, and on steeper slopes; they thicken at the base of slopes or where barriers (e.g. ancient Maya terraces) retard their movement downslope (Beach 1998; Dunning and Beach 2000; Turner 1983). In contrast, *bajos* capture much of the sediment coming off slopes, and over time clay-rich vertisols form, typically underlain by thick (> 1m) gypsum-speckled clays largely devoid of organic matter and nutrients. The vertisols swell during the rainy season and shrink during the dry season, giving rise to deep, slicken-sided cavities and to gilgai activity (Bloom *et al.* 1983) that creates small, irregular humping of the soils on the surface.

The climate, including hurricanes and fire, and soils support two basic forest types highly modified by past human activity: upland (locally, *bosque mediano* or *selva mediana)* and *bajo* (Ch. 4; Fig. 2.3) (Lundell 1934; Miranda 1959; Miranda and Hernández 1963; Pennington and Sarukhán 1968; Rzedowski 1981; Shepherd 1975). These forests are distinguished less by their species composition than by the relative abundance of the dominant tree species and the lower crown height and biomass and higher density of *bajo* forests (Esquivel and Pérez-Salicrup 2000; Pérez-Salicrup 2001*a*; Pérez-Salicrup 2001*b*; Pérez-Salicrup and Foster 2000). Both forest types contain deciduous species that lose foliage during the dry season (Ch. 5). Seasonal leaf fall is especially acute in the drier northwestern portion of the study region where the upland forests are labeled 'subcaducifolia' (Martinez Salas and Galindo-Lea 2000).

Where inundation is most profound (deep and prolonged), the *bajo* forest gives way to sedges, reeds, and grasses (*tular*). *Tulares* typically border or cover extensive portions of *lagunas* during the dry season. In contrast, the 'Savanna Central', located in the center of the region, is a large grassland of some 7.2km$^2$ that retains standing water and significant soil moisture throughout much of the year.[4]

[4] Further south in northeastern El Petén, Guatemala, and in Belize, large stands of pine, coinciding with outwashes from the Maya Mountains and with savannas along the Libertad Arch, and palms can be found. Such vegetation complexes do not exist in the study region.

**Fig. 2.3.** The forest landscape. Looking southward from the Maya site of Calakmul.

## 2. *The Maya Entry and Impact*

There is little evidence to suggest that the physical conditions of the region were significantly different from those described above when the ancient Maya peoples entered the region previous to 3000 BP (but see Dahlin 1983). Geochemical evidence from lake cores to the northeast of the region suggests that the arrival of the Maya to this tropical lowland realm coincided with the onset of an extended dry epoch (Curtis, Hodell and Brenner 1996; Curtis *et al.* 1998; Hodell, Curtis, and Brenner 1995; Hodell, Brenner, and Curtis 2000; Rosenmeier *et al.* 2002). The paleo-record, however, does not suggest a major alteration in the predominant vegetation cover of the region at this time; the area was forested and probably changing in concert with the climate.[5] Indeed, Leyden (1984) argued that the Holocene mesic forest of Petén

[5] Cowgill and Hutchinson's (1963) geochemical analysis of several pits dug into the Bajo de Santa Fe, adjacent to Tikal, Petén, led to the emphatically stated conclusion that climatic oscillations notwithstanding, this feature had been a *bajo* or wetland (not a lake) throughout ancient Maya occupation. They also concluded that the *bajo* had incurred significant sedimentation from Maya use of surrounding terrain. Likewise, work at Cobweb Swamp in northern Belize (Alcala-Herrera *et al.* 1993) indicates that throughout large-scale Maya occupation (*c.* 3400 BP to 500 BP) that swamp was likely a freshwater lagoon, filled in by peat after 500 BP. This finding raises issues about significant climatic desiccation during late Maya occupation. Indeed, the entire problem of climatic desiccation and the end of the Classic Period Maya in the area requires substantial attention.

replaced a more xeric vegetation during Pleistocene, and that this forest remained to the present, save for the Maya impress on it—an impress that would overwhelm much of the lowlands environment (e.g. Islebe *et al.* 1996).

For more than two millennia, the Maya grew in numbers, power, and impact on the land. Ultimately they reached their material apex during the Late and Terminal Classic Periods (AD 550–800 and 800–1000, respectively in this region) when they dominated the landscape with large city-states and settlements of various sizes and farmsteads scattered across the 'rural' hinterlands (Adams 1981; Thomas 1981). While the precise population figures will always be debated, the overall population curve is well recognized. Starting from a small number, the Maya rose in a great wave, ascending to region-wide densities approaching 150 people/km$^2$ by AD 800 (Turner 1986). To be sure, subregional variations in population existed, and these fluctuated through time. Nevertheless, the sheer demand placed by this immense population on sources of food, fiber, and wood fuel (e.g. for cooking, heat, and construction) was substantial and required significant landscape transformation and manipulation to secure production over the long term. This environmental impact is registered in the paleo-environmental record for the region and throughout the Maya lowlands; a plethora of indicators demonstrate an increasingly human-disturbed environment, including major increases in grasses and non-arboreal weedy species and significant declines in arboreal pollen that dominated the pre-Maya pollen record (Binford *et al.* 1987; Rice 1993, 1996; Turner, Klepeis, and Schneider forthcoming).

Initially, the Maya utilized the uplands most intensively. This rapid and severe deforestation contributed to the massive soil loss registered in wetland sediments (Beach 1998). Through time, however, the Maya adjusted their cultivation practices, learning to terrace (Turner 1974, 1983) and in doing so, reducing significantly the rates of soil loss (Beach 1998). By the Middle to Late Classic, if not long before, *in situ* tree cover was replaced by an open landscape of semi-permanent cultivated lands, successional growth, managed forests, and large expanses of orchard-gardens (house gardens, settlement gardens) (Dunning 1992; Dunning and Beach 2000; Fedick 1996; Gómez-Pompa 1987; Gómez-Pompa, Salvador Flores, and Sosa 1987; McKillop 1994; Turner and Miksicek 1984).

Today the region is littered with an astonishing density of relics indicative of the intensity of Maya upland occupation and land use: large numbers of settlements (Adams 1981; Thomas 1981), rock-walled terraces, stone-alignments of various kinds, house mounds, and impressive temple sites (Turner 1974, 1983). Indeed, it is difficult to traverse the landscape, especially from the central portion of region eastward, without encountering some Maya relic every few hundred meters or less.

Eventually, even some of the wetlands on the eastern side of the region succumbed to Maya pressures. The large *bajos* Morocoy and Nicolas Bravo displayed evidence of large and small canals and earthen platforms, indicative of wetland cultivation and perhaps dating to the Late Classic Period.[6] Some of these platforms actually supported small-stone structures (Gliessman *et al.* 1985). Unfortunately, modern agricultural development projects obliterated the majority of these features for mechanized rice cultivation. With this exception, however, no other evidence of large-scale wetland uses by the Maya has been uncovered in the region (Pope and Dahlin 1989, 1993; Turner 1993).

By the apex of the Classic Period there is every reason to believe that the uplands had been significantly denuded, with forests reduced to managed stands (of undetermined size) scattered among an otherwise settled and cultivated landscape, but surely including various stages of successional growth.[7] Individual households, regardless of their location, were surrounded by small orchard-gardens, but the large size of the population necessitated large areas of open land devoted to the production of maize—the Maya staple. Compared to these conditions, there is little evidence to indicate how or if the seasonal wetlands, other than Morocoy and Nicolas Bravo, were systematically used in any way that required significant forest loss. Indeed, given the species composition of the upland and *bajo* forests, Pérez-Salicrup (2001*b*; Ch. 4) hypothesizes that through the intense Maya period that *bajos* may have served as refugia for plant species characteristic of both the uplands and lowlands. Once the Maya collapse and land abandonment took place, the extant species in the *bajo* may have served as sources to restock the uplands.

[6] These features were photographed and reported by Siemens (1983), Harrison (1977, 1990), and Turner (1974) in the early 1970s, and were briefly examined on the ground by Gliessman and his colleagues (1985) in the late 1970s as the *bajo* Nicolas Bravo was being converted to a large-scale rice project. These projects reduced the vertical dimensions of the ancient planting surfaces (wetland fields) that were visible on the ground during the initial construction of the rice projects. The adjacent Maya ditches, long filled in, are still visible from the air during the dry season, apparently owing to soil moisture differences between the ditches and wetland fields.

[7] As discussed in Chs. 4, 6, and elsewhere, bracken fern appears to be a serious invader of some kinds of cropped lands, creating serious problems for sustain cultivation (Vitkay and Pérez-Salicrup 2000). That this invasive species created problems for the ancient Maya is not known.

The collapse of the Maya population was as swift as it was severe.[8] The paleoecological and archaeological records indicate that between AD 800 and 1000, the region was simply abandoned from active use and never again occupied to any significant extent until the second half of the twentieth century (Turner 1986; Whitmore and Turner 1992, 2001).[9] Several small Maya settlements faded in and out of the scene, especially during the Caste Wars of the mid-nineteenth century when many Maya attempted to escape Mexican rule in northern Yucatán by fleeing to the region (Jones 1983, 1989; Whitmore and Turner 1992). This occupation was so sparse that its landscape impacts were ephemeral at most (Klepeis 2000). Indeed, when Cortés attempted to cross from Tabasco into El Petén in 1524–5, just skirting the southern part of our study region, he found little but forest: '. . . there was no open country but jungles instead; the roads were hidden trails, many of them covered by jungle growth. There were no corn milpas, and there was very little food of any kind' (Cerwin 1963: 52–3).

## 3. *The Maya Legacy on the Forest*

Cortés had encountered a vast area with a 500–600-year-old forest cover, which would subsequently remain largely undisturbed for another 400–500 years. Yet, what was encountered by the Spaniards, as well as subsequent botanical scholars in the early part of the twentieth century, was a Mayanized forest, at least in terms of species abundance and distribution and registered in several ways (Pérez-Salicrup 2001*b*). The first involves the abundance of ramón (*Brosimum alicastrum* L.), the Maya breadfruit, which exists throughout the forest and extends well into the central Petén. Ramón's abundance and the fact that its fruit is edible once led to specula-

[8] The cause(s) of the collapse and depopulation of the ancient Maya remain strongly debated and are not reiterated here. It is noteworthy, however, that the case for severe climatic desiccation has been made recently (Hodell, Brenner, and Curtis 2000; Rosenmeier *et al.* 2002), contrary to a much older case for increased mesic conditions (e.g. Huntington 1917). See n. 5 above.

[9] Recent evidence indicates that an early Postclassic population returned to less disturbed sections of the Río de la Pasión in the southwestern Petén (Johnston, Breckenridge, and Hansen 2001). This find raises the question posed by Culbert (1988) concerning the need to examine more thoroughly the 'rural' settlements of the ancient Maya in regard to occupation dynamics. Much of the southern Yucatán peninsular region has been examined in this way in regard to terraces and their adjacent housemounds. This work revealed no significant Postclassic occupation (Turner 1974, 1983).

tion that current stands of the species are vestiges of past Maya ramón orchards. Lambert and Arnason (1981), however, demonstrate that ramón grows well on highly disturbed soils, typically those in which limestone is exposed on the surface, and its abundance on natural and Maya-made exposures is similar. The correlation between ramón and ruins, therefore, may well reflect the species edaphic affinities for decayed limestone structures and plazas around ancient settlement (Fig. 2.4).

Secondly, Lundell (1933, 1938), Miranda (1959), and other botanists described the region's forests as especially abundant in economically useful species, including, among many species, mamey zapote, allspice, guava, and chicle. Lundell (1933, 1938) speculated that this abundance was also the vestige of ancient Maya practices, either forest management (presumably small patches) or tree cropping. The archaeological record indicates significant Maya consumption of fruits, nuts, and other items taken from

**Fig. 2.4.** Ramón (*Brosimum alicastrum* L.) growing from ruins at Calakmul.
*Source*: Whitmore and Turner 2002.

tree species, suggesting the role of orchard-gardens in Maya production (Turner and Miksicek 1984). Gomez-Pompa and colleagues (1987) came to a similar conclusion based on the abundance of small stonewall enclosures replete in Maya forests and their possible links to the Maya *pet kot*, or protected forest or tree areas.

## 4. *Implications for the SYPR Project*

The southern Yucatán peninsular region is a karstic upland dominated by a seasonal tropical forest. Significantly denuded by the ancient Maya, the forest returned with the abandonment of the region about AD 900–1000 and was disturbed minimally from its 'regrowth' until the middle of the twentieth century.

- The forest encountered after the near millennium of sparse occupation and use, however, registered a structure created in part by the ancient Maya, especially in terms of species abundance.

Understanding modern impacts on the region's forests must begin with this recognition.

Beyond this recognition, perhaps the most critical environmental characteristics for understanding land change in regard to cultivation and forest regeneration are those of the mollisols and the rainfall gradient.

- The troporendolls that dominate the upland are quality soils for the tropics in general, but are shallow. Minimal work has been conducted in regard to their capacity to sustain various kinds of cropping strategies, and virtually none of this work had been undertaken in the region. This paucity of work begs various questions raised subsequently in this book about nutrient dynamics (Ch. 5), cropping practices (Chs. 9 and 10), and invasive species (Ch. 6).
- The north–south rainfall gradient, giving rise to forests occupying Am and Aw climates to the north and south respectively, raise a series of questions about nutrient dynamics and the ability of forests to recover from cutting and burning.

These and other issues are returned to later in this work.

## REFERENCES


Adams, R. E. W. (1981). 'Settlement Patterns of the Central Yucatan and Southern Campeche Regions', in Ashmore, W. (ed.), *Lowland Maya Settlement Patterns*. Albuquerque, University of New Mexico Press, 211–57.

Alcala-Herrera, J. A., J. S. Jacob, M. L. Mahain Castillo, and R. W. Neck (1994). 'Holocene Palaeosalinity in a Maya Wetland, Belize, Inferred from the Microfaunal Assemblage'. *Quarternary Research* 41: 121–30.

Beach, T. (1998). 'Soil Catenas, Tropical Deforestation, and Ancient and Contemporary Soil Erosion in the Petén, Guatemala'. *Physical Geography* 19: 378–405.

Binford, M. W., M. Brenner, T. J. Whitmore, A. Higuera-Gundy, E. S. Deevey, and B. W. Leyden (1987). 'Ecosystems, Paleoecology and Human Disturbance in Subtropical and Tropical America'. *Quaternary Science Reviews* 6: 115–28.

Bloom, P. R., M. Pohl, C. Buttleman, F. Wiseman, A. Covich, C. Mikscek, J. Ball, and J. Stein (1983). 'Prehistoric Maya Wetland Agriculture and the Alluvial Soils near San Antonio Rio Hondo, Belize'. *Nature* 301: 417–19.

Cerwin, H. (1963). *Bernal Diaz—Historian of the Conquest*. Norman, Okla., University of Oklahoma Press.

Cowgill, U. M., and G. E. Hutchinson (1963). 'El Bajo de Santa Fé'. *Transactions of the American Philosophical Society* NS 53: 3–50.

Culbert, T. P. (1988). 'The Collapse of Classic Maya Civilization', in T. P. Culbert and D. R. Rice (eds.), *Precolumbian Population History in the Maya Lowlands*. Albuquerque, NM, University of New Mexico Press, 103–22.

Curtis, J. H., D. A. Hodell, and M. Brenner (1996). 'Climate Variability on the Yucatan Peninsula (Mexico) during the Past 3500 Years, and Implications for Maya Cultural Evolution'. *Quarternary Research* 46: 37–47.

Curtis, J. H., M. Brenner, D. A. Hodell, R. A. Balsar, G. A. Islebe, and H. Hooghiemstra (1998). 'A Multi-Proxy Study of Holocene Environmental Change in the Maya Lowlands of Petén, Guatemala'. *Journal of Paleolimnology* 19: 139–59.

Dahlin, B. H. (1983). 'Climate and Prehistory in the Yucatán Peninsula'. *Climate Change* 5: 245–63.

Dunning, N. P. (1992). *Lords of the Hills: Ancient Maya Settlement in the Puuc Region, Yucatán, Mexico*. Madison, Wis., Prehistory Press.

Dunning, N. P., and T. Beach (2000). 'Stability and Instability in Prehispanic Maya Landscapes', in D. L. Lentz (ed.), *Imperfect Balance: Landscape Transformations in the Precolumbian Americas*. New York, Columbia University Press, 186–202.

Esquivel, R., and D. R. Pérez-Salicrup (2000). 'Presence of *Hypsiphylla grandella* in Plantations of *Cedrella odorata* and *Swietenia macrophylla* in Southeastern Campeche, Mexico', in M. J. Wade (ed.), *Evolution 2000*. Bloomington. CD on Indiana University Conferences.

Fedick, S. L. (ed.) (1996). *The Managed Mosaic: Ancient Maya Agriculture and Resource Use*. Salt Lake City, University of Utah Press.

Finch, W., Jr. (1965). *The Karst of Yucatan*. Urbana-Champaign, University of Illinois.

Folan, W. J. (1983). 'Physical Geography of the Yucatan Peninsula', in W. J. Folan, E. R. Kintz, and L. A. Fletcher (eds.), *Coba: A Classic Maya Metropolis*. New York, Academic Press, 21–48.

Garcia, E. (1970). *Carta de Climas*. México DF, Instituto de Geografía, Universidad Nacional Autonóma de México.

Gliessman, S. R., B. L. Turner II, F. J. Rosado May, and M. F. Amador (1985). 'Ancient Raised-Field Agriculture in the Maya Lowlands of Southeastern Mexico', in J. P. Darch (ed.), *Drained Field Agriculture in Central and South America*. Oxford, British Archaeological Reports International Series 189: 11–30.

Gómez-Pompa, A. (1987). 'On Maya Silviculture'. *Mexican Studies/Estudios Mexicanos* 3: 1–17.

Gómez-Pompa, A., J. Salvador Flores, and V. Sosa (1987). 'The "Pet Kot": A Man-Made Tropical Forest of the Maya'. *Interciencia* 12: 10–15.

Harrison, P. D. (1977). 'The Rise of the *Bajos* and the Fall of the Maya', in N. Hammond (ed.), *Social Processes in Maya Prehistory: Studies in the Memory of Sir Eric Thompson*. London, Academic Press, 469–508.

——(1990). 'The Revolution in Ancient Maya Subsistence', in F. S. Clancy and P. D. Harrison (eds.), *Vision and Revision in Maya Studies*. Albuquerque, University of New Mexico Press, 99–113.

Hodell, D. A., M. Brenner, and J. H. Curtis (2000). 'Climate Change in the Northern American Tropics and Subtropics since the Last Ice Age: Implications for Environment and Culture', in D. Lentz (ed.), *Imperfect Balance: Landscape Transformations in the Precolumbian Americas*. New York, Columbia University Press, 13–38.

Hodell, D. A., J. H. Curtis, and M. Brenner (1995). 'Possible Role of Climate in the Collapse of Classic Maya Civilization'. *Nature* 375: 391–4.

Huntington, E. (1917). 'Maya Civilization and Climate Change', in *Proceedings of the 19th Congress of Americanists*. Washington DC, 150–64.

Islebe, G. A., H. Hooghiemstra, M. Brenner, J. H. Curtis, and D. A. Hodell (1996). 'A Holocene Vegetation History from Lowland Guatemala'. *The Holocene* 6: 265–71.

Jennings, J. N. (1985). *Karst Geomorphology*. Oxford, Basil Blackwell.

Johnston, K. J., A. J. Breckenridge, and B. C. Hansen (2001). 'Paleoecological Evidence of an Early Postclassic Occupation in the Southwestern Maya Lowlands: Laguna Las Pozas, Guatemala'. *Latin American Antiquity* 12: 149–66.

Jones, G. D. (1983). 'The Last Maya Frontier of Colonial Yucatan', in M. J. MacLeod and R. Wasserstrom (eds.), *Southeastern Mesoamerica. Essays on the History of Ethnic Relations*. Lincoln, University of Nebraska Press, 64–91.

——(1989). *Maya Resistance to Spanish Rule: Time and History on a Colonial Frontier*. Albuquerque, University of New Mexico Press.

Klepeis, P. (2000). 'Deforesting the Once Deforested: Land Transformation in Southeastern Mexico', Ph. D. Dissertation, Clark University, Worcester, MA.

Lambert, J. D. H., and T. Arnason (1981). '*Ramón* and Maya Ruins: An Ecological, Not Economic, Relation'. *Science* 216: 298–9.

Leyden, B. W. (1984). 'Guatemalan Forest Synthesis after Pleistocene Aridity'. *Proceedings, National Academy of Sciences, USA* 81: 4856–5859.

Lopez Ramos, E. (1973). 'Estudio Geológico de la Peninsula de Yucatán'. *Boletín de la Asociación Mexicana de Geologos Petroleros* 25: 23–76.

Lundell, C. L. (1933). 'The Agriculture of the Maya'. *Southwest Review* 19: 65–77.

——(1934). 'Preliminary Sketch of the Phytogeography of the Yucatan Peninsula'. *Contribution to American Archaeology* 12: 436. Washington DC, Carnegie Institute of Washington.

——(1938). 'Plants Probably Utilized by the Old Empire Maya of Petén and Adjacent Lowlands'. *Papers of the Michigan Academy of Science, Arts and Letters* 24: 37–56.

McKillop, H. (1994). 'Ancient Maya Tree Cropping: A Viable Subsistence Adaptation for the Island Maya'. *Ancient Mesoamerica* 5: 129–40.

Martinez Salas, E., and C. Galindo-Lea (2000). *La Vegetación de Calakmul, Campeche, México: Clasificasión, Descripción y Distribución*. Herbario Nacional de México (MEXU), Instituto de Biología-Universidad Nacional Autónoma de México and Center for Conservation Biology, Stanford University.

Miranda, F. (1959). 'Estudios Acerca de la Vegetación', in E. Beltran (ed.), *Los Recursos Naturales del Sureste y su Aprovechamiento*. México DF, Instituto Mexicano Recursos Naturales Renovables, 215–71.

Miranda, F., and E. Hernández X. (1963). 'Los tipos de vegetación de México y su clasificación'. *Boletín de la Sociedad Botánica de México* 28: 29–179.

Pennington, T. D., and J. Sarukhán (1968). *Manual para la Indentificacíon de Campo se los Principales Árboles Tropicales de México*. México DF, Instituto Nacional de Investigaciones Forestales, Secretaria de Agricultura y Ganaderia.

Pérez-Salicrup, D. R. (2001*a*). 'Forest Types in the Southern Yucatan Peninsula: Effects of Long- and Short-Term Land Use on Regional Diversity', in K. N. Ganeshaiah, R. Shaanker, and K. S. Bawa (eds.), *Tropical Ecosystems: Structure, Diversity and Human Welfare. Proceedings of the International Conference on Tropical Ecosystems*. New Delhi, Oxford-IBH, 142.

——(2001*b*). 'Forest Types and Forest and Recovery after Long-and Short-Term Land Use in the Southern Yucatan, Mexico', in A. Gómez-Pompa (ed.), *21st Symposium in Plant Biology: Lowland Maya Area: Three Millennia at the Human–Wildlife Interface*. Riverside, Calif., University of California, 26.

Pérez-Salicrup, D. R., and D. F. Foster (2000). 'Forest Types and Forest Succession in Southern Yucatan, Mexico', in M. J. Wade (ed.), *Evolution 2000*. Bloomington, CD on Indiana University Conferences.

Pope, K. O., and B. H. Dahlin (1989). 'Ancient Maya Wetland Agriculture: New Insights from Ecological and Remote Sensing Research'. *Journal of Field Archaeology* 16: 87–106.

——(1993). 'Radar Detection and Ecology of Ancient Maya Canal Systems—Reply to Adams *et al.*'. *Journal of Field Archaeology* 20: 379–83.

Rice, D. S. (1993). 'Eight-Century Physical Geography, Environment and Natural Resources in the Maya Lowlands', in J. Sabloff and J. Henderson (eds.), *Lowland Maya Civilization in the Eight Century A.D.* Washington DC, Dumbarton Oaks, 11–63.

Rice, D. (1996). 'Paleolimnological Analysis in the Central Petén, Guatemala, in S. Fedick (ed.), *The Managed Mosaic: Ancient Maya Agriculture and Resource Use*. Salt Lake City, University of Utah Press, 193–235.

Rosenmeier, M. G., D. A. Hodell, M. Brenner, J. H. Curtis, and T. P. Guilderson (2002). 'A 4000-year Lacustrine Record of Environmental Change'. *Quaternary Research* 57: 183–90.

Rzedowski, J. (1981). *Vegetación de México*. México DF, Editorial Limusa.

Scarborough, V. L., and G. G. Gallopin (1991). 'A Water Storage Adaptation in the Maya Lowlands'. *Science* 251: 658–62.

Shepherd, J. D. (1975). 'Phytosociology of a seasonal forest in southeastern Mexico'. MA Thesis. University of Wisconsin-Madison.

Siemens, A. H. (1983). 'Wetland Agriculture in Pre-Hispanic America'. *Geographical Review* 73: 166–81.

Thomas, P. M., Jr (1981). *Prehistoric Maya Settlement Patterns at Becan, Campeche, Mexico.* New Orleans, Tulane University.

Turner, B. L., II (1974). 'Prehistoric Intensive Agriculture in the Mayan Lowlands'. *Science* 185: 118–24.

——(1983). *Once Beneath the Forest: Prehistoric Terracing in the Río Bec Region of the Maya Lowlands.* Boulder, Westview Press.

——(1986). *Population Reconstruction of the Central Maya Lowlands: 1000 B.C. to Present*. Technical Paper 2. Millennial Long Waves of Human Occupance Project, Clark University.

——(1993). 'Rethinking the "New Orthodoxy": Interpreting Ancient Maya Agriculture and Environment', in K. Mathewson (ed.), *Culture, Form, and Place: Essays in Cultural and Historical Geography, Geoscience and Man 32*. Baton Rouge, Geoscience Publications, Louisiana State University, 57–88.

Turner, B. L., II, and C. H. Miksicek (1984). 'Economic Plant Species Associated with Prehistoric Agriculture in the Maya Lowlands'. *Economic Botany* 38: 179–93.

Turner, B. L., II, P. Klepeis, and L. Schneider (forthcoming). 'Three Millennia in the Southern Yucatán Peninsula: Implications for Occupancy, Use and "Carry Capacity" ', in A. Gómez Pompa, M. Allen, S. Fedick, and J. Jimenez Osorino (eds.), *Lowland Maya Area: Three Millennia at the Human Wildlife Interface*. New York, Haworth Press.

Vitkay, K., and D. R. Pérez-Salicrup (2000). 'Effect of *Sabal mexicana* Palms on *Pteridium aquilinum* Invaded Fields in the Southern Yucatan', in M. J. Wade (ed.), *Evolution 2000*. Bloomington, CD on Indiana University Conferences.

Weidie, A. E., and W. C. Ward (eds.) (1976). *Carbonate Rocks and Hydrogeology of the Yucatán Peninsula, Mexico*. New Orleans, New Orleans Geological Society.

Whitmore, T. M., and B. L. Turner II (1992). 'Landscapes of Cultivation in Mesoamerica on the Eve of the Conquest'. *Annals of the Association of American Geographers* 82: 402–25.

——(2001). *Cultivated Landscapes of Middle America on the Eve of Conquest*. Oxford, Oxford University Press.

Wilhelmy, H. (1981). *Welt und Umwelt der Maya: Aufstieg und Untergang einer Hochkultur*. Munich, R. Piper.

Wilson, E. M. (1980). 'Physical Geography of the Yucatan Peninsula', in E. H. Moseley and E. D. Terry (eds.), *Yucatan: A World Apart*. University, Ala., University of Alabama Press, 5–40.

# 3

# Forest Extraction to Theme Parks: The Modern History of Land Change

*Peter Klepeis*

Modern-day deforestation in the southern Yucatán peninsular region began in earnest in the late 1960s. The composition of the region's forest and options for land uses, however, were partly shaped by eighty years of activity leading up to the 1960s, just as it was by the ancient Maya over a millennium ago (Ch. 2). Most of the modern impacts began in the twentieth century and are traced here through three major episodes of use and occupation of the region: *forest extraction*, 1880–1983; *big projects and forest clearing*, 1975–82; and *land-use diversification, conservation, and tourism*, 1983 to the present. Each episode corresponds to different visions of how the region should be used and to different human–environment conditions shaping the kind, location, and magnitude of land change. Understanding these changing conditions underpins all other assessments of the SYPR project.

The episode of *forest extraction* spans the bulk of the modern history of the region. It began in the late nineteenth century and ended with the demise of parastatal logging companies in the 1970s and early 1980s, due primarily to the depletion of reserves of mahogany and Spanish cedar throughout the region. Before this episode fully expired, a new one, that of *big projects and forest clearing* began, marked by large-scale rice and cattle schemes undertaken in the mid to late 1970s and early 1980s. This episode accelerated the road construction that began in the latter part of the 1960s, and it witnessed expanded settlement linked to colonization programs. The Mexican debt crisis of 1982 brought this episode to an abrupt halt, triggering the search for a new alternative to developing the frontier. This search, made in the context of neoliberal economic reforms, led to the establishment of the Calakmul Biosphere Reserve in 1989 and other, more recent initiatives, defining the most recent episode of *land-use diversification, conservation, and tourism*.

## 1. *Forest Extraction (1880–1983)*

From the collapse of the Classic Maya civilization to the twentieth century, the occupation of the region was sparse (Turner 1990), the forest serving as a refuge during the colonial period for those Maya fleeing Spanish domination along the coasts and in the north, especially during the Caste War of the middle nineteenth century, when the northern Maya revolted against Mexico (Jones 1989). Prior to modern forestry, a few trading routes cut across the region linking northern Yucatán to the Bahía de Chetumal and El Petén of Guatemala (Fremont 1861), but the impact of travelers as well as refugees on the forest was negligible, at least at the regional scale.

Modern attention to the forests of the region began in earnest at the end of the nineteenth century when national and international entrepreneurs opened zones abutting the region for specialized logging, but intensive forest extraction in the region proper awaited the advent of chewing gum and the demand for chicle, a species and product in which the region's forests were rich. Intensive logging in the region awaited the mid-twentieth century and triggered the first pronounced modern occupation of this frontier region. While minor logging continues in the present, the zenith of timber extraction took place between the mid-1940s and 1960s (Table 3.1).

**Table 3.1.** Forestry periods for the greater southern Yucatán region

| Period | Approximate start year | Approximate end year |
|---|---|---|
| Logwood | mid-1500s | early 1900s |
| Chicle | late 1880s | present |
| Boom period | 1903 | 1945 |
| Selective logging | 1600s | Present |
| First period | 1890s | 1914 |
| Second 'boom' period | 1940s | 1955 |
| Gradual decline in logging | 1955 | 1980s |
| Social forestry | 1983 | present |

## 1.1. *Logwood*

The opening of the study region to capitalist forest extraction in the late 1800s was due primarily to four factors. (1) The coastal and riverine wetlands abutting the region to the southeast and southwest contained logwood—*palo de tinte* or *palo negro* (*Haematoxylon campechianum* L.)—used for the extraction of a red dye. (2) The upland forests were rich in mahogany (*Swietenia macrophylla* King) and other hardwoods, and the stocks of these species were depleted elsewhere in Mexico by the late nineteenth century (de los Santos Valadez 1987; de Vos 1988; Ponce Jiménez 1990). (3) Policies instituted by President Porfirio Díaz (1876–1911) opened Mexico's frontier locations to development, encouraging foreign capital investments (Chenaut 1989; Galletti 1994; Konrad 1991; Macías Zapata 1996). And, (4) the Cruzob (meaning rebel) or Santa Cruz Maya rebellion was repressed, and the violence linked to the Caste War finally subsided in 1901, leading to a period of relative political stability throughout the Yucatán Peninsula (Macías Zapata 1996).

Trade in dye extracted from logwood was lucrative in the US and European markets, especially between 1870 and 1895, when a synthetic substitute was discovered (Villalobos Gonzalez 1993). Its extraction declined precipitously after the turn of the century (Konrad 1991) and was replaced by a shift to the logging of mahogany and cedar (*Cedrela odorata* L.), which was intensive in the northwest and southeast corners of the study region—in large part because the construction of small-gauge rails allowed transportation of logs from these areas—and within 60km of Río Hondo (Snook 1998) (Fig. 3.1). Absent rail and waterways, efficient transportation did not exist, restricting the distance that logs could be hauled.[1]

## 1.2. *Chicle*

Before logging could be profitable in the interior of the southern Yucatán peninsular region, forest industries refocused on a new commodity, the resin from the *sapodilla* or *chicozapote* tree (*Manilkara zapota* L. Van Royen, formerly, *Achras sapota* L. (Lundell 1933; Villaseñor 1958)). Not only did a

[1] Mules and oxen were used to drag logs for short distances over trails until the early twentieth century, when narrow-gauge rail and crawling tractors enabled logs to be moved upwards of 30–60km (Snook 1998). Access to the interior of the Yucatán Peninsula awaited the mid-twentieth century when semi-permanent roads were built and rubber-tired vehicles introduced.

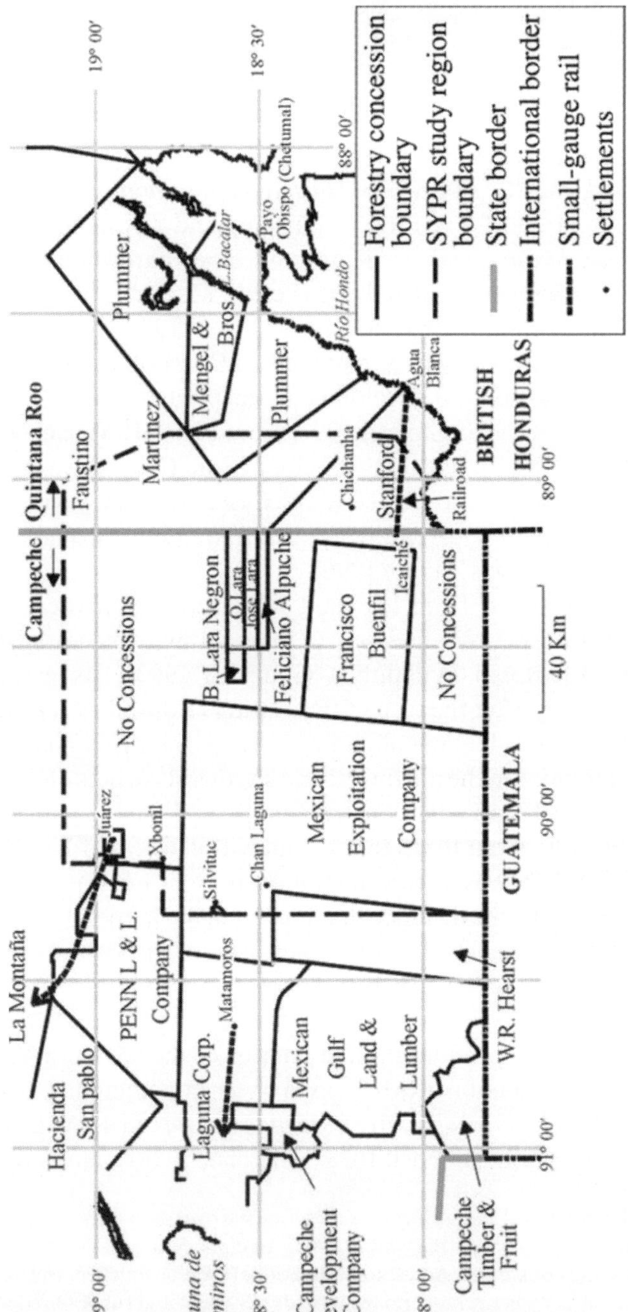


**Fig. 3.1.** Forest concessions for southern half of the Yucatán peninsula, *c.*1920.
*Source*: Graham, Gomez, and Gutierrez, 1920.

strong international market exist for chicle resin, the collapse of the henequen (*Agave sislana* Perrine) economy in the north and the need to pacify rebel Maya in the southeast created an economic and political incentive for Mexico's promotion of the chicle industry. The federal government thus sought to assist Mexican business and co-opt the rebels into the forestry industry. This last objective would dampen the chances of potential Maya linkages with the British by establishing a Mexican presence along the border with British Honduras (Konrad 1991). President Díaz awarded forest extraction concessions to mostly US and, perhaps ironically, British interests, usually through Mexican intermediaries (Villalobos González 1993). By the boom period of chicle extraction (1903–45), the position of foreign and Mexican companies in the zone was well established, leading to the rapid exploitation of chicle resources (Konrad 1991; Ponce Jiménez 1990).

The chicle boom attracted workers from outside the region. Between 1920 and 1930 over 6,000 *chicleros* (seasonal workers extracting chicle) were brought from British Honduras, other parts of Campeche, Veracruz, and Yucatán (Barrera de Jorgenson 1993). Until the development of permanent settlements in the region in the late 1920s and 1930s, most provisions were transported from elsewhere. As *ejidos* (land grants to twenty or more peasants giving them usufruct rights, see Ch. 7) and other permanent settlements sprung up, *chicleros* brought their families to the region to establish permanent residence. Homesteading meant that swidden or *milpa* cultivation (Ch. 9) surrounded the settlements, and this activity, not chicle extraction *per se*, began the first modern 'deforestation'.[2]

### 1.3. *Hardwoods*

The post-war discovery of a synthetic substitute for chicle had catastrophic consequences on the chicle market (Barrera de Jorgenson 1993; Jiménez, M. 1951; Konrad 1987; Ponce Jiménez 1990). Government officials and industry refocused attention on the abundant stands of mahogany and cedar in the region, which were now accessible by way of improved trucks. The Mexican government created parastatal forest companies to tackle this commerce (Fig. 3.2). Caobas Mexicanas/Impulsora Forestal Peninsular set up operations in Zoh Laguna in the late 1940s, beginning the logging boom of the southern Yucatán peninsular region (Acopa and Boege 1998; Snook 1998). Soon *ejidos* with sawmills were set up at Silvituc, Gustavo Diáz Ordaz, and Nicolas Bravo (the center of the *ejido* Laguna Om).

[2] Deforestation is used here in reference to the clearing of trees greater than 20 years of age.

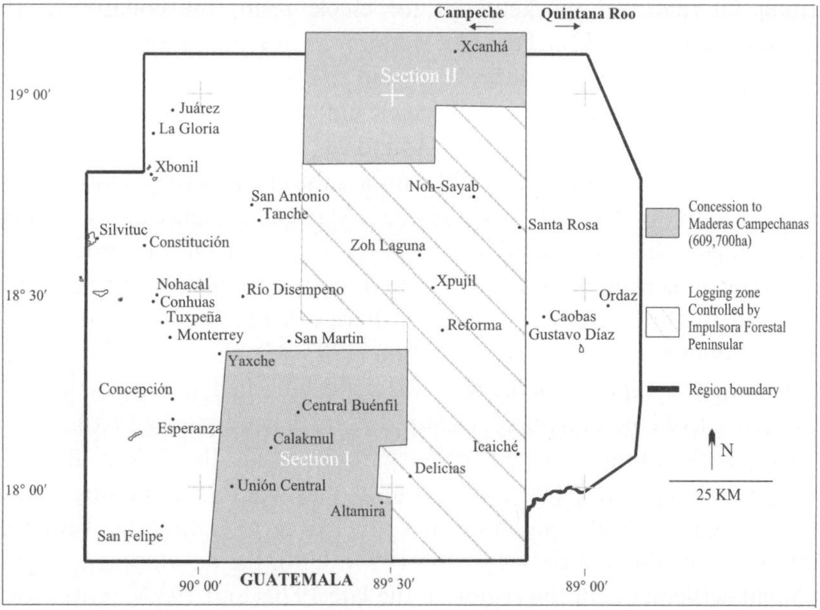


**Fig. 3.2.** Forest concession boundary for Unidad Industrial Maderas Campechanas and the approximate zone of operations of Impulsora Forestal Peninsular.
*Source*: Argüelles Suárez and Armijo 1999.

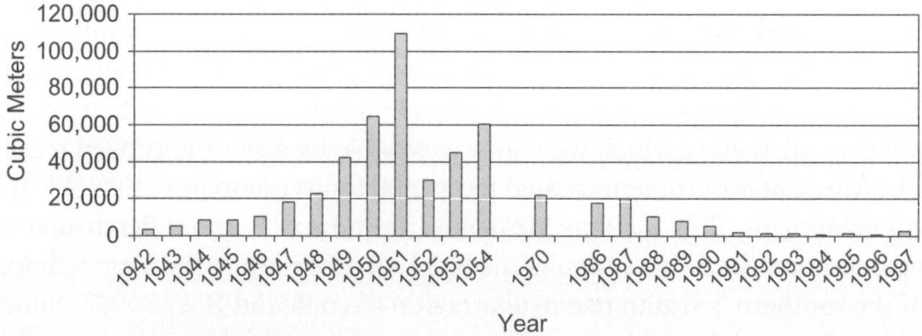


**Fig. 3.3.** Volume of mahogany logged in Campeche for select years (most of timber extracted from study region; no data for years 1955–69 and 1971–85).
*Sources*: SAG 1972; SEMARNAP 1998; Villaseñor 1958.

Intensive logging followed throughout the next decade, resulting in a marked depletion of stands of mahogany and cedar (see Ch. 4) (Fig. 3.3). The activity by Caobas Mexicanas/Impulsora Forestal Peninsular would continue until the late 1970s, but the most intensive logging undertaken after the 1950s boom was by another parastatal company, MIQRO (Maderas Industrializadas de Quintana Roo), which focused on the northernmost and southernmost sections of the Quintana Roo portion of the study region. Despite the impacts of Hurricane Janet, which disturbed much of the forests in southeastern Quintana Roo in 1955, intensive logging continued in the east through the 1960s.

A large, if ephemeral, road network established by the parastatals exposed the region to potential colonization beyond loggers and *chicleros*. Moreover, forestry workers as *ejidatarios* began to demand rights to the forest themselves (Flachsenberg and Galletti 1998). These pressures and government support of agricultural programs provided the parastatal companies with the incentive to extract as much timber as possible before the concessions ceased (Argüelles Suárez 1999; Argüelles Suárez and Armijo 1995). Hardwoods were depleted and replanting requirements were largely not followed.

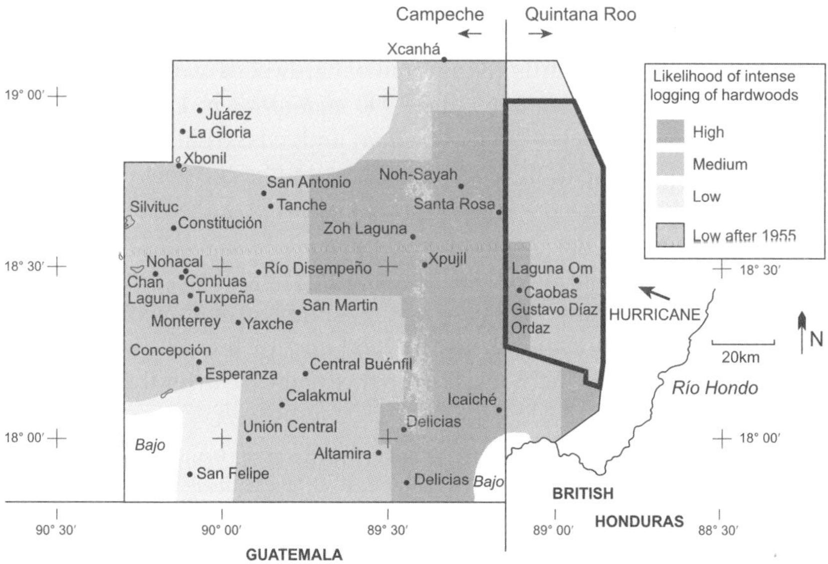


**Fig. 3.4.** Estimated intensity of the logging of hardwoods.
*Source*: Klepeis and Turner 2001.

Logging was apparently more intensive in the east-central part of the region (Fig. 3.4), and least active in the northwestern part, dominated by the drier, 'subcaducifolia' forest (Ch. 2). Logging roads identified in aerial photographs taken in 1969 indicate that the most intensive logging at that time took place in the southeastern portion of the study region and east of the road that crosscuts the study region north–south through Xpujil (Fig. 1.1), an area controlled by Impulsora Forestal Peninsular (Fig. 3.2). Furthermore, the need to open a sawmill at Gustavo Díaz Ordaz (San Antonio de Soda) in 1950 indicates the focus on logging in the southeastern quadrant of the region. The 1969 aerial photographs, archival evidence, concession maps, and maps circa 1980 showing logging roads suggest that the northern and southern parts of the study region, west of the north–south road, were not logged as heavily as those in the east, although records for the southwestern portion of the region are slim. Although Lundell (1934) indicates that the region contained much mahogany and Spanish cedar in the early part of the twentieth century, the southwestern portion was one of the few areas from which sizeable mahogany was being extracted in the late 1990s. This area may have been bypassed in the logging boom period due to its isolation. Impulsora Forestal Peninsular—the force behind the region's most intensive logging—was not in control of this area and the *ejidos*, such as Silvituc, that held forest *ampliaciones* (land grants to existing *ejidos* to be used for forestry only) in this zone did not possess the capital to create the infrastructure necessary to extract timber from the area.

By the mid-1960s onwards, the amount of hardwoods extracted in the east began to decline. Between 1965 and 1990 *ejidatarios* in Laguna Om estimated that approximately 42,000 mahogany and cedar trunks were extracted from the forest reserve in the *ejido* (Uc Chan and Sosa Romero 1997), and other data indicate that the volume of mahogany extracted from the three forestry *ejidos* within the Quintana Roo portion of the study area probably surpassed no more than 100,000m$^3$ between 1960 and 1990 (Table 3.2).

Overall, a conservative estimate of the volume of mahogany extracted from the entire study region between 1900 and 1990 is 420,400m$^3$ (some 156,000 trees), and probably well over this mark.[3] Currently, mahogany

[3] Various estimates and inventories were used to arrive at the figures provided here for mahogany logged in the study region. For example, Rodríguez Caballero (1944) found that approximately 3.0 trees (8.14m$^3$) were marketable per hectare and that only 0.12 trees (0.32m$^3$)/ha was adequate for international markets. An estimate for the study region based on these data should be taken cautiously, however, since forest inventories were inadequate and unable to capture a sense of the density of mahogany for the region. An average density of 0.12 trees/ha—the measure used here—may be conservative.

**Table 3.2.** Estimates of mahogany extraction

| Period | Zone | Mahogany ($m^3$) |
|---|---|---|
| | Entire study region | |
| 1900–30 | | 70,400 |
| 1930–60 | | 250,000 |
| | Forestry *ejidos* in region within Quintana Roo: | |
| 1959–67 | Caoba | 7,000 |
| 1959–67 | Tres Garantias | 11,800 |
| 1965–90 | Laguna Om* | 31,500 |
| 1960–90 | Quintana Roo portion of study region | 100,000 |

*Assumes that the percentage of cedar logged in comparison to mahogany was usually around 25%, as indicated by Villaseñor (1958).

stands worth cutting are so rare in *mediano* or upland forest that the logging of mahogany is restricted to *bajos* where the species are shorter and generally of poor market quality (Ch. 3; Turner *et al.* 2001). Various attempts are underway to plant small mahogany stands, but the seedlings will take years to mature. Finally, recent attempts by Plan Piloto Forestal (Pilot Forestry Plan, a social forestry initiative that started in 1983; see Ch. 7) and *ejido*-led initiatives to market alternative tree species (e.g. granadillo (*Platymiscium yucatanum* Standl.) and siricote (*Cordia dodecandra* DC.)) have had marginal success.

## 2. *Big Projects and Forest Clearing (1975–1982)*

The devastating Hurricane Janet hit the southern Yucatán in 1955, revealing the isolation of the region as well as that of Chetumal, the territorial capital of Quintana Roo situated on the southeastern coastline (Fig. 1.1). Coinciding with federal aims to develop and populate Quintana Roo sufficiently to make it a state, the first paved road between Escárcega (on the main highway from México DF to Mérida, Yucatán) and Chetumal was largely completed in the late 1960s and fully paved by 1972 (Vogeler 1974). The highway opened the southern Yucatán peninsular region directly to the rest of Mexico and to new kinds of development.

Fueled by an international oil boom, Mexico devised an ambitious settlement scheme for the region, one that would open lands to land-stressed peasants from elsewhere in the country and provide rice, cattle, and maize for commercial distribution. This last activity was supported partly by the

Programa Nacional de Desmonte (National Deforestation Program), which started in 1975 and gave smallholders subsidies and loans to convert upland forest to pasture and cropland (CIQRO 1980; Fuentes Aguilar 1980). Survey respondents for the SYPR project (see Ch. 8) indicate that the amount of land cleared under this program was considerable. For example, one *ejidatario* who normally cultivated around 5ha of maize per year cleared some 30ha of *mediano* after receiving credit from the government to do so.[4] Earlier programs apparently had similar consequences. Vogeler (1974, 1976) reported that some swidden farmers in Xpujil cultivated as much as 19.2ha as early as 1969, apparently in order to sell maize to government-sponsored programs.

In addition to individual household cultivation, large-scale, mechanized projects were begun in the wetlands adjacent to Lago Silvituc in the west and within the large *bajos* of the eastern part of the region (Figs. 1.1, 3.5). Linked directly to government programs, over 5,000ha of forest were cleared in the connected Bajo Nicolas Bravo and Bajo Morocoy alone between 1975 and 1982 (Cortina Villar, Macario Mendoza, and Ogneva-Himmelberger 1999), and a total in excess of 10,000ha of *bajo* forest was cleared in the Quintana Roo portion of the study region over the same period. These projects and new *ejidos* spawned a dramatic increase in the regional population from approximately 2,174 persons in 1960 to over 35,000 persons in 1995, although population densities remained sparse, with no more than a few persons/km$^2$ (INEGI 1991*a*, 1991*b*, 1996*a*, 1996*b*; SIC-DGE 1963*a*, 1963*b*).

The '*bajo*' experiments in rice cultivation and *bajo* and upland efforts in cattle quickly failed owing to a number of problems, including faulty design and mismanagement. The rice projects sought to use the annual inundation in large *bajos* for crop production, relying solely on rainfall and natural drainage. The annual flux in the amount and timing of the rains proved huge barriers to production, however, and by the time that canals were cut for drainage and wells drilled to provide water for periods of shortfalls, declines in soil fertility, increased weeds, and the Mexican debt crisis made the government re-evaluate sustained investment in the venture. Experiments in cattle production on *ejido* lands suffered from inadequate

[4] This *ejidatario*, Pedro Macario Mendoza, would subsequently become an ecologist in El Colegio de la Frontera Sur—Unidad Chetumal (ECOSUR-UC), a collaborating institution on the SYPR project. Indeed, Macario Mendoza worked on this project. He still cultivates his swiddens.

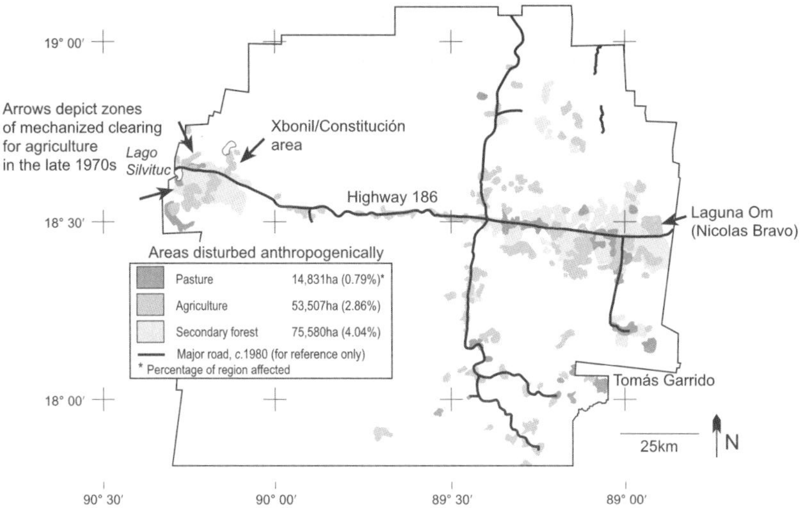


**Fig. 3.5.** Land-use/cover change in the mid-1970s.
*Source*: after INEGI 1985.

dry season water sources and disease (Ch. 2), but a few *ranchos* (small, privately owned ranches) persist, although none apparently are fiscally successful.[5] The highly touted, large projects left impacts far beyond their economic losses. The large loss of *bajo* forests has remained mostly intact today, converted to new cattle experiments; although selected *bajos* have been allowed to revert to forest (e.g. that in Tomás Garrido, southern Quintana Roo). Much of the upland forest taken reverted to successional growth, and smaller amounts of forest were lost to infrastructure (e.g. roads, *ejidal* centers) (INEGI 1985; Fig. 3.5).

Taken together, the large-scale projects and individual smallholders accelerated the rates of deforestation through the 1970s, especially along the east–west axis of the new highway. By the end of the 1970s, about 7.7 per cent of the region had been disturbed anthropogenically, and by 1982, the end of the episode in question, this figure rose to 8–10 per cent (Cortina Villar, Macario Mendoza, and Ogneva-Himmelberger 1999; INEGI 1985; Fig. 3.5). Cumulatively, subsistence swidden disturbed more forest area by

[5] A few *ranchos* were established in the region, especially east of Xpujil, with the construction of Highway 186 in the late 1960s. How these lands were obtained by absentee landlords and how they function is poorly understood. The SYPR project did not pursue investigation of this land use owing to limitations on research labor and expenses, and the small amount of land occupied by *ranchos* in the region compared to *ejidos*.

far than did rice or pasture, but the locations of cuts were dispersed and much of the land placed in long rotational fallows (Ch. 9).

Much of the deforested area succumbed to exotic and invasive species, raising questions about longer-term environmental impacts (Turner, Klepeis, and Schneider forthcoming). Aleman (*Paspalum dimidiatum* L.) and Estrella de Africa (*Cynodon plectostachyus* [K. Schum.] Pilg.) were introduced as pasture grasses, while Johnson grass (*Sorgum halapense* [L.] Pers.) was unintentionally introduced, apparently through sugarcane seed used to the southeast of the study region. The former two species were planted for pasture and the last unintentionally. An endemic species, bracken fern (*Pteridium aquilinum* [L.] Kuhn) invaded some swidden fields as early as the 1970s (Turner 1983). The area under bracken fern almost doubled between the mid-1980s and the mid-1990s, now covering 9,200ha of the study region (Chs. 2, 6; Turner *et al.* 2001).

The 1970–82 boom–bust period of federal investment in settlement and agricultural projects was as profound as it was short-lived. Not only did various schemes and settlements lead to considerable deforestation and land-cover change, the region was physically opened to the rest of the country for occupation. Once opened, landless *campesinos* have continued to enter the region to gain land, both legally and illegally.

### 3. *Land-Use Diversification, Conservation, and Tourism (1983–2000)*

The 'command' economy of the large projects gave way to neoliberal reforms empowering the smallholders (*ejidatarios*) with greater control over land-use decisions than at any other time in the region's modern history. This control, with the assistance of non-governmental organizations (NGOs) and various government programs, has led to a more diversified household economy, including land-use options. Forest extractive reserves, agroforestry, production of special market crops (e.g. chili, tomatoes, and oranges), small-scale pasture for livestock, beekeeping, and off-farm income constitute various experiments and opportunities that have expanded the range of possible land and livelihood strategies for *ejidatarios* (Chs. 7–8). Spontaneous colonization by land-hungry settlers continues, but by 1990 most available land was occupied, and some migrants began to set up homesteads illegally on government lands, especially in the southern parts of the region (Fig. 3.6)

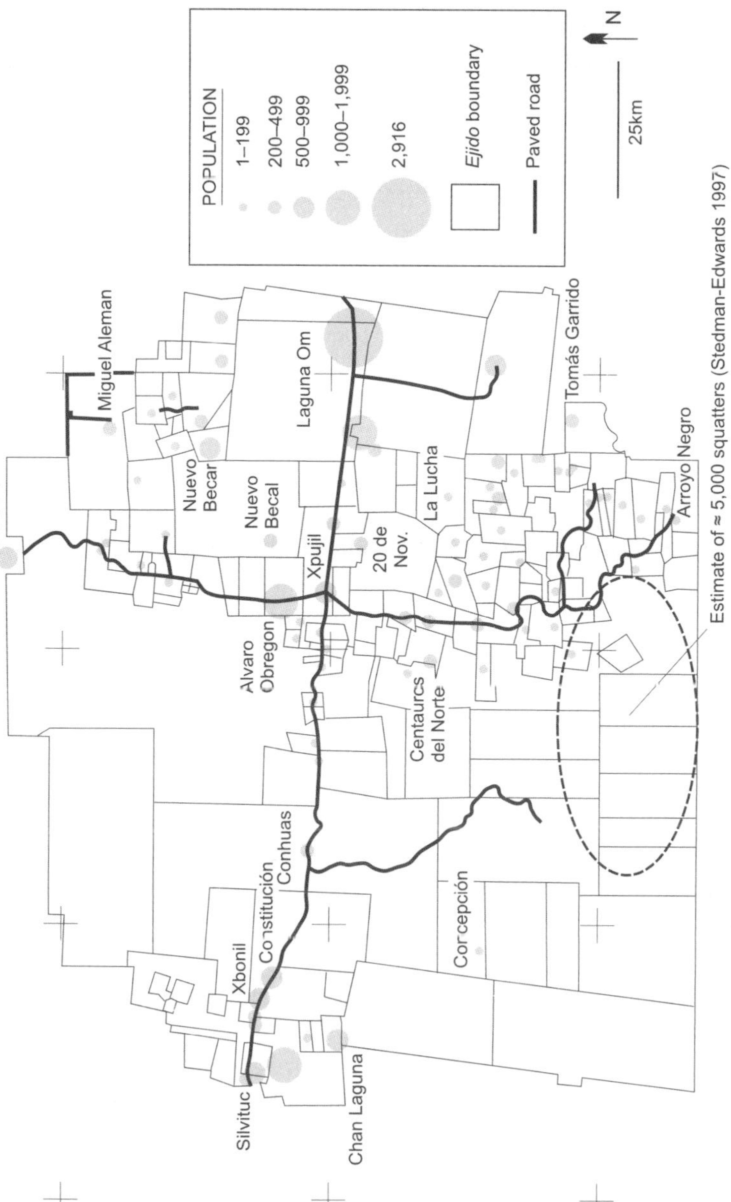


**Fig. 3.6.** Population centers, 1990, and *ejido* boundaries.
*Source*: INEGI 1991*a*, *b*.

Despite economic liberalization, culminating in the potential for ejidos to divest into private properties, federal and external controls continue to play a significant role in land use and land cover. Responding to various interests, a large biosphere reserve (Calakmul, 7,232km$^2$) was created in 1989 and registered with UNESCO's Man and the Biosphere Program in 1993 (Fig. 3.7; Acopa and Boege 1998). Consistent with the chronic mismanagement of centralized activities in the region, this reserve carries with it a series of problems, including the potential to displace smallholders whose lands fall within the reserve, while leading to the intensification of human activity outside it. Forty-four *ejidos* have portions of their land within the buffer zone of the reserve (including *ejidos* and *ampliaciones*, lands designated for forestry use by *ejidos* located elsewhere; see Ch. 7), and another twenty-one have holdings within the nuclear zone, an area in which minimal, if any, human activity is supposed to take place. With no formal management plan in existence throughout the 1990s and with no clear guidelines to govern their use of land resources, local smallholders continue to use the land as they did before 1989 (Stedman-Edwards 1997). Indeed, there was a brief period of intensive forest extraction within the reserve's boundaries immediately following its creation, perhaps a response by *campesinos* to impending restrictions to access (Martens 1999). Pending effective enforcement, most *ejidatarios* would probably ignore reserve land-use rules because of confusion over the idea of a biosphere reserve, their pressing economic needs, and, in some cases, an unwillingness to embrace the concept of an externally defined conservation program (Haenn 2000*a*). After all, many *ejidatarios* were vehemently opposed to the establishment of the reserve, some being informed of it after the fact (Haenn 2000*b*).

The Calakmul Biosphere Reserve is only one, if potentially powerful, cog in the new national vision for the region, however. The larger vision is to use 'nature' itself as a commodity, including the Maya ruins beneath the forest.[6] The development scheme itself is called El Mundo Maya (The Maya World), an international development project proposed in the 1980s (Garrett 1989) and begun in 1993. The program does not represent one single policy but a common goal for Belize, El Salvador, Guatemala, Honduras, and Mexico to develop the Maya region economically via archaeo-ecotourism. The Calakmul Biosphere Reserve, of course, offers a

[6] As noted in Ch. 2, the region is replete with Maya ruins of all kinds and sizes. Those promising to hold relics of value have largely been looted, registered by small holes dug to the center of pyramids and palaces. Looting remains a fixture in the region.

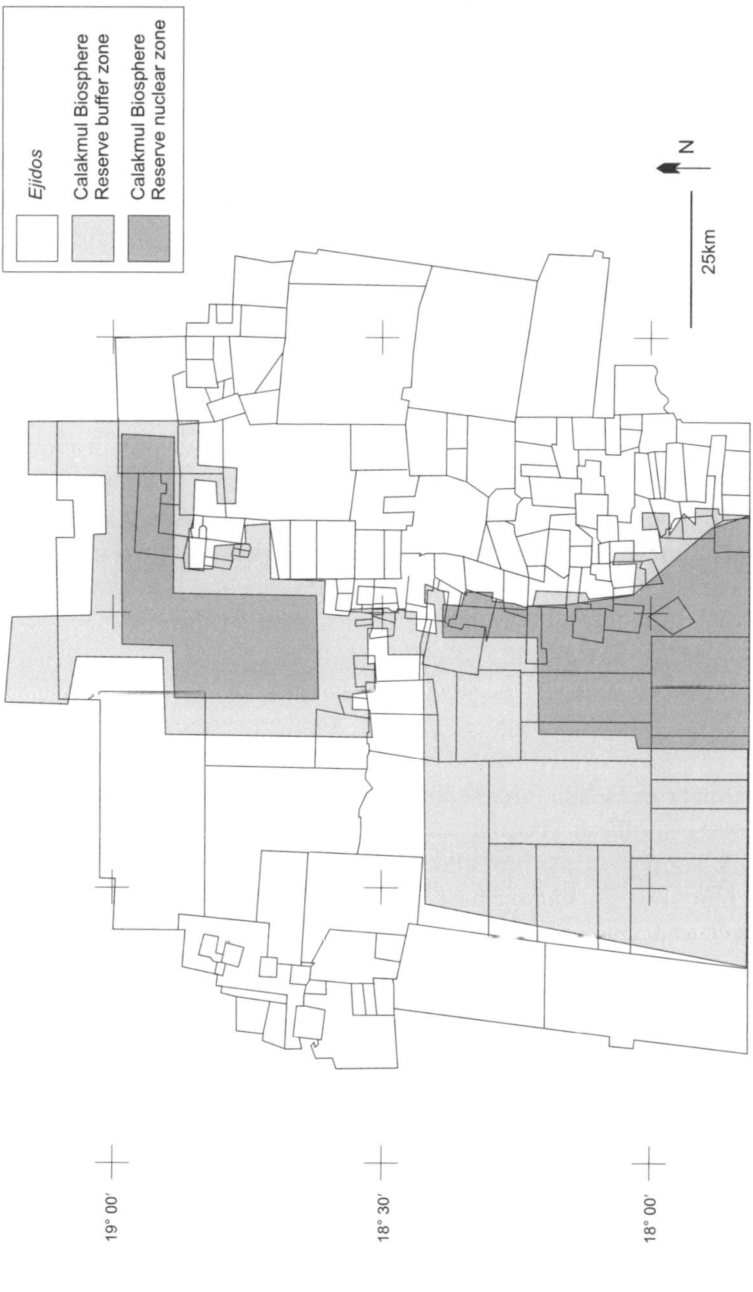


**Fig. 3.7.** Calakmul Biosphere Reserve and *ejido* boundaries.

huge forest conservation zone around which El Mundo Maya can and has built (e.g. the new hotel, Ecovilla, near Xpujil). And, the centrality of the southern Yucatán peninsular region in relation to reserves and parks in Guatemala and Belize to the south and nature reserves, such as Río Lagartos and San Felipe to the north, have placed the region within the design framework of the Corredor Biologico Sian Ka'an–Calakmul, which is part of an international program—Mesoamerican Biological Corridor—seeking to facilitate the movement of biota (Kaiser 2001; Ch. 7).

To facilitate access to the region by tourists, the government has funded substantial road improvements, new road construction, and electrification. Regional infrastructure improvements, in turn, facilitate the growth of *ejidos* as well as the range of choice in land-use activities for *ejidatarios*. With growth in human occupation, land clearance for swidden increases (focused on subsistence), as does the ability to access and produce for markets, imperfect as they are, exemplified most dramatically by chili production (Ch. 10). These changes and those noted above have made the 1990s to the present a time in which the region has witnessed its most diversified use since the ancient Maya, more than a millennium ago, and if they continue to gain momentum, promise to sustain deforestation further.

## 4. *Episodes and Forces of Land Change: Summary*

Both proximate and distal forces combine to precipitate land changes anywhere, and the southern Yucatán peninsular region is no exception. Given the region's frontier status throughout the last century, it is not surprising that dominant forces of change have been distal in kind—lying beyond the region itself and involving federal policies that change the structure of the regional political economy. Beginning with logwood extraction on the region's southeastern and southwestern peripheries, the episode of *forest extraction* developed in the final decades of the nineteenth century. Extraction then shifted resource focus, first to chicle and second to hardwoods, and areal focus, to the region's interior in the twentieth century. Once these resources were mined and Mexico was awash in oil-driven development in the 1970s, regional emphasis switched to the episode of *big projects and forest clearing*, the projects of which incurred significant financial loss for the government and the *ejidatarios* who bought into them through bank loans. The current episode of *land-use diversification, conservation, and tourism* is driven in part by archaeo-ecotourism that remains unproven economically, although the massive development of the northeastern coast of Quintana Roo (including Cancún) may hold clues for the future.

Federal decisions opened the region to settlement by way of migration, but once in place, residents, primarily *ejidatarios*, have become the proximate agents of change as they interpret (and bend) subsidies and programs to their own ends and as they undertake their own land experiments under neoliberal reforms (Klepeis and Vance 2003). Foremost among their efforts has been the development of commercial chili cultivation, which is currently the fastest growing land use in the region. As well, NGOs have pressed for various kinds of small-scale production activities that are forest-friendly in their planning. The decision to adopt or experiment with these systems rests with *ejidos* and households, although they may be offered various inducements to undertake the efforts. The long-term viability of these efforts on land diversification, however, resides in markets for the intended produce, and these markets are currently thin.

The central messages provided by analysis of the forces driving modern land changes in the region are clear: the kind and amount of land-cover changes are connected to episodes in the regional political economy that generate different human–environment conditions and which, in turn, are tied to Mexican policy. Throughout most of its modern history, the region operated under federal command controls. The final two decades of the twentieth century witnessed a shift to neoliberal policies, investing in local land-use decisions as they might operate under the umbrella of the biosphere and nature reserves and El Mundo Maya.

From this history, some key points must be considered in assessments of current conditions operating in the southern Yucatán peninsular region:

- Forest extraction in the first 60 years of the last century severely mined the region of its mahogany and Spanish cedar, affecting species composition of the upland forest.
- Major development activities begun in the late 1960s triggered the region's first significant deforestation since ancient Maya times, the pace of which was accelerated by large project land clearance.
- After opening the region to large-scale occupation, much of it was placed under a biosphere reserve and an international scheme to promote archaeo-ecotourism, such that from the 1990s onwards a tension has existed between the need to clear land for cultivation and to conserve forests under the reserve and tourism schemes.

- The different impacts of these changing human–environment conditions are central to all assessments of deforestation in the region, especially those that seek to explain and project rates and spatial patterns of deforestation.

## REFERENCES

Acopa, D., and E. Boege (1998). 'The Maya Forest in Campeche, Mexico: Experiences in Forest Management at Calakmul', in R. B. Primack, D. Bray, H. A. Galletti, and I. Ponciano (eds.), *Timber, Tourists, and Temples: Conservation and Development in the Maya Forest of Belize, Guatemala, and Mexico*. Washington DC, Island Press, 81–97.

Argüelles Suárez, A. (1999). *Diagnóstico de las Poblaciones de Caoba en México*. Chetumal, PROARCA/CAPAS.

Argüelles Suárez, A., and N. Armijo Canto (1995). *Utilización y Conservación de los Recursos Forestales en Quintana Roo, Problemática y Perspectivas del Manejo Forestal*. Chetumal, Union Nacional de Organizaciones de Foresteria Comunal, AC.

Barrera de Jorgenson, A. (1993). 'Chicle Extraction and Forest Conservation in Quintana Roo, Mexico', MA Thesis, University of Florida.

Chenaut, V. (1989). *Migrantes y Aventureros en la Frontera Sur*. Mexico, Secretaria de Educación Publica.

CIQRO (Centro de Investigación de Quintana Roo) (1980). *Quintana Roo: Organización Espacial*. Puerto Morelos, Quintana Roo, CIQRO.

Cortina Villar S., P. Macario Mendoza, and Y. Ogneva-Himmelberger (1999). 'Cambios en el uso del suelo y deforestacion en el sur de los estados de Campeche y Quintana Roo, Mexico'. *Boletín del Instituto de Geografía de la UNAM: Investigaciones Geográficas* 38: 41–56.

de los Santos Valadez, M. (1987). 'Inventarios Forestales en el Estado de Quintana Roo', in H. G. Lund, M. Caballero-Deloya, and R. Villarreal-Canton (eds.), *Land and Resource Evaluation for National Planning in the Tropics: Proceedings of the International Conference and Workshop*. Washington DC, Department of Agriculture, Forest Service, 223–7.

de Vos, J. (1988). *Oro Verde: La Conquista de la Selva Lacandona por Los Madereros Tabasquenos 1822–1949*. México DF, Fondo de Cultura Economica.

Flachsenberg, H., and H. A. Galletti (1998). 'Forest Management in Quintana Roo, Mexico', in R. B. Primack, D. Bray, H. A. Galletti, and I. Ponciano (eds.), *Timber, Tourists, and Temples: Conservation and Development in the*

*Maya Forest of Belize, Guatemala and Mexico*. Washington DC, Island Press, 47–60.

Fremont, H. (1861). Plano del Estado de Campeche. Campeche, Agencia Ministeria de Fomento.

Fuentes Aguilar, L. (1980). 'Desmontes y Colonizacion'. Paper presented at the Symposium, *Memorias de Simposio Quintana Roo-Problemática y Perspectivas*. Cancun, Quintana Roo. October. Unpublished.

Galletti, H. A. (1994). 'Las Actividades Forestales y su Desarrollo Histórico', in *Estudio Integral de la Frontera México-Belize*. Chetumal, CIQRO, 109–71.

Garrett, W. E. (1989). 'La Ruta Maya'. *National Geographic* 176: 424–79.

Graham C. G., M. R. Gómez, and J. Gutiérrez L. (1920). *Plano General del Estado de Campeche*. Campeche, Gobierno del Estado de Campeche.

Haenn, N. (2000*a*). 'Biodiversity is Diversity in Use: Community-based Conservation in the Calakmul Biosphere Reserve', *The Nature Conservancy's Working Paper Series América Verde* 7.

——(2000*b*). 'Calakmul's Hidden History: Constructing a Culture Region and a Research Site', in *Conference Abstracts of the Annual Meeting of the Latin American Studies Association*. Miami, Florida. Unpublished.

INEGI (1985). Carta de Uso del Suelo y Vegetación Escala 1:250,000. Instituto Nacional de Estadística Geografía e Informática (INEGI), Aguascalientes.

——(1991*a*). *XI Censo General de Poblacion y Vivienda, 1990. Resultados definitivos. Datos por localidad. Quintana Roo*. Aguascalientes, Instituto Nacional de Estadística, Geografía e Informática.

——(1991*b*). *XI Censo General de Poblacion y Vivienda, 1990. Resultados definitivos. Datos por localidad. Campeche*. Aguascalientes, Instituto Nacional de Estadística, Geografía e Informática.

——(1996*a*). *Conte de Poblacion y Vivienda 95: Campeche*. Aguascalientes, Instituto Nacional de Estadística, Geografía e Informática.

——(1996*b*). *Conte de Poblacion y Vivienda 95: Quintana Roo*. Aguascalientes, Instituto Nacional de Estadística, Geografía e Informática.

Jiménez M., L. G. (1951). *El Chicle: Su Explotación Forestal y Industrial*. México DF, Imprenta Manual Casas C.

Jones, G. D. (1989). *Maya Resistance to Spanish Rule: Time and History on a Colonial Frontier*. Albuquerque, University of New Mexico Press.

Kaiser, J. (2001). 'Bold Corridor Project Confronts Political Reality'. *Science* 293: 2192–4.

Klepeis, P., and B. L. Turner II (2001). 'Integrated Land History and Global Change Science: The Example of the Southern Yucatán Peninsular Region'. *Land Use Policy* 18: 27–39.

Klepeis, P., and C. Vance (2003). 'Neoliberal Policy and Deforestation in Southeastern Mexico: An Assessment of the PROCAMPO Program' *Economic Geography* 79: 221–40.

Konrad, H. W. (1987). 'Capitalismo y trabajo en los bosques de las tierras bajas tropicales Mexicanas: El Caso de la Industria del Chicle'. *Historia de México* 36: 465–505.

Konrad, H. W. (1991). 'Capitalism on the Tropical-Forest Frontier: Quintana Roo, 1880s to 1930', in J. T. Brannon and J. Gilbert (eds.), *Land, Labor, and Capital in Modern Yucatan: Essays in Regional History and Political Economy*. Tuscaloosa, University of Alabama Press, 143–71.

Lundell, C. L. (1933). *Chicle Exploration in the Sapodilla Forest of the Yucatan Peninsula*. Carnegie Institute of Washington, Publication 436. Washington DC, Carnegie Institute.

—— (1934). Preliminary Sketch of the Phytogeography of the Yucatan Peninsula. *Contribution to American Archaeology*. 12/436. Washington DC, Carnegie Institute of Washington.

Macías Zapata, G. A. (1996). 'Economía y politica entre los Mayas Icaiches de Quintana Roo, 1893–1980', in *Investigaciónes Antropologicas Recientes*. Institut für Ethnologie der Universität Bern, 83–91.

Martens, Ron (1999). 'Participación y Capacitación en el Mundo Maya'. <http://www2.planeta.com/mader/ecotravel/mexico/yucatan/ronmartens/1.html> last accessed 7 April 2002.

Ponce Jiménez, M. P. (1990). *La Montaña Chiclera Campeche: Vida Cotidiana y Trabajo (1900–1950)*. México DF, Centro de Investigaciones y Estudios Superiores en Antropológica Social.

Rodríguez Caballero, R. (1944). La explotación de los montes de Caoba en el territorio de Quintana Roo. M. S. Thesis. Ingeniero Agrónomo, Chapingo, México.

SAG (1972). *Anuario de la Produccion Forestal de México, año de 1970. Campeche*. México DF, Secretaria de Agricultura y Ganaderia.

SEMARNAP (1998.). *Informe*. Campeche, Secretaria de Medio Ambiente, Recursos Naturales y Pesca.

SIC-DGE (1963*a*). *VIII Censo General de Población, 1960. Estado de Campeche*. México DF, Secretaria de Industria y Comercio-Dirección General de Estadística.

—— (1963*b*). *VIII Censo General de Población, 1960. Estado de Quintana Roo*. México DF, Secretaria de Industria y Comercio-Dirección General de Estadística.

Snook, L. K. (1998). 'Sustaining Harvests of Mahogany (*Swietenia macrophylla* King) from Mexico's Yucatán Forests: Past, Present, and Future', in R. B. Primack, D. Bray, H. A. Galletti, and I. Ponciano (eds.), *Timber, Tourists, and Temples: Conservation and Development in the Maya Forest of Belize, Guatemala, and Mexico*. Washington DC, Island Press, 61–80.

Stedman-Edwards, P. (1997). 'Socioeconomic Root Causes of Biodiversity Loss: The Case of Calakmul, Mexico'. México DF, World Wildlife Fund. Unpublished.

Turner, B. L., II (1983). *Once Beneath the Forest: Prehistoric Terracing in the Río Bec Region of the Maya Lowlands*. Boulder, Westview Press.

——(1990). 'The Rise and Fall of Population and Agriculture in the Central Maya Lowlands: 300 BC to Present', in L. F. Newman (ed.), *Hunger in History: Food Shortage, Poverty, and Deprivation*. Cambridge, Basil Blackwell, 178–211.

Turner, B. L., II, P. Klepeis, and L. S. Schneider 2003. 'Three Millennia in the Southern Yucatán Peninsular Region: Implications for Occupancy, Use, and "Carrying Capacity"', in A. Gómez-Pompa, M. Allen, S. Fedick, and J. Jimenez-Osornio (eds.), *Lowland Maya Area: Three Millennia at the Human–Wildland Interface*. New York, Haworth Press, 361–87.

Turner, B. L., II, S. C. Villar, D. Foster, J. Geoghegan, E. Keys, P. Klepeis, D. Lawrence, P. M. Mendoza, S. Manson, Y. Ogneva-Himmelberger, D. P. Salicrup, R. Roy Chowdhury, B. Savitsky, L. Schneider, B. Schmook, and C. Vance (2001). 'Deforestation and Agricultural Change in the Southern Yucatán Peninsular Region: Integrative Land Change for Global Change Studies'. *Forestry, Ecosystems, and Management* 154: 343–70.

Uc Chan, L., and J. Sosa Romero (1997). 'Memoria del Taller de Organización y Manejo de los Recursos Forestales, Ejido Laguna Om, Quintana Roo'. Organización de Ejidos Forestales de Quintana Roo, 'Chaktemal', S. C. Chetumal, Quintana Roo. Unpublished.

Villalobos González, M. H. (1993). 'Las concesiones forestales en Quintana Roo a fines del Porfiriato'. *Relaciones 53: Estudios de Historia Y Sociedad* 14: 87–112.

Villaseñor, A. R. (1958). 'Los Bosques y su explotación', in E. Beltran (ed.), *Los Recursos Naturales del Sureste y su Aprovechamiento*. México DF, Instituto Mexicano de Recursos Naturales Renovables, AC, 273–326.

Vogeler, I. (1974). 'The Cultural Ecological Setting of Southeastern Campeche', in R. E. W. Adams (compiler), *Preliminary Report on the Archaeology of the Rio Bec Region*. New Orleans, Middle American Research Institute, Tulane University, 31: 110–12.

——(1976). 'The Dependency Model Applied to a Mexican Tropical Frontier Region'. *Journal of Tropical Geography* 43: 64–8.

# PART II

## *Land-Cover Characteristics and Change: Forests, Forest Cover, and Nutrient Dynamics*

The southern Yucatán peninsular region maintains the largest expanse of seasonal or dry tropical forest remaining in Middle America. Despite the importance of this forest for chicle (gum) and tropical hardwoods during the early and mid-1900s and the attention given to it by various botanists during the same period, surprisingly little quantitative data and analysis of the forests or forested landscapes has been undertaken. The SYPR project seeks to remedy this lacuna, in part because the character of the forest and its dynamics are essential for determining local and regional impacts of its current uses.

These qualities are examined through primary research reported here. The 'reality' of two basic forests—upland and *bajo* forests—that dominate the region is established. These types are characterized more so by differences in species abundance and crown height than by difference in species *per se*. For the most part, trees in the upland forest grow taller than those in the *bajo*, where seasonal inundation apparently plays some role in producing a shorter stature. Nutrient cycling in the upland forests, land of choice for swidden cultivation, is affected by the strong north–south precipitation gradient and, apparently by the history of past swidden activity. Significant declines in litter production are experienced in successional forests having incurred more than two cycles of swidden cultivation. Yet, after about twenty-five years, these forests are indistinguishable from 'old growth' forest in terms of species and their abundance, absent mahogany and Spanish cedar which has been more-or-less logged out. Indeed, the critical age of the forest in terms of tree species diversity and carbon storage appears to be about twenty-five years.

Can imagery analysis separate upland forests at this critical age? Phase 1 of the SYPR project did not seek this critical threshold, as it was not known until the end of that phase. Landsat TM imagery analysis, assisted by ancillary data, is used to separate upland from *bajo* forest, with modest confusion where their characteristics overlap: shallow slopes grading into poljes or *bajos* that incur annual flux in the amount and length of inundation. Classification has also shown considerable success in separating various land-covers and associated uses, given sufficient training sites, including successional growth. Thus it would appear that the second phase of the project may well be able to separate forests in terms of the critical, 25-year age. Beyond this, the imagery analysis indicates that significant deforestation has been underway in the region, especially on its eastern and western flanks, and along an axis skirting the southeastern border of the Calakmul Biosphere Reserve. From the late 1960s into the late 1990s, annual deforestation rates ranged from about 0.3 per cent to 0.4 per cent, most of the deforested lands put into some swidden fallow cycle.

*The Editors*

# 4

# Forest Types and their Implications

*Diego Pérez-Salicrup*

The southern Yucatán peninsular region contains the largest and most rapidly disappearing continuous tract of tropical forest in Mexico (Flores and Espejel Carvajal 1994; Delfín Gonzales, Parra, and Echazarreta 1995; Acopa and Boege 1998). Vegetation in the region is a mosaic of forest types with different structural appearances (Flores and Espejel Carvajal 1994; Hernández-Xolocotzi 1959; Miranda 1958) that primarily reflect variation in environmental and edaphic conditions (Ibarra-Manríquez 1996). However, the structure and tree composition of forests in the region, as elsewhere in the central Maya lowlands, has been and remains strongly influenced by human activity (Ch. 2). In spite of the abundance of botanical work throughout the Yucatán peninsula, little attention has been devoted to characterizing the forests in this frontier region quantitatively, and the variation and distribution of forests remain poorly documented. Yet, it is precisely this kind of documentation that is required for integrated land studies of the kind that the SYPR project is undertaking (Turner *et al.* 2001).

## 1. *Forests in the Region*

Since the third decade of the twentieth century, botanical interest has focused on the flora of the Yucatán Peninsula, especially that located in the historically more accessible portion of the peninsula (Ibarra-Manríquez 1996). Early twentieth-century studies (Lundell 1938; Standley 1930) led to a broad classification of the primary vegetation as deciduous tropical forests (Miranda 1958), or evergreen tropical forests (Rzedowski 1981), controlled in distribution by the northwest to southeast precipitation gradient (Fig. 4.1), distinctive dry season, and karstic terrain (Ch. 2). Today, the entire region is appropriately labeled a seasonally dry tropical forest (Bullock, Mooney, and Medina 1995). During the rainy season (May–October) most species have their canopies fully displayed and light is a limiting factor in the forest understory (Martínez-Ramos 1985, 1994). For the remainder of the year, monthly precipitation usually does not exceed 100mm. During the lowest rainfall months (February–April), water may become limiting and considerable defoliation takes place, especially in the north and west.

Other factors controlling forest structure and composition include topography, twentieth-century land-use history, and hurricanes (Brokaw and Walker 1991; Cooper-Ellis *et al.* 1999). At a broad scale, the higher, interior elevations in the Zoh Laguna plateau (term used to designate hill lands on some older maps) reach 300m amsl or more, while the eastern and western margins of the region descend to about 100m amsl. Across the entire region at all elevations karstic sinks or *bajos* capture and retain surface water during the rainy season and form local wetlands characterized by fine-grained soils. The impacts of twentieth-century land use on vegetation have been highly concentrated in the central and eastern portions of the region, and with the exception of one or two large *bajo* complexes, are focused on rolling or hill terrain (Ch. 3). In addition, the Yucatán Peninsula is highly exposed to hurricanes that may exert profound impacts on forest structure and composition (Boose, Foster, and Hall 2003; Foster, Fluet, and Boose 1999; Konrad 1996; Sánchez and Islebe 1999; Whigham, Lynch, and Dickinson 1998). However, the frequency and intensity of hurricanes is relatively low and uniform in the southern Yucatán peninsular region relative to elsewhere in the peninsula, undoubtedly diminishing the role of wind disturbance in controlling the distribution of forest types over the long term (Boose, Foster, and Hall 2003; Foster 1988).

Considerable uncertainty exists in the description of forests in this region, with a range of forest types from five to eighteen, depending on the

**Fig. 4.1.** Forest study locations and rainfall gradient. Sampling of forest plots was stratified into six locations (black dots) that represent regional gradients in elevation(from 100m in the margins to in excess of 250m in the center), rainfall (see climate diagrams), and 20th century land use (more intense use in the east than in the west). Climograms show mean annual precipitation (MAP), and rainfall seasonality at five locations within the study region.

*Source*: Comisión Nacional del Agua (1998). 'Datos Mensual de Precipitación y Temperatura, 1980–1998: Estados de Campeche y Quintana Roo'. Mérida. Unpublished.

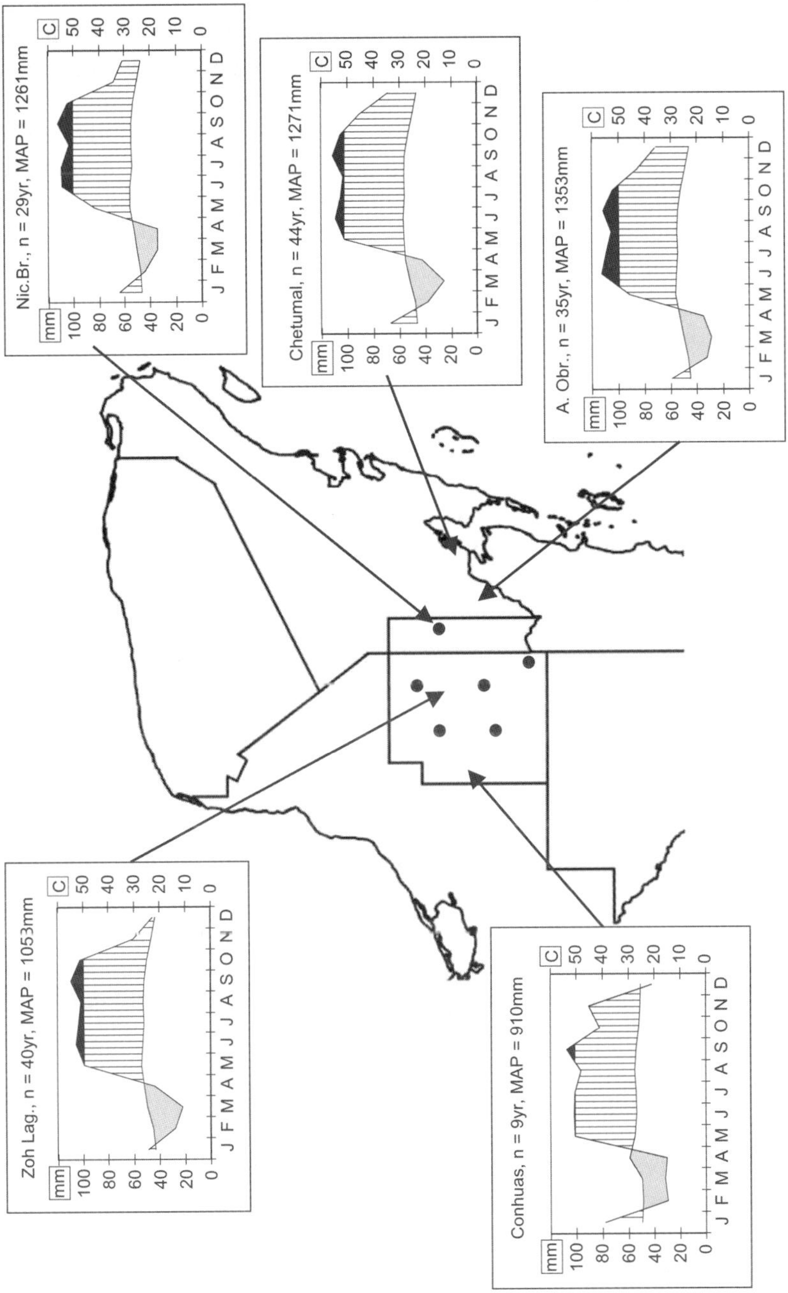
Nic.Br., n = 29yr, MAP = 1261mm
Chetumal, n = 44yr, MAP = 1271mm
A. Obr., n = 35yr, MAP = 1353mm
Zoh Lag., n = 40yr, MAP = 1053mm
Conhuas, n = 9yr, MAP = 910mm

particular classification (e.g. Flores and Espejel Carvajal 1994; Lundell 1938; Miranda 1958; Olmsted *et al.* 1995; Rzedowski 1981; Shepherd 1975). All classifications, however, agree in placing two forest types at opposite extremes: mid-statured forests (MSF, *bosque mediano* or upland forest), and low-statured forests (LSF, *bosque bajo* or *bajo* forest). These two basic forest types were recognized independently by botanists working in the south of the Yucatán peninsula in both Mexico and Belize, during the late 1950s. Their studies also agreed in identifying the depth and degree of flooding of soils as the major factors controlling the distinct structural variation between these forests (Miranda 1958; Wright *et al.* 1959). Mid-statured forests grow on the tops and sides of the gently rolling hills across the landscape (Fig. 2.1), where well-drained soils consisting of loams and clays dominate. In contrast, low-statured forests grow on low-lying terrain, mostly sinks and depressions infilled with deep clays that waterlog during the rainy season and harden during the dry season. Nutrient absorption by plants is difficult in these saturated soils, despite the fact that the soils, may be richer in nutrient concentration than the upland sites (Schulze and Whitacre 1999).

In addition to these distinct forest types, Miranda (1958), Hernández-Xocolotzi (1959), and others have also recognized the importance of the successional forest stands that comprise the cycle of shifting or swidden agriculture. Swidden, practiced throughout the neotropics since pre-European arrival (Barrera, Gómez-Pompa, and Vázquez-Yanes 1977; Horn and Kennedy 2001; Kennedy and Horn 1997), gives rise to a dynamic mosaic of vegetation where stands of different ages maintain a high regional species diversity (Barrera, Gómez-Pompa, and Vázquez-Yanes 1977). Owing to poor soil drainage in LSF, most swidden cultivation in the region is conducted in MSF. Hence, forest stands of different successional age primarily represent stages of regeneration of MSF following its clearance, burning, and cultivation for crops over a few-year period (Ch. 5).

## 3. *Ordination and Description of Forest Types*

In regions where forest types are clearly defined, it is possible to evaluate the relative impact of natural events versus human history in shaping forest structure and composition (Foster 1992; Foster *et al.* 1992, 1997; McLachlan, Foster, and Menalled 2000). The use of different and inconsistent systems of classification of the vegetation of the southern Yucatán peninsular region, however, makes this assessment difficult. A first step towards understanding the impact of land-use history on the region's forest, therefore, is to develop a quantitative analysis of forest vegetation based on field sampling and multivariate statistical analysis, including ordination. Rather than creating an entirely new forest classification, however, a simpler and more direct approach was taken in this study. Since there is general agreement that MSF and LSF are the opposite ends of a gradient of forests in the region, this study began by evaluating these two forest types in detail: specifically, examining how distinct they actually are in terms of structure and tree species composition.

### 3.1. *Field Methods*

A network of over 150 circular 500m$^2$ permanent plots was established in six subregions across the southern Yucatán area that encompass gradients of rainfall and elevation, and different histories of twentieth-century land use (Fig. 4.1). Within each subregion, plots were located in mature LSF, MSF, or in successional forest stands. Discrimination between these forest types in the field was accomplished by visual inspection, and corroboration with local farmers on their specific histories. Plots in LSF and MSF were considered mature when *ejidatarios* confirmed that the area had not been used for swidden agriculture. It is recognized, however, that most forests across the study region have experienced some type of human use, such as timber or chicle extraction. Sites with obvious signs of recent disturbance were avoided. All plots were established at a randomly selected location within a large patch area of a given forest type. In the case of successional stands, interviews with *ejidatarios* were used to estimate the number of years since last agricultural cultivation. Successional forest stands were classified into six categories of age from last harvest: 0–5, 5–10, 10–15, 15–20, 20–25, and 25–30 years.

In each plot, all trees ≥ 10cm dbh (diameter at breast height, 1.3m or above buttresses) were marked, identified, and measured. Tree diversity was used as a surrogate for biodiversity, although it is recognized that not all taxonomic groups correlate with trees in their response to human land use. Each tree was classified by its epiphyte load into one of four classes: no epiphytes; at least one epiphyte present; more than 10 epiphytes but less than 25 per cent of the tree covered with epiphytes; and more than 25 per cent of the tree with epiphytes. All lianas (climbing plants) ≥ 2cm dbh, all standing dead trees ≥ 5cm dbh, and all tree stumps were counted and measured. Canopy height was classified in 21 equally interspersed points within each plot into one of the following categories: 0–2, 2–5, 5–10, 10–15, 15–20, 20–25, 25–30, and > 30 m in height. A $100m^2$ subplot located at the center of each main plot was used to count, identify, and measure all trees 5–10cm in dbh, and measure the dbh in all lianas 0.2–2cm dbh and all saplings and shrubs > 2.0m tall.

Once all trees were measured and identified, NMS (Non-metric Multidimensional Scaling) ordinations and MRPP (Multi-Response Permutation Procedure) tests were used to evaluate whether LSF, MSF, and successional stands were distinguishable in terms of tree species presence, tree species abundances, and tree species basal areas. In addition, an NMS ordination and an MRPP test were used to evaluate whether LSF, MSF, and successional plots could be distinguished based on structural features. Analyses were conducted with the program PC-Ord version 4 (McCune and Mefford 1999).

## 3.2. *Results*

Ordination of vegetation plots based on tree species presence, density, and basal area yielded very similar results (Fig. 4.2). Patterns in the ordination were confirmed with significant MRPP tests. Three major vegetation patterns emerge from this analysis. The most important is that LSF is clearly separated from MSF based on their tree species composition (Fig. 4.2). Overlap between these two forest types, however, supports the notion that these types occupy a vegetation continuum along an environmental gradient, apparently determined by local topography and soils, especially soil moisture (Schulze and Whitacre 1999). A second notable pattern is that successional forests stands form a temporal gradient to mature MSF. And finally, successional forest stands converge rapidly on the tree species composition of MSF; after 20–30 years species composition in successional plots is no longer distinguishable from plots in mature MSF.

### 3.2.1. *Comparison between LSF and MSF*

Species area curves for each forest type indicate that the rate at which new tree species are added to the sample as area increases is slightly higher for MSF plots than for LSF plots (no. of tree species in MSF = $(\text{area})^{0.438}$, $r^2 = 0.96$, $F = 1026.86$, $p < 0.001$; no. of tree species in LSF = $(\text{area})^{0.437}$, $r^2 = 0.96$, $F = 1086.44$, $p < 0.001$; difference in slopes, df = 1, 93, $F = 12.706$, $p < 0.001$), as predicted by several authors (e.g. Miranda 1958). This result

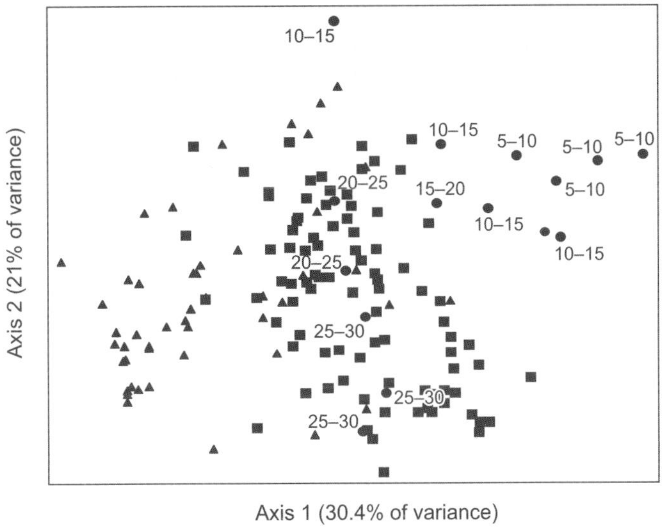


**Fig. 4.2.** Two main axes of a non-metric multi-dimensional scaling ordination of forest plots.
Triangles = low-statured forest; squares = medium-statured forest; circles = successional forest. Numbers adjacent to circles indicate approximate number of years of succession in SF plots.

is somewhat misleading, however. Structurally, trees in LSF have smaller diameters than trees in MSF, and there are less tree stems (≥ 10 cm dbh) within a given LSF plot than in an MSF plot of the same size ($df = 1, 135$; $F = 7.021$, $p = 0.009$). Consequently, it is important to construct species accumulation curves based on the mean number of individuals sampled per plot, instcad of arca (Condit *ct al.* 1996). Treated this way, there are no significant differences between the two forest types in species accumulation curves or total number of tree species (no. of tree species in MSF or LSF = $(\text{area})^{0.437}$ $r^2 = 0.96$, $F = 2168.0$; difference in slopes, $df = 1, 93$; $F = 0.089$, $p = 0.766$).

The two forest types have few species that are exclusive to either type. Of the total number of species found through the study, 7 per cent were endemics to LSF, and 11 per cent were endemics to MSF. These numbers are slightly biased by the fact that more plots were established in MSF. Holding the number of plots for each forest type constant, about 9 per cent of the total pool of species is endemic to each type, while 82 per cent of the tree species are found in both forest types.

An inspection of the Relative Importance Value (RIV or IV) for all tree species shows that few species share high importance values in both forest types, but tree species with a high RIV in one forest type have lower values for the other (Table 4.1). A few of these species are worth noting. *Haematoxylum campechianum* L. ranks first in IV in LSF. This species was the center of timber exploitation during Colonial times (Ch. 3), when a dye extracted from its wood was of high value, giving the tree its common name *tinto* or *palo de Campeche*. Its exploitation, however, centered on coastal lands or inland waterways; evidently, in the remote southern Yucatán peninsular region, this species was not exploited extensively (Justo Sierra 1998). *H. campechianum*, and *Bucida buceras* L. grow on temporarily or permanently waterlogged sites, hence they are rarely found in MSF. *Manilkara zapota* (L.) P. Royen, which ranks second and third in IV in LSF and MSF respectively, is the dominant tree species throughout the region. This tree is the source of the resin for natural chewing gum, which became the most important non-timber forest product in Mexico until artificial chewing gum developed in the mid-1900s (Ch. 3). Despite five decades of continuous exploitation of its sap, it is interesting to note that *M. zapota* remains dominant. *Brosimum alicastrum* S.W., locally known as *ramón*, is the second most dominant species in terms of IV in MSF. Several authors have noted that *B. alicastrum* is particularly abundant near Maya ruins, apparently as a result of its capacity to germinate and grow in exposed calcareous substrate found in ruins (Lambert and Arnason 1982; Peters 1983). Because Maya ruins are far more common in MSF (see below) and exposed calcareous rock rarely exists in the deep soils common to LSF, it is not surprising that *B. alicastrum* is preferentially abundant in MSF. Finally, *Bursera simaruba* (L.) Sarg. is the tree with the highest IV in MSF, but ranks sixth in LSF. This tree is also one of the first tree species found in early successional plots. Hence, *B. simaruba* should receive more attention as a long-lived pioneer that may potentially affect the course of upland forest succession.

Timber extraction in the region, especially in the second half of the twentieth century (Ch. 3), has had an important impact on at least two other species: *Swietenia macrophylla* King, and *Cedrella odorata* L., which are among the most valuable timber tree species in the neotropics. According to Rodríguez Caballero (1944) and Miranda (1958), these species, particularly *S. macrophylla*, were relatively abundant in MSF at the beginning of the timber exploitation era. Today, neither species is common or dominant (Turner *et al.* 2001: 357). Furthermore, both have a higher IV in LSF than in MSF, apparently reflecting heavier logging in the more accessible MSF.

**Table 4.1.** Ten tree species with high relative importance value (RIV) in low-statured forest and medium-statured forest

| Low-statured forests | | Mid-statured forests | |
|---|---|---|---|
| Species | RIV | Species | RIV |
| *Haematoxylum campechianum* | 29.85 | *Manilkara zapota* | 16.64 |
| *Manilkara zapota* | 20.78 | *Bursera simaruba* | 16.08 |
| *Metopium brownei* | 18.60 | *Thouinia canescens* | 15.91 |
| *Coccoloba cozumelensis* | 9.88 | *Brosimum alicastrum* | 13.39 |
| *Bucida buceras* | 7.76 | *Ampelocera hottlei* | 12.03 |
| *Gymnopodium floribundum* | 5.92 | *Pouteria unilocularis* | 11.38 |
| *Gymnanthes lucida* | 5.87 | *Krugiodendron ferreum* | 10.94 |
| *Bursera simaruba* | 5.22 | *Eugenia buxifolia* | 6.21 |
| *Jatropha gaumeri* | 4.36 | *Lonchocarpus yucatanensis* | 6.06 |
| *Byrsonima bucidaefolia* | 4.35 | *Gymnopodium floribundum* | 4.75 |

*Note*: RIV = [relative abundance + relative frequency + relative basal area]/3 × 100. Species are ranked by descending RIV.

In terms of forest structure, LSF and MSF are clearly different. Trees in LSF are thinner and shorter, and carry higher epiphyte loads. In contrast, MSF have larger, taller trees, and are also more abundant in lianas. Interestingly, both forest types have the same tree basal area because LSF have smaller, but more trees than MSF. The most frequent canopy height category in MSF was 15–20m, while in LSF it was 5–10m. Finally, 6.5 per cent of forest canopy in LSF was 0–2m, while only 1.2 per cent of forest canopy in MSF fell into that category. Hence, the forest canopy in MSF is higher and less open than the forest canopy in LSF.

## 3.3. *Succession following Swidden Cultivation*

Understanding forest succession following agricultural abandonment in the tropics has become increasingly important for interpretation of landscape pattern and as the basis for management and restoration strategies (Aide *et al.* 1996; Finegan 1992; Purata 1986; Uhl 1982). In landscapes dominated by shifting agriculture, the maintenance of species composition at a regional scale depends on the dynamic mosaic of forest stands of different successional stages (Hernández-Xolocotzi 1959; Quintana *et al.* 1996; Uhl 1987). Importantly, either the cessation of human activities or the termination of succession can lead to a decline in regional diversity or an individual species' extinction.

Studies on forest succession after human land uses in the tropics have concentrated on abandoned pasture regeneration (Aide *et al.* 1996; Butler and Chazdon 1998; Dupuy and Chazdon 1998; Zahawi and Augspurger 1999) and on wet tropical forests (Dewalt, Schnitzer, and Denslow 2000; Holl 1999; Hughes, Kauffman, and Jaramillo 2000; Nicotra, Chazdon, and Iriarte 1999), and have paid less attention to the slash-and-burn dynamics on seasonally dry tropical forests. While pasture development and abandonment are unquestionably important processes affecting forests in the tropics of the Western Hemisphere (Bushbacher 1986), swidden or shifting agriculture is geographically more extensive and has affected tropical forests in this hemisphere for historically longer periods of time than have pastures (Gómez-Pompa 1987; Horn and Kennedy 2001; Kennedy and Horn 1997; Uhl 1987). Also, although seasonally dry tropical forests cover 42 per cent of all tropical world vegetation (Murphy and Lugo 1986, 1995), fewer studies have been conducted on this vegetation type compared to their wetter counterparts (Mooney, Bullock, and Medina 1995; Gentry 1995).

Contemporary swidden or *milpa* cultivation became prevalent in the southern Yucatán in the middle of the last century (Chs. 2, 3). Although mechanization in land clearance and agriculture has increased recently (Chs. 9, 10), swidden remains the dominant land use. Furthermore, some forest stands that were cleared for agriculture more than thirty years ago have been allowed to lie fallow, offering a unique opportunity to follow the course of forest regeneration in this seasonally dry tropical forest.

Results indicate that forest succession during fallow periods takes place relatively rapidly (Fig. 4.2). After twenty to thirty years of agricultural abandonment, successional stands have a tree-species composition that is no longer distinguishable from that of mature MSF. This rate of species recovery has been observed for other sites in the tropics (Aide *et al.* 1996), and may be the result of two important factors. First, much of the region had not experienced cultivation in over 1,000 years until the middle of the twentieth century (Ch. 2); thus, conditions for the re-establishment of tree species are ideal, in part because soil nutrient depletion may be low and because agricultural fields still occupy small openings embedded in a large matrix of mature forest. Second, these forests experienced a range of extensive and intensive human disturbances in ancient Maya times (Ch. 2) and, therefore, the pool of available trees may be comprised of species that are resilient to human disturbance. This last observation seems to be particularly important, since most of the species that become established early in the successional cycle remain dominant even in mature MSF, such as *B. simaruba* and *Dendropanax arboreum* (L.) Decne. and Planch. It is possible that tree species that are negatively affected by the conditions created during the shifting agricultural cycle might have become locally extinct or quite uncommon since the time of the Maya. On the other hand, it is important to recognize that in this study mature forests were defined as sites with no agricultural use in their recent (last fifty years at least) history, but other land uses, such as timber and non-timber forest product extraction might have given mature forests a more successional character.

Successional stands go through different structural changes during forest regeneration, conforming consistently with the patterns suggested by Gómez-Pompa and Vázquez-Yanes (1985). First, *Cecropia peltata* L., *Acacia* spp., *Lysiloma latisiliquum* (L.) Benth., *B. simaruba*, *Croton* spp. and *Coccoloba* spp. establish in the first five years, creating dense stands with an irregular and open canopy 2–5m high. Over the next ten years, the stand passes through a thinning phase, in which many of the small trees die, and later successional species take over, although some of the early established species remain. The most notable example is *B. simaruba*. After fifteen years of succession, a 500m$^2$ plot contains 12–15 tree species, almost the total number found in a similar mature forest. After twenty to thirty years, successional stands resemble mature forests in terms of composition, but with a higher density due to the smaller size of individual trees. Other structural features that 20–30-year-old successional stands are lacking in comparison with mature MSF include dead standing trees and large diameter lianas.

Recent, local land-use changes may potentially affect tree composition at the regional level. Mechanized agriculture, which appears to be associated with more-or-less permanent cropping of jalapeño (*Capsicum annuum* L.) rotated with other crops (Ch. 10) is expanding at the expense of mature and successional forest. As cropping intensifies, specific seed sources are diminished at the regional scale, potentially affecting regeneration once the land is abandoned (Holl 1999; McClanahan 1986). Secondly, as plots are used more frequently and intensively, they become dominated by early successional tree species and weedy species for longer periods of time (Mesquita 2000; Steininger 2000), potentially affecting the regeneration and presence of late successional tree species as part of the landscape mosaic (Peña-Claros 2001). And, as indicated in Ch. 5, intensive agriculture may be associated with significant declines in soil nutrient status.

Finally, as a consequence of the increasing scale and intensity of land use, the integrity of forests in the region may be affected by impeded or arrested succession (Lugo 1997; Niering and Goodwin 1974; Putz and Canham

1992). Across the southern Yucatán succession may be slowed or halted in a herbaceous and shrub stage due to the establishment and aggressive growth of *Pteridium aquilinum* (L.) Kuhn and *Viguiera dentata* (Cav.) Spreng. (Ch. 6) in cultivated fields exposed to repeated fires and agricultural cycles (Fig. 6.9). These species are pyrophilic, so swidden creates a positive feedback loop in which these species become dominant and inhibit forest regeneration. As swidden and associated agricultural practices increase across the region the extent of disturbed land supporting these species increases as well. It is costly in labor and capital to eradicate these species once established in large plots and consequently the land enters into an expanding phase of degradation. Succession is impeded on old agricultural areas, which become abandoned from productive use and farmers are forced to cut an equivalent amount of new farmland from forest.

## 4. *A Hypothesis from Forest Research*

As detailed in Ch. 2, the forests of the southern Yucatán peninsular region were substantially cleared and otherwise affected by diverse but large-scale Maya activities for several thousands of years up to about AD 1000. As yet, there is little evidence that most *bajos* were used for cultivation (Ch. 2), a finding supported in the forest surveys undertaken in this study. Eighty-five per cent of plots established in MSF revealed Maya artefacts (e.g. stone walls, mounds), whereas similar structures appeared in only 2 per cent of the plots in LSF. Consequently it appears that Maya deforestation concentrated on the rolling upland areas. The apparent magnitude of this deforestation raised an important question: Where was the regeneration stock for new forests when the region was abandoned by the Maya from active use? A possible answer rests in the results found in this research. Similarities between the LSF and MSF forests suggest that the forests growing on former Maya agricultural lands across the uplands may well have regenerated from Low Statured Forest left in *bajos*. This suggestion stands in opposition to ideas put forward elsewhere (Green 1973), which considers LSF solely as a successional stage of MSF. It is consistent, however, with the history of reforestation of many other regions following human abandonment; forest regeneration starts from species that grow in sites that are marginal for human activities, and which harbor a large pool of available species. It is possible that *bajos* played this role in the southern Yucatán leading to the high tree species affinity between MSF and LSF.

## 5. *Conclusions and Observations from the Forest*

As extensive as the forests of the southern Yucatán peninsula are, evidence of human impress are widespread. As demonstrated in Chs. 2 and 3, the ancient landscape was highly altered by the Maya, abandoned and largely unoccupied from about AD 900–1000 to the 1900s, and progressively occupied and altered from the middle of the last century to the present. Of the modern uses, forest extraction appears to have had the first, large impacts on the region's forests. The extraction of *S. macrophylla* and *C. odorata* must have had several impacts on forest structure and tree-species composition (Argüeyes 1991; Flachsenberg and Galletti 1998; Snook 1998). The large-diameter individuals of these species appear to have been logged from this forest, comparing current diameters to those described in the past (Snook 1998). In addition, the extraction of these species and the sap of *M. zapota* must have been accompanied by many other impacts including hunting and widespread changes in the relative importance of many wildlife species, most notably large predators and browsers.

These more recent changes notwithstanding, the overwhelming majority of the region is covered with the two basic forest types identified here, LSF and MSF, and successional growth largely taken from MSF. Of these, MSF is the forest type that is more likely to undergo changes due to agriculture, because it grows in soils that are more adequate for that land use. This explains why successional growth, including invasive species, has increased as a proportion of the landscape over the last forty years or more, especially east of the north–south roadway bisecting the region (Fig. 1.1) and increasingly in the southeast quadrant of the region (Ch. 6). The creation of the Calakmul Biosphere Reserve, inasmuch as it can protect forest areas from other direct uses, promises to yield many insights into long-term forest dynamics in this region. Because three of the six regions of study are located within the limits of the Biosphere reserve, it will be possible in the near future to compare the growth and dynamics of forests within the reserve with those exposed to other land uses outside it. In particular, it will be interesting to compare tree-fall dynamics, basal area accumulation, and the response of plots to known human and natural disturbances.

## REFERENCES

Acopa, D., and E. Boege (1998). 'The Maya Forest in Campeche, México: Experiences in Forest Management at Calakmul', in R. B. Primack, D. Bray, H. A. Galleti, and I. Ponciano (eds.), *Timber, Tourists and Temples, Conservation and Development in the Maya Forest of Belize, Guatemala, and Mexico*. Washington DC, Island Press, 81–97.

Aide, T. M., J. K. Zimmerman, M. Rosario, and H. Marcano (1996). 'Forest Recovery in Abandoned Cattle Pastures Along an Elevational Gradient in Northeastern Puerto Rico'. *Biotropica* 28: 537–48.

Argüeyes, L. A. (1991). 'Plan de manejo forestal para el bosque tropical de la empresa ejidal Noh-Bec'. M. S. Thesis, Universidad Autónoma de Chapingo, Mexico.

Barrera, A., A. Gómez-Pompa, and C. Vázquez-Yanes (1977). 'El manejo de las Selvas por los Mayas: sus Implicaciones Silvícolas y Agrícolas'. *Biotica* 2: 47–61.

Boose, E. R., D. R. Foster, B. P. A., and B. Hall (2003). 'Geographical and Historical Variation in Hurricanes across the Yucatán Peninsula', in A. Gómez-Pompa, M. F. Allen, S. Fedick, and J. J. Jiménez-Osornio (eds.), *Lowland Maya Area: Three Millennia at the Human-Wildland Interface*. Oxford, Haworth Press, 193–213.

Brokaw, N. V. L., and L. R. Walker (1991). 'Summary of the Effects of Caribbean Hurricanes on Vegetation'. *Biotropica* 23: 442–7.

Bullock, S. H., H. A. Mooney, and E. Medina (eds.) (1995). *Seasonally Dry Tropical Forests*. Cambridge, Cambridge University Press.

Bushbacher, R. J. (1986). 'Tropical Deforestation and Pasture Development'. *Bioscience* 36: 22–8.

Butler, B. J., and R. L. Chazdon (1998). 'Species Richness, Spatial Variation, and Abundance of the Soil Seed Bank of a Secondary Tropical Rain Forest'. *Biotropica* 30: 214–22.

Condit, R., S. P. Hubbell, J. V. Lafrankie, R. Sukumar, N. Manokaran, R. Foster, and P. Ashton (1996). 'Species-Area and Species-Individual Relationships for Tropical Trees: A Comparison of Three 50-ha Plots'. *Journal of Ecology* 84: 549–62.

Cooper-Ellis, S., D. R. Foster, G. Carlton, and A. Lezberg (1999). 'Forest Response to Catastrophic Wind: Results from an Experimental Hurricane'. *Ecology* 80: 2683–96.

Delfín Gonzalez, H., T. V. Parra, and G. C. Echazarreta (eds.) (1995). *Conocimiento y manejo de las selvas de la Península de Yucatán*. Mérida, Yucatán, Universidad Autónoma de Yucatán.

Dewalt, S. J., S. A. Schnitzer, and J. S. Denslow (2000). 'Density and Diversity of Lianas along a Chronosequence in a Central Panamanian Lowland Forest'. *Journal of Tropical Ecology* 16: 1–9.

Dupuy, J. M., and R. L. Chazdon (1998). 'Long-Term Effect of Forest Regrowth and Selective Logging on the Seed Bank of Tropical Forests in NE Costa Rica'. *Biotropica* 30: 223–37.

Finegan, B. (1992). 'The Management Potential of Neotropical Secondary Lowland Rain Forest'. *Forest Ecology and Management* 47: 295–321.

Flachsenberg, H., and H. Galletti (1998). 'Forest Management in Quintana Roo, Mexico', in R. B. Primack, D. Bray, H. A. Galletti, and I. Ponciano (eds.), *Timber, Tourists, and Temples: Conservation and Development in the Maya Forest of Belize, Guatemala, and Mexico*. Washington DC, Island Press. 47–60.

Flores, J. S., and I. Espejel Carvajal (1994). *Tipos de vegetación de la Península de Yucatán*: Mérida, Yucatán, Universidad Autónoma de Yucatán.

Foster, D. R. (1988). 'Species and Stand Response to Catastrophic Wind in Central New England, U.S.A'. *Journal of Ecology* 76: 135–51.

Foster, D. R. (1992). 'Land-Use History (1730–1990) and Vegetation Dynamics in Central New England, USA'. *Journal of Ecology* 80: 753–72.

Foster, D. R., M. Fluet, and E. R. Boose (1999). 'Human or Natural Disturbance: Landscape-Scale Dynamics of the Tropical Forests of Puerto Rico'. *Ecological Applications* 9: 555–72.

Foster, D. R., T. Zebryk, P. Schoonmaker, and A. Lezberg (1992). 'Post-Settlement History of Human Land-Use and Vegetation Dynamics of a *Tsuga canadensis* (Hemlock) Woodlot in Central New England'. *Journal of Ecology* 80: 773–86.

Foster, D. R., J. D. Aber, J. M. Melillo, R. D. Bowden, and F. A. Bazzaz (1997). 'Forest Response to Disturbance and Anthropogenic Stress'. *Bioscience* 47: 437–45.

Gentry, A. H. (1995). 'Diversity and Floristic Composition of Neotropical Dry Forests', in S. H. Bullock, H. A. Mooney, and E. Medina (eds.), *Seasonally Dry Tropical Forests*. Cambridge, Cambridge University Press, 146–94.

Gómez-Pompa, A. (1987). 'On Maya Silviculture'. *Mexican Studies* 3: 1–17.

Gómez-Pompa, A., and C. Vázquez-Yanes (1985). 'Estudios sobre la regeneración de selvas en regiones calido-húmedas de Mexico', in A. Gómez-Pompa, and S. Del Amo (eds.), *Investigaciones sobre la regeneración de selvas altas en Veracruz, Mexico*. México DF, Alhambra, 1–25.

Green, E. (1973). 'Location Analysis of Prehistoric Maya Sites in Northern British Honduras'. *American Antiquity* 38: 279–93.

Hernández-Xolocotzi, E. (1959). 'La agricultura en la Península de Yucatán', in E. Beltrán (ed.), *Los Recursos Naturales del Sureste y su Aprovechamiento*. México DF, Instituto Mexicano de Recursos Naturales Renovables AC, 3–57.

Holl, K. D. (1999). 'Factors Limiting Tropical Rain Forest Regeneration in Abandoned Pasture: Seed Rain, Seed Germination, Microclimate, and Soil'. *Biotropica* 31: 229–42.

Horn, S. P., and L. M. Kennedy (2001). 'Pollen Evidence of Maize Cultivation 2700 B.P. at La Selva Biological Station, Costa Rica'. *Biotropica* 33: 191–6.

Hughes, R. F., J. B. Kauffman, and V. J. Jaramillo (2000). 'Ecosystem-Scale Impacts of Deforestation and Land Use in a Humid Tropical Region of Mexico'. *Ecological Applications* 10: 515–27.

Ibarra-Manríquez, G. (1996). 'Biogeografía de los árboles nativos de la península de Yucatán: un enfoque para evaluar su grado de conservación'. Doctoral Thesis, Universidad Nacional Autónoma de México.

Justo Sierra, C. (1998). *Breve história de Campeche*. (1st edn.). México DF, Fondo de Cultura Económica/El Colegio de Mexico.

Kennedy, L. M., and S. P. Horn (1997). 'Prehistoric Maize Cultivation at the La Selva Biological Station, Costa Rica'. *Biotropica* 29: 368–70.

Konrad, H. W. (1996). 'Caribbean Tropical Storms'. *Revista Mexicana del Caribe* 1: 98–130.

Lambert, J. D. H., and T. Arnason (1982). 'Ramon and Maya Ruins: An Ecological Not an Economic Reason'. *Science* 216: 298–9.

Lugo, A. E. (1997). 'The Apparent Paradox of Reestablishing Species Richness on Degraded Lands with Tree Monocultures'. *Forest Ecology and Management* 99: 9–19.

Lundell, C. L. (1938). 'The 1938 Botanical Expedition to Yucatán and Quintana Roo, Mexico'. *Carnegie Institution of Washington Year Book, 1937–1938* 37: 7–11.

McClanahan, R. R. (1986). 'The Effect of a Seed Source on Primary Succession in a Forest Ecosystem'. *Vegetatio* 65: 175–18.

McCune, B., and M. J. Mefford (1999). *Multivariate Analysis of Ecological Data Version 4.01*. Gleneden Beach, Oreg., MjM Software.

McLachlan, J. S., D. R. Foster, and F. Menalled (2000). 'Anthropogenic Ties to Late-Successional Structure and Composition in Four New England Hemlock Stands'. *Ecology* 81: 717–33.

Martínez-Ramos, M. (1985). 'Claros, ciclos vitales en los árboles tropicales y la regeneración natural de las selvas altas perenifolias', in A. Gómez-Pompa and S. Del Amo (eds.), *Investigaciónes sobre la regeneración de selvas altas en Veracruz, México*. México DF, Alhambra, 191–239.

——(1994). 'Regeneración natural y diversidad de especies arbóreas en selvas humedas'. *Boletín de la Sociedad Botánica de México* 54: 179–224.

Mesquita, R. C. G. (2000). 'Management of Advanced Regeneration in Secondary Forests of the Brazilian Amazon'. *Forest Ecology and Management* 130: 131–40.

Miranda, F. (1958). 'Estudios acerca de la vegetación', in E. Beltrán (ed.), *Los Recursos Naturales del Sureste y su Aprovechamiento*. México DF, Instituto Mexicano de Recursos Naturales Renovables AC, 215–71.

Mooney, H. A., S. H. Bullock, and E. Medina (1995). 'Introduction', in S. H. Bullock, H. A. Mooney, and E. Medina (eds.), *Seasonally Dry Tropical Forests*. Cambridge, Cambridge University Press, 1–8.

Murphy, P. G., and A. E. Lugo (1986). 'Ecology of Tropical Dry Forests'. *Annual Review of Ecology and Systematics* 17: 67–88.

—— (1995). 'Dry Forests of Central America and the Caribbean', in S. H. Bullock, H. A. Mooney, and E. Medina (eds.), *Seasonally Dry Tropical Forests*. Cambridge, Cambridge University Press, 9–34.

Nicotra, A. B., R. L. Chazdon, and S. V. B. Iriarte (1999). 'Spatial Heterogeneity of Light and Woody Seedling Regeneration in Tropical Wet Forests'. *Ecology* 80: 1908–26.

Niering, W. A., and R. H. Goodwin (1974). 'Creation of Relatively Stable Shrublands with Herbicides: Arresting "Succession" on Rights-of-Way and Pastureland'. *Ecology* 55: 784–95.

Olmsted, I., R. Durán García, J. A. González-Iturbe, J. Granados Castellanos, J. C. Trejo Torres, D. Zizumbo V., G. Campos Ríos, and G. Ibarra-Mariquez (1995). 'Diagnostico del conocimiento y manejo de las selvas de la peninsula de Yucatán', in G. H. Delfín, T. V. Parra, and G. C. Echazarreta (eds.), *Conocimiento y manejo de las selvas de la Península de Yucatán*. Mérida, Yucatán, Universidad Autónoma de Yucatán, 139–62.

Peña-Claros, M. (2001). *Secondary Forest Succession, Processes Affecting the Regeneration of Bolivian Tree Species*. Utrecht, Promab Scientific Series 3.

Peters, C. M. (1983). 'Observations on Maya Subsistence and Ecology of a Tropical Tree'. *American Antiquity* 48: 610–15.

Purata, S. E. (1986). 'Floristic and Structural Changes during Old-Field Succession in the Mexican Tropics in Relation to Site History and Species Availability'. *Journal of Tropical Ecology* 2: 257–76.

Putz, F. E., and C. D. Canham (1992). 'Mechanisms of Arrested Succession in Shrublands: Root and Shoot Competition between Shrubs and Tree Seedlings'. *Forest Ecology and Management* 49: 267–75.

Quintana-Ascencio, P. F., M. González-Espinosa, N. Ramírez-Marcial, G. Domínguez-Vázquez, and M. Martínez-Icó (1996). 'Soil Seed Banks and Regeneration of Tropical Rain Forest from Milpa Fields at the Selva Lacandona, Chiapas, Mexico'. *Biotropica* 28: 192–209.

Rodríguez Caballero, R. (1944). 'La explotación de los montes de Caoba (*Swietenia macrophylla* King) en el territorio de Quintana Roo'. Ingeniero Agrónomo, Chapingo, México, M. S. Thesis.

Rzedowski, J. (1981). *Vegetación de México*. México DF, Limusa.

Sánchez, S. O., and G. A. Islebe (1999). 'Hurricane Gilbert and Structural Changes in a Tropical Forest in South-Eastern Mexico'. *Global Ecology and Biogeography* 8: 29–38.

Schulze, M. D., and D. F. Whitacre (1999). 'A Classification and Ordination of the Tree Community of Tikal National Park, Petén, Guatemala'. *Bulletin of the Florida Museum of Natural History* 41: 169–297.

Shepherd, J. D. (1975). 'The Phytosociology of a Tropical Seasonal Forest'. Master's dissertation, University of Wisconsin-Madison.

Snook, L. K. (1998). 'Sustaining Harvests of Mahogany (*Swietenia macrophylla* King) from Mexico's Yucatán Forests: Past, Present, and Future', in R. B. Primack, D. Bray, H. A. Galletti, and I. Ponciano (eds.), *Timber, Tourists, and Temples: Conservation and Development in the Maya Forest of Belize, Guatemala, and Mexico*. Washington DC, Island Press, 61–80.

Standley, P. C. (1930). 'Flora of Yucatan'. *Field Museum of Natural History Botanical Series* 3: 157–89.

Steininger, M. K. (2000). 'Secondary Forest Structure and Biomass Following Short and Extended Land-Use in Central and Southern Amazonia'. *Journal of Tropical Ecology* 16: 689–708.

Turner, B. L. II., S. Cortina Villar, D. Foster, J. Geoghegan, E. Keys, P. Klepeis, D. Lawrence, P. Macario Mendoza, S. Manson, Y. Ogneva-Himmelberger, A. B. Plotkin, D. R. Pérez-Salicrup, R. Roy Chowdhury, B. Savitsky, L. Schneider, B. Schmook, and C. Vance (2001). 'Deforestation in the Southern Yucatán Peninsular Region: An Integrative Approach'. *Forest Ecology and Management* 154: 353–70.

Uhl, C. (1982). 'Recovery Following Disturbances of Different Intensities in the Amazon Rain Forest of Venezuela'. *Interciencia* 7: 19–24.

—— (1987). 'Factors Controlling Succession Following Slash-and-Burn Agriculture in Amazonia'. *Journal of Ecology* 75: 377–407.

Whigham, D. F., J. F. Lynch, and M. B. Dickinson (1998). 'Dynamics and Ecology of Natural and Managed Forests in Quintana Roo', in R. B. Primack, D. Bray, H. A. Galleti, and I. Ponciano (eds.), *Timber, Tourists and Temples: Conservation and Development in the Maya Forest of Belize, Guatemala, and Mexico*. Washington DC, Island Press. 267–81.

Wright, A. C. S., D. H. Romney, R. H. Arbuckle, and V. E. Vial (1959). *Land in British Honduras*. London, Her Majesty's Stationery Office.

Zahawi, R. A., and C. K. Augspurger (1999). 'Early Plant Succession in Abandoned Pastures in Ecuador'. *Biotropica* 31: 538–52.

# 5

# Recovery of Nutrient Cycling and Ecosystem Properties Following Swidden Cultivation: Regional and Stand-Level Constraints

*Deborah Lawrence and David R. Foster*

## 1. *Tropical Dry Forest Dynamics: Mexico and the Southern Yucatán Peninsular Region*

The total area of agricultural systems in tropical Mexico increased by 64 per cent from 1977 to 1992—a mean annual deforestation rate of 1.9 per cent (Cairns *et al.* 2000). In all likelihood, this rate has continued for the past ten years. Dry tropical forest covers 8 per cent of Mexico and is subject to conversion for agricultural use (Trejo and Dirzo 2000). Because the southern Yucatán contains the largest contiguous block of dry tropical forest in Mexico and Central America, understanding the biogeochemical consequences of land-use change there is important for effective national and international conservation and development efforts.

Over the past four decades the southern Yucatán peninsular region has undergone an increasing amount and intensity of land use (Chs. 3, 9, 10).

Many thanks to the farmers of El Refugio, Nicolas Bravo, and Arroyo Negro who worked with us. We would also like to thank Juan Mendez-Diaz for essential assistance and leadership in the field. Thanks to numerous students for excellent help in the field and the lab: Fausto Bolom, Jakara Hubbard, Larissa Read, Jessica Sisco, Heidi Wasson, and Tana Wood. Without the logistical support of Pedro Macario Mendoza of ECOSUR, this project would not have been possible. For funding, we thank the sponsors noted in Ch. 1 and the University of Virginia.

These land uses, many focused on swidden practices, alter the structure and function of forested lands and often generate new feedbacks in terms of subsequent human use. Consequently, a major goal in assessing regional environmental change is to understand how biogeochemical processes respond to land-use change, emphasizing the potential of a human-dominated landscape to sustain continued human use. One of the greatest challenges in these studies is to untangle the effects of environmentally induced variation from, for example, climate, geology, or natural disturbance, from that induced by human activity. In the SYPR project the approach to this challenge has been to investigate variation in ecosystem processes in several study sites across the dominant environmental gradients while focusing on the influence of local, human-controlled factors within a given area.

In the southern Yucatán peninsular region annual precipitation increases by more than 50 per cent over a distance of 120km. Median annual precipitation varies from about 900mm in the northern part of the study area to about 1,400mm in the southern part. This dramatic gradient overlies a seasonal pattern shared by all sites regardless of their total annual precipitation. Rainfall is highly variable, with a pronounced dry period lasting from four to six months, depending on latitude (Fig. 5.1). The range in precipitation observed in the study area encompasses approximately 50 per cent of the variation in precipitation of dry tropical forests worldwide (Murphy and Lugo 1986). Thus, the gradient represents a broad sampling of the environment of the region and provides data relevant to many other dry forests.

Studying several sites along this gradient makes it possible to explore the manner in which interannual variability or moderate climate differences affect ecosystem function in intact forests and those recovering from past land use. The major stand-level factors determining the rate and dynamics of ecosystem processes are forest age, cultivation history, and management (land-use type, landowner preferences). A mosaic of agriculture, secondary forest, and mature forest laid upon a dry to moist tropical forest gradient is the matrix embodying the consequences of land-use change in the region.

Because swidden (locally, *milpa*) agriculture is the predominant cause of deforestation in the region (Ch. 9), forest recovery following this kind of disturbance was the central focus of investigations. The main objective was to assess how nutrient cycling, productivity, and biomass change during the process of secondary forest succession. Plot-based studies were placed into a broader context by testing for the influence of added local and regional

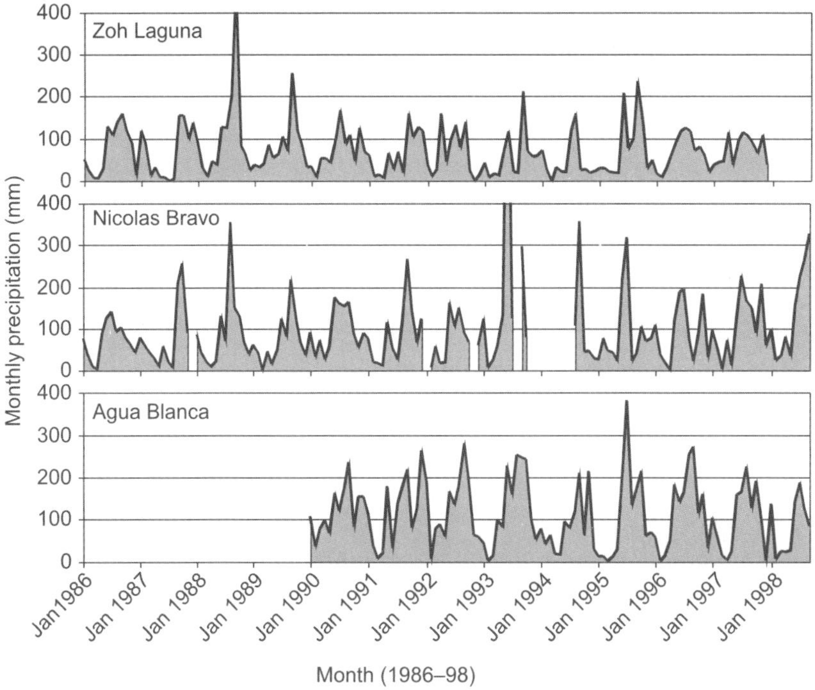


**Fig. 5.1.** Monthly precipitation for the period 1986–98.
For the three villages nearest to the principal study sites: Zoh Laguna (*c*.900mm/yr), 20km south of El Refugio, Nicolas Bravo (*c*.1,100mm/yr), and Agua Blanca (*c*.1,400mm/yr), 70km east-northeast of Arroyo (courtesy of Instituto Nacional de Estadística, Geografía e Informática, Nationale Environmentales Geographico, INEGI, Mexico). Zero precipitation values indicate missing data. Data from extreme storm events in 1988 and 1993 were clipped. All dates refer to calendar years.

drivers. Specifically, the influence of cultivation history (including time since abandonment and cultivation intensity) and geographical location (along established environmental gradients) on patterns of forest recovery was evaluated. This approach enables an understanding of the multiple constraints on the trajectory of individual patches on the landscape. Owing to both stochastic and deterministic processes, the landscape under swidden cultivation consists of patches at various stages of recovery. The two-tiered conceptual approach, investigating local sites within a regional framework, also makes it possible to characterize the structure and function of the entire landscape at any one point in time, and to project it into the future, through integration with predictive models (Chs. 12, 13). Before making projections, it is necessary to understand current constraints on forest recovery in the region. With that goal in mind, the specific questions addressed in this chapter are:

1. What are the regional environmental influences on forest processes at the scale of 10s to 100s of kilometers? Do soils and precipitation vary predictably in time and space?
2. How do forest processes vary along these environmental gradients?
3. Within a given region, how do they vary as a function of forest age?
4. How does cultivation history influence the rate of recovery?

## 2. *Methods*

### 2.1. *Study Sites*

Fieldwork was concentrated in three study sites positioned 60–120km apart in order to sample the regional precipitation gradient. Thirteen plots were established in El Refugio, the northernmost and the driest site and at Nicolas Bravo, an area of intermediate rainfall to the south and east. Further south in the wettest part of the region, in the village of Arroyo Negro, ten plots were sampled (Fig. 1.1). In all three areas, samples included both mature forest stands (never cleared for agriculture in recent history) and secondary forest fallows regenerating after swidden cultivation of maize. None of the stands had received inputs of fertilizer, pesticide, or herbicide. The secondary forest stands ranged in age from 2 to 25 years since the last harvest at the plot (Table 5.1). The exact age of the mature forests is impossible to characterize due to the absence of long-term historical data or tree ring chronologies. Evaluation of biomass suggests, however, that they are at least 50 years old (Read and Lawrence 2003a). Despite the absence of stumps or other evidence of past disturbance some or all of the forests may have experienced selective timber extraction in the last 40–100 years (Ch. 3; Klepeis and Turner 2001).

Secondary forest sites were selected from the database compiled by the socio-economic research team. Information on the farmers and details on the history of their parcels made it possible to control many of the variables that confound studies of recovery following deforestation. These factors, interesting in their own right, include differences in prior land-use type,

**Table 5.1.** Location, ownership, and age in 1998 of parcels included in nutrient cycling studies

| Location | Landowner | Forest age (years) |
|---|---|---|
| El Refugio (ER) | 1 | 2 |
| 18° 49' N | 1 | 5 |
| 89° 23' W | 1 | 12 |
| | 2 | 4 |
| | 2 | 8 |
| | 2 | 12 |
| | 3 | 3 |
| | 3 | 10 |
| | 3 | 12 |
| | 4 | 8 |
| | 5 | mature |
| | 6 | mature |
| | Reserva ER | mature |
| Nicolas Bravo (NB) | 7 | 3 |
| 18° 27' N | 7 | 6 |
| 88° 56' W | 7 | 25 |
| | 8 | 8 |
| | 6 | 18 |
| | 9 | 5 |
| | 9 | *c.*25 |
| | 10 | 5 |
| | 10 | 16 |
| | 10 | 24 |
| | Reserva NB1 | mature |
| | Reserva NB2 | mature |
| | Reserva NB4 | mature |
| Arroyo Negro (AN) | 11 | 4 |
| 17° 53' N | 11 | 8 |
| 89° 17' W | 11 | 18 |
| | 12 | 5 |
| | 12 | 9 |
| | 12 | 15 |
| | 13 | 5 |
| | 13 | 7 |
| | 14 | mature |
| | 15 | mature |

intensity of cultivation, and use of chemical inputs or machinery (e.g. Bushbacher, Uhl, and Serrao 1988; Fernandez and Sanford 1995; Hughes, Kauffman, and Jaramillo 1999). In each study site, several parcels were located with young, middle-aged, and old fallow plots. Although clearing and cultivation practices may differ by owner, within a farmer's parcel it is possible to minimize management differences. Thus, a major effort was made in the establishment of chronosequences to sample the effects of forest age, rather than the effects of contrasting management approaches. In this sense, the true sampling unit was the farmer, with the various parcels owned by each farmer representing one trajectory of forest recovery. This approach, exploiting the relative uniformity of practice within a farm, is unusual and was facilitated by the cross-disciplinary nature of our project.

## 2.2. *Regional Environmental Gradients in Soil*

At each site, composite soil samples were collected at four depths: 0–15cm, 15–30cm, 30–60cm and 90–100cm. For this study, analyses are presented for surface soils (0–15cm) only. Sieved, air-dried samples were analyzed for physical and chemical properties by Brookside Laboratories (New Knoxville, Ohio). Mature forest soils were compared in order to assess regional soil differences, while avoiding the confounding influences of shifting cultivation. To determine changes in soil properties as a function of forest age, sites were divided into three secondary forest age classes (2–5 y.o., 6–10 y.o., and 12–25 y.o.) plus mature forests.

## 2.3. *Productivity, Biomass, and Nutrient Cycling*

One 500m$^2$ plot was established in each stand. Litter was collected in four 1m$^2$ litter traps arrayed across each plot on a monthly basis from November 1998 through January 2000. Fine litter (leaves, reproductive material, bark and wood < 2cm in diameter) was separated from coarse litter (bark and wood > 2cm in diameter), dried at < 70 °C and weighed. The dry mass of fine litter was used as an index of productivity whereas the nutrient content (N and P) of the fine litter was used to gauge the effects of land-use change on nutrient cycling across the sites (cf. Odum 1969; Proctor *et al.* 1983; Vitousek 1984). Above-ground biomass, including living vegetation and detritus, was sampled in each plot. To estimate the biomass of detritus on the forest floor, all fine and coarse litter was sampled in four 1m$^2$ areas, one adjacent to each of the litter traps. The diameters of all stems in the plot with diameter at breast height (dbh) > 10cm were measured, and a nested plot design was used to sample smaller stems (see Read and Lawrence 2003a). On a systematically sampled subset of trees from all plots, measurements of height were taken, and regressions were created to estimate heights for all measured trees. Then, standing live biomass was estimated from dbh, height, and wood density measures, using established regressions (cf. Martinez-Yrizar *et al.* 1992).

## 3. *Results and Discussion*

### 3.1. *Effects of Seasonal Variation in Rainfall*

As in the dry forests of western Mexico and northern Yucatán, and in the wetter forests of Guatemala, Costa Rica, and Panama to the south, litter production is strongly tied to the timing of precipitation (Frankie. Baker, and Opler 1974; Kunkel Westphal and Kunkel 1979; Martinez-Yrizar and Sarukhan 1990; Whigham *et al.* 1990; Wieder and Wright 1995). In the southern Yucatán peninsular region, both the production and nutrient concentration of litter responded to seasonal drought from December to May. Fine litter production was relatively constant throughout most of the year, at 25–30g/m$^2$/mo. After several months of extremely dry weather, however, litterfall increased fivefold, apparently due to water stress (Holbrook, Whitbeck, and Mooney 1995; Reich and Borchert 1984). Thirty per cent of total annual litter production fell during the peak, between March and April, and roughly 60 per cent fell from January to May (Fig. 5.2; also Lawrence n.d.).

Litter nutrient concentrations reached their minima during peak litterfall, presumably because forest trees reabsorbed essential nutrients prior to massive leaf loss (Chapin 1980; Killingbeck 1996) (Fig. 5.3). On average, litter P concentrations were reduced 47 per cent from the maximum observed in a given site; in contrast, litter N concentrations were reduced by only 33 per cent (Read and Lawrence 2003b.). Lower nutrient concentrations in dry season litter have been reported in both wetter and drier forests (McGrath *et al.* 2001; Swift, Russell-Smith, and Perfect 1981; Wieder and Wright 1995). Despite reductions in nutrient concentration, the dramatic increase in litter mass during the dry season is likely to produce a pulse of nutrients that become available with the onset of the rains in June, lasting for weeks to months (Campo, Jaramillo, and Maass 1998; Lawrence n.d.; McGrath *et al.* 2001; Singh *et al.* 1989; Swift, Russell-Smith, and Perfect 1981). The entire region is likely to exhibit a similar pattern of highly seasonal soil nutrient dynamics.

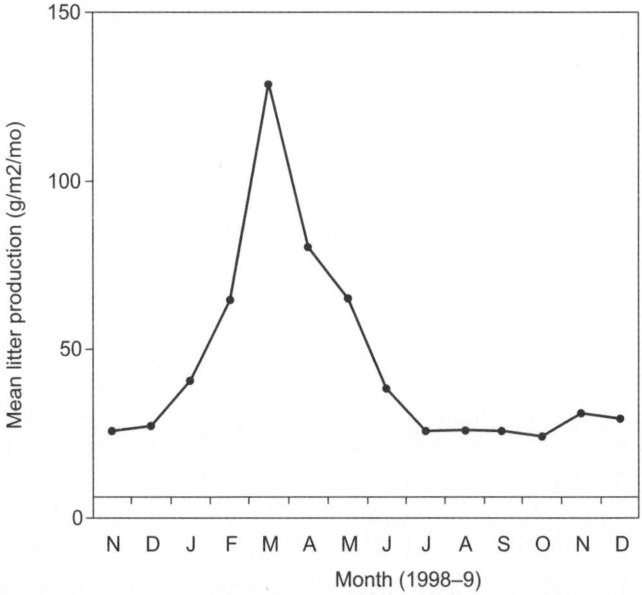


**Fig. 5.2.** Seasonal pattern in fine litter production (leaves, twigs, and wood <2cm in diameter). Mean ± standard error for all 36 plots.

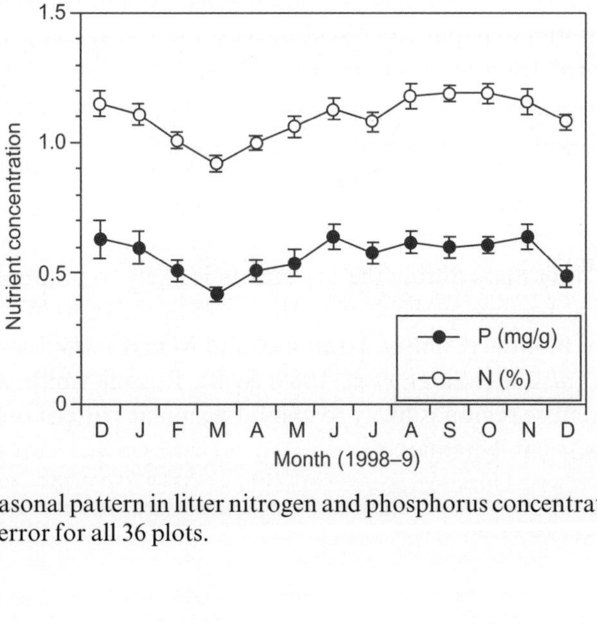


**Fig. 5.3.** Seasonal pattern in litter nitrogen and phosphorus concentrations. Mean ± standard error for all 36 plots.

### 3.3. *Soil Gradients*

Although the study region is relatively homogeneous in topography and parent material, variability is observed in the physical and chemical properties of soils in mature forest (Table 5.2). Some of this variation may be the result of autogenic processes occurring as a result of interactions between vegetation and soils through geologic time. Some may result from the historic practices of colonists or from ancient Mayan activity. An important part of the variation, especially in soil organic matter, is most likely the result of changes in ecosystem processes along the regional precipitation gradient. In particular, the southern sites tended to have a higher organic matter content, as expected if litter inputs were higher due to greater moisture availability resulting from higher annual precipitation (cf. Vitousek 1984). Other aspects of soil variation, such as the higher percentage of sand-sized particles at Arroyo Negro may relate to subtle differences in topographic position of the plots sampled. Over time, smaller particles tend to move downslope, leaving a coarser soil texture at the top of a catena relative to the lower portions (Brady 1990). Despite moderate topography and elevation (100–300m amsl) throughout the region, mature sites in Arroyo Negro were somewhat higher and closer to the ridge top.[1] Sites in El Refugio and Nicolas Bravo were lower on the slope. Local variation in soil moisture consistent with differences in soil texture was proposed to explain differences in ridge-top and mid-slope plots in western Mexico (Martinez-Yrizar and Sarukhan 1990).

As a result of parent material rich in calcium carbonate, soil pH was generally quite high across the region (7.4–7.7). The calcium content of soils in Nicolas Bravo, however, was significantly lower than that of the other two sites. This difference was reflected in a lower pH and lower concentration of total exchangeable cations. Significantly lower levels of soluble sulfur and significantly higher concentrations of detrimental and potentially toxic metals (aluminum, copper, iron, and manganese) suggest that overall fertility may be lower in Nicolas Bravo. Perhaps this lower fertility reflects the longer history of agriculture there. Unlike the other study sites which were

[1] Arroyo Negro, however, is located near the nexus of the borders of Belize and El Petén, Guatemala, as well as adjacent to the ephemeral stream, Arroyo Negro. Belize and Petén hold patches (elongated strips) of land containing granitic outwash, usually sands, apparently from the Maya Mountains to the south (Ch. 2). As an alternative explanation, the sands found in this study may be related to the same source or to downstream outwash from the sand strips to the south.

**Table 5.2.** Characteristics of mature forest soils in the three focal study areas (ER, n = 3; NB, n = 3; AN, n = 5 2)

| | El Refugio | Nicolas Bravo | Arroyo Negro | P-value |
|---|---|---|---|---|
| *Physical properties* | | | | |
| Clay (%) | 23±4 | 19±6 | 14±1 | 0.51 |
| Silt (%) | 25±4 | 33±4 | 27±3 | 0.39 |
| Sand (%) | 51±2 [a] | 47±2[a] | 59±2 [b] | 0.022 |
| Organic matter (%) | 11.4±1.1[a] | 14.5±0.6[ab] | 21.8±5.6[b] | 0.077 |
| pH | 7.7±0.0[a] | 7.4±0.1[b] | 7.7±0.1[ab] | 0.068 |
| TEC (meq/g) | 134±18[a] | 54±10[b] | 139±4[a] | 0.012 |
| *Chemical properties (ppm)* | | | | |
| Sulfur | 58±4[a] | 42±3[b] | 70±7[a] | 0.014 |
| Calcium | 25,700±3,600[a] | 9,800±2,000[b] | 26,900±800[a] | 0.011 |
| Magnesium | 495±21 | 469±43 | 489±54 | 0.87 |
| Potassium | 384±67[ab] | 441±45[a] | 197±14[b] | 0.071 |
| Sodium | 25±3 | 23±2 | 29±7 | 0.64 |
| Manganese | 30±5[a] | 83±29[b] | 18±3[a] | 0.023 |
| Boron | 2.0±0.2 | 2.0±-.3 | 2.6±0.4 | 0.32 |
| Iron | 6.7±1.3[a] | 23±6.5[b] | 3.5±0.5[a] | 0.053 |
| Copper | 1.7±0.1[a] | 2.4±0.1[b] | 1.5±0.1[a] | 0.003 |
| Zinc | 1.2±0.0 | 1.5±0.4 | 1.6±0.2 | 0.63 |
| Aluminum | 30±7[a] | 164±55[b] | 17±2[a] | 0.066 |

*Note*: P value indicates significance of separate analyses of variance (ANOVAs) by region. Significant differences among regions indicated by different letters.

colonized 20–30 years ago, Nicolas Bravo has been settled since the 1950s. This long history is correlated with large areas of the invasive bracken fern whose dominance may be facilitated by nutrient-poor soils (Chs. 4, 6; Suazo 1998).

Potassium levels were significantly higher in Nicolas Bravo, however, suggesting that turnover of organic matter is either more rapid or that more material is cycling through the forest-floor litter layer and subject to leaching. This result is not incongruent with soil fertility limitations on productivity—cycling through the organic layer is all the more important when the mineral soil is depleted in essential nutrients (Chapin 1981; Jordan and Herrera 1981). The most parsimonious explanation for enrichment in metals and loss of calcium may be topography. Nicolas Bravo is at a slightly lower elevation, on a flatter part of the study area away from the moderate spine of the peninsula. Thus, it is relatively older geologically.[2] Weathering

[2] Personal communication from Grant Goodell, Research Professor, Environmental Sciences, University of Virginia.

has been occurring longer in this area than in the more hilly areas we sampled in the northwest (ER) and southwest (AN). Following its release due to weathering, calcium may have been lost from the system through leaching over time (Brady 1990). However, ancient Mayan land use cannot be ruled out as having had disproportionate and negative effects on the soils around Nicolas Bravo. Ruins are extremely abundant from Nicolas Bravo west to Becan (Turner 1974). Yet, El Refugio is also potentially in a zone of intense Mayan activity in the past. It seems most likely that the intensity of recent historic land use best explains the difference among sites.

### 3.4. *Regional Environmental Effects on Ecosystem Function*

Regional differences in both total annual precipitation and soil characteristics may contribute to differences in structure and function among the study sites. To isolate regional influence on forest processes from the effects of forest age and cultivation history, mature forests were examined separately. Forest-floor biomass increased with increasing precipitation (Fig. 5.4b, solid bars). This result suggests a concomitant increase in litter production. If litter production were constant, enhanced decomposition with higher precipitation would lead to lower forest-floor biomass. In fact, mean litter production was greater in the south (Fig. 5.4b, hatched bars). Due to small sample size, however, especially in Arroyo Negro where one of two sites demonstrated surprisingly low biomass and production, the relationship between region and annual litter production was not statistically significant. Martinez-Yrizar and Saruhkan (1990) found enhanced litter production in the moister of their two study sites in Chamela, Mexico. Their results are consistent with our finding that differences in moisture availability can lead to differences in litter production.

Despite significant differences in rainfall and some differences in soil properties, above-ground biomass did not vary significantly among the regions (Fig. 5.4a). The highest values came from Nicolas Bravo, where soils were apparently less fertile. Lower than expected values in Arroyo Negro may be due to unquantified impacts of recent logging (see Ch. 3). Mature forests in Arroyo Negro had fewer big trees (> 10cm dbh) and lower basal area than the other sites, suggesting that some trees may have been extracted in recent history (Read and Lawrence 2003a). On the other hand, high biomass in Nicolas Bravo may be due to an overabundance of lianas—perhaps another legacy of historic disturbance, human or natural (Boose, Foster, and Fluet 1994; Gerwing and Lopes Farias 2000). The mean

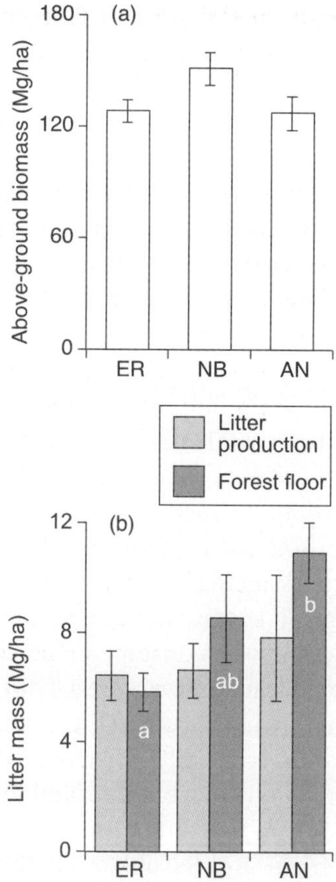


**Fig. 5.4.** Biomass, forest floor, and litter production by study location. Above-ground live biomass (>1cm dbh)) (a); forest floor biomass (b, dark bars); and annual fine litter production (b, light bars). ER = El Refugio, NB = Nicolas Bravo, AN = Arroyo Negro. Mean ± standard error for all plots within a region. Statistically significant differences between regions indicated by different letters.

contribution of lianas to total biomass was 10.2 per cent in Nicolas Bravo v. 5.6 per cent in El Refugio and 4.8 per cent in Arroyo Negro.

Phosphorus but not nitrogen dynamics followed the trend in annual precipitation. Total annual inputs of phosphorus were greater in the wettest region, Arroyo Negro (Fig. 5.5a), presumably as a result of both greater productivity and higher P concentrations in the litter (Read and Lawrence 2003b). Nitrogen inputs increased slightly with increasing precipitation, but the trend was not significant (Fig. 5.5b). This modest trend reflects only differences in total productivity, as litter concentrations did not vary significantly among regions (Read and Lawrence 2003b). The response of forest P cycling to regional precipitation gradients and the lack of response in the N cycle suggest that forests in the region are more limited by phosphorus than by nitrogen (ibid.). Furthermore, the manifestation of nutrient limitation depends on water availability. This may explain why Campo and colleagues (2001) reported that phosphorus did not appear to be limiting in the very dry forests of Chamela, although small amounts of phosphorus were retained in the system. The indication of phosphorus limitation came out of a regional analysis, where more P, but not more N, was observed to cycle in the wettest region. N cycling appears less sensitive to increased precipitation than P cycling.

## 3.5. *Effects of Forest Age on Biomass and Nutrient Cycling*

Forest age had a greater impact on ecosystem function than did regional environmental gradients, considering the range of variation induced. Considering plots from all sites together, above-ground live biomass increased significantly as a function of forest age with the largest difference between mature forests and the oldest secondary forests (Fig. 5.6a). Although total woody basal area had recovered to 63 per cent of mature

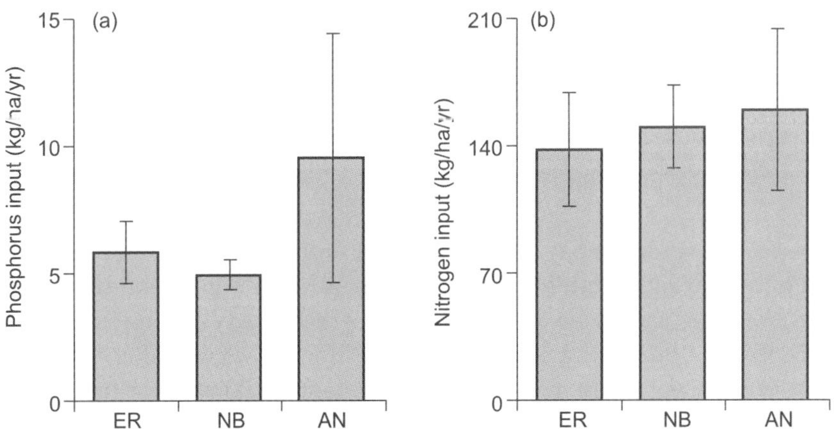


**Fig. 5.5.** Total annual phosphorus and nitrogen inputs deposited in fine litter in each study location. Mean ± standard error for all plots within a region.

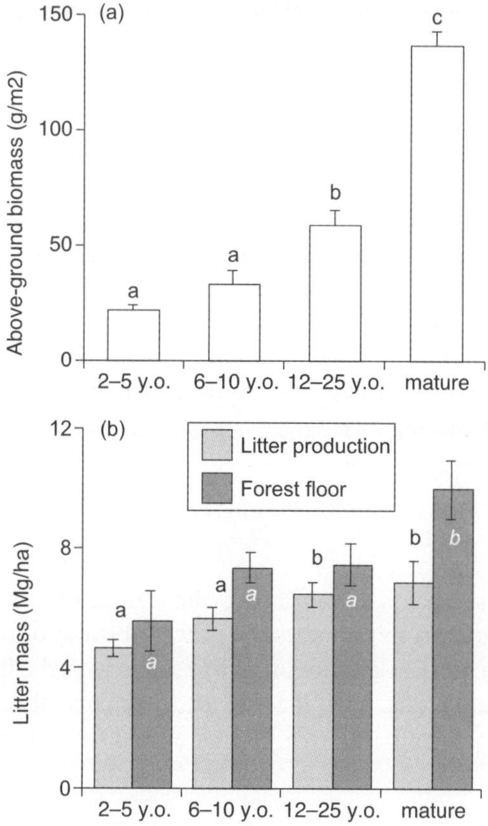


**Fig. 5.6.** Biomass, forest floor and litter production as a function of forest age. Above-ground live biomass (>1cm dbh)) (a); forest floor biomass (b, dark bars); and annual fine litter production (b, light bars) across all sites. Mean ± standard error for all plots within an age class. Statistically significant differences between age classes indicated by different letters.

forest levels by 25 years, live biomass reached only 40 per cent of mature forest levels (58 v. 136Mg/ha). The mean biomass of the youngest age categories did not differ significantly from each other (21 v. 33Mg/ha), but their biomass was significantly lower than that of older secondary and mature forests.

Both litter production and forest-floor biomass increased consistently as a function of forest age, with less dramatic a difference between old secondary and mature forests (Fig. 5.6b). With only 15 per cent of the above-ground biomass, the youngest forests produced almost 70 per cent as much fine litter as mature forests. These results are similar to those reported by Ewell (1976) in Guatemala, and match patterns of productivity for successional forests in the temperate zone as well (Odum 1969). The increase in forest-floor biomass was more rapid than expected given increases in litter production, suggesting that decomposition slowed with forest age. Experimental results confirm that decomposition rates decline in older forests in response to lower litter quality rather than a decline in the suitability of the environment (Xuluc-Tolosa *et al.* 2003). The change in litter quality seems to be related to phosphorus content: litter phosphorus concentration declined with age, but no significant change in litter nitrogen concentration was found (Read and Lawrence 2003b).

Whereas the amount of phosphorus, but not nitrogen, responded to increased precipitation, mean phosphorus inputs across all regions did not increase as a function of forest age, whereas nitrogen inputs did so (Fig. 5.7). This result suggests that phosphorus cycling changes markedly during the process of forest recovery. Older forests produced significantly greater amounts of litter (Fig. 5.6b), while cycling the same amount of phosphorus as younger forests. As a result, litter P concentration declined and P-use efficiency increased with age (ibid.). In contrast, the pool of nitrogen cycled through litter increased in step with greater litter production in older forests. Litter N concentration did not increase significantly, although there was a tendency for slightly higher concentrations in older forests. Taken together, these data suggest that N-use efficiency declined, if it changed at all, as a function of forest age.

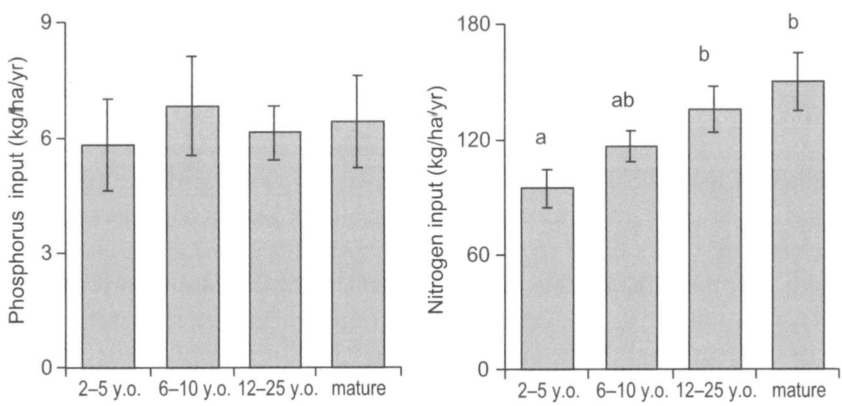


**Fig. 5.7.** Total annual phosphorus and nitrogen inputs deposited in fine litter as a function of forest age. Mean ± standard error for all plots within an age class. Statistically significant differences between age classes indicated by different letters.

Evaluation of changes in nutrient cycling during forest recovery depends on the nutrient examined. Furthermore, understanding nutrient constraints on plant growth and ecosystem processes depends on considering the interaction between water- and nutrient-limitation. In the above analysis of nutrient dynamics as a function of forest age, regional precipitation gradients were disregarded. Water was considered to be broadly limiting. Under this scenario, productivity seemed to be limited by nitrogen availability: increased nitrogen inputs correlated with increased litter production. All forests cycled similar amounts of phosphorus regardless of age. Perhaps this level represents the maximum phosphorus pool accessible given the soil–climate regime. More P cannot enter the system unless fixed biogeochemical constraints are lifted. In a sense, P may be limiting, but we cannot determine the limitation by comparing P return as a function of forest age because recovery time is not the only issue. Although the potential exists for N uptake through biological fixation, there are no additional pools to tap to alleviate P limitation through time. Increasing production depends, then, on increasing nitrogen availability through time as the forest and its soils recover from cultivation. Only when water limitation is alleviated and soil biological activity is enhanced (as it appears to be in Arroyo Negro), can additional phosphorus enter the system (see Fig. 5.5). This phosphorus yielded gains in litter production relative to the drier, more P-starved sites (Fig. 5.4), suggesting that phosphorus may become limiting when water is no longer the primary limiting factor.

Changes in soil properties with forest age corroborate the suggestion that nitrogen availability increases during recovery from shifting cultivation (e.g. Popenoe 1959). N accumulation would allow N litter concentrations to remain stable while productivity increases. Although data on the N and P content of the soils is lacking, soil organic matter did increase significantly with age (Table 5.3). In the top 15cm of soil, organic matter increased from 10–11 per cent in secondary forests under age 10 to 15 per cent in mature forest soils. Older secondary forest soils demonstrated intermediate levels. Most probably, this increase in soil organic matter represents an increase in litter inputs and a slowing of mineralization rates in the soil. This increase parallels the increase in forest-floor mass with age attributed to added inputs and slowed decomposition. Data on potassium concentrations support this interpretation: K concentration was lowest in mature forest soils, where turnover rates of organic matter are expected to be slowest due to a decline in litter quality.

**Table 5.3.** Soil variation as a function of forest age (2–5 y.o., n = 10; 6–10 y.o., n = 8; 12–25 y.o., n = 10; mature forest, n = 8) across all the three study sites

| | 2–5 y.o. | 6–10 y.o. | 12–25 y.o. | Mature | P-value |
|---|---|---|---|---|---|
| *Physical properties* | | | | | |
| Clay (%) | 28±4 | 34±5 | 28±2 | 20±3 | 0.074 |
| Silt (%) | 34±1 | 30±5 | 38±3 | 29±2 | 0.22 |
| Sand (%) | 38±4[a] | 36±5[a] | 34±3[a] | 52±2[b] | 0.011 |
| Organic matter (%) | 11.0±0.3[a] | 10.1±0.7[a] | 12.3±0.7[ab] | 15.2±1.9[b] | 0.009 |
| pH | 7.7±0.1[a] | 7.5±0.1[ab] | 7.4±0.1[b] | 7.6±0.1[ab] | 0.070 |
| TEC (meq/g) | 94±15 | 98±14 | 68±11 | 105±16 | 0.27 |
| *Chemical properties (ppm)* | | | | | |
| Sulfur | 54±5 | 55±6 | 43±5 | 55±5 | 0.25 |
| Calcium | 17,600±3,000 | 18,400±2,800 | 12,400±2,300 | 20,000±3,300 | 0.26 |
| Magnesium | 523±24 | 579±47 | 539±31 | 483±19 | 0.26 |
| Potassium | 616±70[a] | 483±43[ab] | 508±35[a] | 359±45[b] | 0.013 |
| Sodium | 26±1 | 31±2 | 28±2 | 25±2 | 0.14 |
| Manganese | 83±19 | 93±34 | 108±15 | 47±12 | 0.28 |
| Boron | 1.9±0.1 | 1.7±0.2 | 1.8±0.1 | 2.1±0.2 | 0.20 |
| Iron | 17±3 | 22±6 | 26±4 | 12±4 | 0.19 |
| Copper | 2.1±0.1 | 2.3±0.3 | 2.3±0.1 | 1.9±0.1 | 0.39 |
| Zinc | 1.5±0.1 | 1.4±0.2 | 1.5±0.2 | 1.4±0.1 | 0.97 |
| Aluminum | 176±54 | 203±68 | 258±68 | 77±31 | 0.16 |

*Note*: P value indicates significance of separate analyses of variance (ANOVAs) by age. Significant differences among age classes are indicated by different letters.

Surprisingly, both decomposition experiments (Xuluc-Tolosa *et al.* 2003) and field observations suggest that P may limit changes in decomposition and mineralization as a function of forest age, although N appears to limit plant productivity. P may also limit N fixation—this effect has been suggested in other studies (e.g. Crews, Farrington, and Vitousek 2000; Uliassi *et al.* 2000). With the potential to constrain decomposition, mineralization, and N fixation, phosphorus may play a critical role in limiting N availability to recovering vegetation. It is important to emphasize the difficulty in identifying either P or N as the primary limiting nutrient in this system, especially if P mediates N limitation. Both nutrients appear to limit key ecosystem processes.

### 3.6. *Effect of Cultivation History*

As indicated by the error bars on Fig. 5.6, production of fine litter varies substantially within a given age class. The major hypothesis to explain this variation is that sites subjected to more intensive cultivation prior to abandonment are less productive. Through collaborative efforts with the social scientists on the research team, it was possible to identify basic components of the cultivation history at each patch. In particular we know when the patch was abandoned (forest age), when the patch was originally cleared, how many years it was under maize cultivation, and how many swidden cycles had occurred there. We examined the latter in this analysis. Because mean litter production did not differ significantly between the 12–25-year-old secondary forests and mature forests (Fig. 5.6b), it was possible to compare patches that had experienced zero, one, and two swidden cycles. Production declined in parcels that had been cleared more than once (Fig. 5.8). The minimum level of production did not decline, but the maximum level did, lowering the mean. Given the importance of apparent feedbacks between litter production, soil organic matter, and nutrient availability to plants, diminished litter production in secondary forests with more intense land-use history may constrain future agricultural productivity in the absence of chemical inputs.

## 4. *Conclusions*

Sampling along natural and human-made gradients allowed the evaluation of both broad environmental drivers of ecosystem processes and stand-

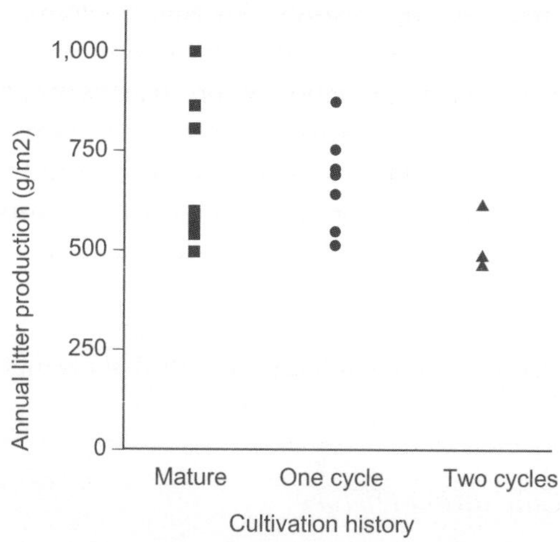


**Fig. 5.8.** Annual fine litter production as a function of the number of prior crop–fallow cycles. Note: only mature forest and sites fallowed 12–25 years included. A cycle refers to one complete crop–fallow period.

level impacts of essential human land-use decisions such as where to clear forest for agriculture and how long to fallow a patch of land. Environmental factors do constrain litter production, forest floor mass, and soil organic matter in mature forest. Accounting for natural gradients in soil and precipitation may be sufficient to explain the variability observed. To understand differences in the live above-ground biomass of mature forests, however, the human history of the area may add critical information, including logging history and decisions about what areas have been left untouched (e.g. upper slopes). All aspects of structure and function in young secondary forests—live above-ground biomass, litter production, forest floor mass, and critical soil properties—are strongly influenced by the age and cultivation history of the stand. The presence of humans on the landscape results in changes that are greater than those induced by natural, environmental variation at the regional scale. For the best understanding of these landscapes, and for scaling up from case studies to regional and global assessments, integrating human impacts on ecosystem processes is vital.

## 5. *Implications*

Preliminary analysis, based on tree basal area only, suggested that above-ground biomass accumulates rapidly during forest regeneration following swidden agriculture, with basal area increasing to 80 per cent of mature forest levels within 25 years (Turner *et al.* 2001). Subsequent analysis of biomass based on allometric equations involving the height, diameter, and wood density of all stems (including lianas and small trees < 5cm dbh) confirms the 40–60 year recovery time to reach mature forest levels (Read and Lawrence 2003a). This analysis supercedes prior reports. The similarity of the results, however, depends on the assumption of a linear increase in biomass beyond 25 years. The initial recovery of biomass (to 40% by 25 years) is slower than the recovery of total basal area (to 63% in this new analysis), because of the influence of height and wood density. If biomass growth is not linear, and the annual biomass increment declines, for example, by 50 per cent after forests achieve a closed canopy, then recovery would be slowed. If canopy closure occurs 12–25 years following abandonment, it may take as long as 55–95 years to reach mature forest levels.

These estimates are somewhat longer than our initial estimates of the time necessary to recover to the biomass level observed currently in mature forests (Turner *et al.* 2001). As noted previously, these mature forests have been heavily logged, removing the largest stems consisting of mahogany and Spanish cedar. If these or other large species were to dominate the forest canopy once again, recovery to an unlogged state might take an additional 10–25 years, assuming an ultimate density of 6 trees/ha with diameters about 100cm (see Rodriguez Caballero 1944; Snook 1998). Thus, we estimate that forests similar in total above-ground biomass to those observed in the nineteenth century might be reached 65–120 years after agricultural abandonment. This estimate is more in line with estimates of recovery time in the wet tropical forests of Los Tuxtlas, Mexico. Those forests may take 70 years or more to reach biomass equivalent to mature forests (Hughes, Kauffman, and Jaramillo 1999).

Efforts to use remotely sensed data to scale up from point samples to regional values will depend critically on the ability to distinguish old secondary forest from mature forest (Ch. 6). Models that predict landscape-scale carbon stocks, for example, will be extremely sensitive to the mapping of field-derived age categories onto image-derived land-cover classes. On the other hand, distinguishing older secondary forest and mature forests may be less critical for determining regional productivity, given the similarity in litter production. Failure to distinguish between young and old secondary forest would have less of an impact but could still introduce error. The amount of classification error that can be safely tolerated depends on the goal of the model: carbon stocks or carbon fluxes. Similar considerations apply to constructing biogeochemical models to explain the response of limiting nutrients to land-use change.

The study of nutrient dynamics in the region provides important clues about the regenerative capacity of the forests and the sustainability of the current and dominant mode of cultivation, the swidden. Any assessment of land use and land-cover change, therefore, must take into account the following critical points.

- Understanding the interaction between human-induced and environmental variation in controlling ecosystem processes requires studies conducted at regional scales.
- Regional precipitation gradients strongly influence nutrient cycling through soil organic matter and litter, but have less effect on plant productivity and biomass.
- Although forest age profoundly influences nutrient cycling (especially of phosphorus), its greatest effect is to alter litter production and biomass.

- Phosphorus cycling responds to both increased precipitation and to increasing forest age; P use efficiency declines with increased precipitation and increases with age.
- Nitrogen cycling appears less constrained by water availability; N use efficiency tends to decline with forest age, but N inputs do not respond dramatically to increased precipitation.
- Declining litter production in secondary forests having experienced more than one cycle of shifting cultivation indicates that cultivation history must be considered in evaluating ecosystem processes at the landscape scale, and in assessing the sustainability of current land-use practices.
- Scaling up estimates of carbon stocks using remote sensing requires effective distinction between old secondary and mature forests; estimates of forest productivity do not depend critically on this distinction.

## REFERENCES

Boose, E. R., D. R. Foster, and M. Fluet (1994). 'Hurricane Impacts to Tropical and Temperate Forest Landscapes'. *Ecological Monographs* 64: 369–400.

Brady, N. C. (1990). *The Nature and Properties of Soils* (10th edn.). New York, Macmillan.

Buschbacher, R., C. Uhl, and E. A. S. Serrao (1988). 'Abandoned Pastures in Eastern Amazonia II: Nutrient Stocks in the Soil and Vegetation'. *Journal of Ecology* 76: 682–99.

Cairns, M. A., P. K. Haggerty, R. Alvarez, B. H. J. De Jong, and I. Olmsted (2000). 'Tropical Mexico's Recent Land-Use Change: A Region's Contribution to the Global Carbon Cycle'. *Ecological Applications* 10: 1426–41.

Campo, J., V. J. Jaramillo, and J. M. Maass (1998). 'Pulses of Soil Phosphorus Availability in a Mexican Tropical Dry Forest: Effects of Seasonality and Level of Wetting'. *Oecologia* 115: 167–72.

Campo, J., M. Maass, V. J. Jaramillo, A. Martinez-Yrizar, and J. Sarukhan (2001). 'Phosphorus Cycling in a Mexican Tropical Dry Forest Ecosystem'. *Biogeochemistry* 53: 161–79.

Chapin, F. S. (1980). 'The Mineral Nutrition of Wild Plants'. *Annual Review of Ecology and Systematics* 11: 233–60.

Crews, T. E., H. Farrington, and P. M. Vitousek (2000). 'Changes in Asymbiotic Heterotrophic Nitrogen Fixation on Leaf Litter of *Metrosideros polymorpha* with Long-Term Ecosystem Development in Hawaii'. *Ecosystems* 3: 386–95.

Ewel, J. J. (1976). 'Litter Fall and Leaf Decomposition in a Tropical Forest Succession in Eastern Guatemala'. *Journal of Ecology* 64: 293–307.

Fernandez, D. N., and R. L. Sanford (1995). 'Effects of Recent Land-Use Practices on Soil Nutrients and Succession under Tropical Wet Forest in Costa Rica'. *Conservation Biology* 9: 915–22.

Frankie, G. W., H. G. Baker, and P. A. Opler (1974). 'Comparative Phenological Studies in Tropical Wet and Dry Forests in the Lowlands of Costa Rica'. *Journal of Ecology* 62: 881–919.

Gerwing, J. H., and D. Lopes Farias (2000). 'Integrating Liana Abundance and Forest Stature into an Estimate of Total Aboveground Biomass for an Eastern Amazonian Forest'. *Journal of Tropical Ecology* 16: 237–335.

Holbrook, N. M., J. L. Whitbeck, and H. A. Mooney (1995). 'Drought Responses of Neotropical Dry Forest Trees', in S. H. Bullock, H. A. Mooney, and E. Medina (eds.), *Seasonally Dry Tropical Forests*. Cambridge, Cambridge University Press, 243–76.

Hughes, R. F., J. B. Kauffman, and V. J. Jaramillo (1999). 'Biomass, Carbon, and Nutrient Dynamics of Secondary Forests in a Humid Tropical Region of Mexico'. *Ecology* 80: 1892–907.

Jordan, C. F., and R. Herrera (1981). 'Tropical Rain Forests: Are Nutrients Really Critical?' *The American Naturalist* 117: 167–80.

Killingbeck, K. T. (1996). 'Nutrients in Senesced Leaves: Keys to the Search of Potential Resorption Proficiency'. *Ecology* 77: 1716–29.

Klepeis, P., and B. L. Turner, II (2001). 'Integrated Land History and Global Change Science: The Example of the Southern Yucatan Peninsular Region Project'. *Land Use Policy* 18: 27–39.

Kunkel Westphal, I., and P. Kunkel (1979). 'Litter Fall in a Guatemalan Primary Forest, with Details of Leaf-Shedding by Some Common Tree Species'. *Journal of Ecology* 67: 665–86.

Lawrence, D. (n.d.). 'Effect of Precipitation Regime and Forest Age on Seasonal Litterfall Patterns in the Dry Tropical Forests of the Southern Yucatan'. Unpublished manuscript.

McGrath, D. A., N. B. Comerford, and M. L. Duryea (2001). 'Litter Dynamics and Monthly Fluctuations in Soil Phosphorus Availability in an Amazonian Agroforest'. *Forest Ecology and Management* 131: 167–81.

Martinez-Yrizar, A., and J. Sarukhan (1990). 'Litterfall Patterns in a Tropical Deciduous Forest in Mexico over a Five-Year Period'. *Journal of Tropical Ecology* 6: 433–44.

Martinez-Yrizar, A., J. Sarukhan, E. Perez-Jimenaz, E. Rincon, J. M. Maass, A. Solis-Magallanes, and L. Cervantes (1992). 'Above-Ground Phytomass of a Tropical Deciduous Forest on the Coast of Jalisco, Mexico'. *Journal of Tropical Ecology* 8: 87–96.

Murphy, P. G., and A. E. Lugo (1986). 'Ecology of Tropical Dry Forest'. *Annual Review of Ecology and Systematics* 17: 67–88.

Odum, E. P. (1969). 'The Strategy of Ecosystem Development'. *Science* 164: 262–70.

Popenoe, H. (1959). 'The Influence of the Shifting Cultivation Cycle on Soil Properties in Central America', in *Proceedings of the Ninth Pacific Science Congress of the Pacific Science Association*, 1957. Bangkok, 72–7.

Proctor, J., J. M. Anderson, S. C. L. Fogden, and H. W. Vallack (1983). 'Ecological Studies in Four Contrasting Rain Forests in Gunung Mulu National Park, Sarawak'. *Journal of Ecology* 71: 261–83.

Read, L., and D. Lawrence (2003a). 'Recovery of Biomass Following Shifting Cultivation in Dry Tropical Forests of the Yucatán'. *Ecological Applications* 13: 85–97.

Read, L., and D. Lawrence (2003b). 'Litter Nutrient Dynamics During Succession in Dry Tropical Forests of the Yucatán: Regional and Seasonal Effects, *Ecosystems* (forthcoming).

Reich, P. B., and R. Borchert (1984). 'Water Stress and Tree Phenology in a Tropical Dry Forest in the Lowlands of Costa Rica'. *Journal of Ecology* 72: 61–74.

Rodriguez Caballero, R. (1944). *La Explotación de los Montes de Caoba en el Territorio de Quintana Roo*. Ingeniero Agronomo: México DF, Chapingo.

Singh, J. S., A. S. Raghubansi, R. S. Singh, and S. C. Srivastava (1989). 'Microbial Biomass Acts as a Source of Plant Nutrients in Dry Tropical Forest and Savanna'. *Nature* 338: 499–500.

Snook, L. K. (1998). 'Sustaining Harvests of Mahogany (*Swietenia macrophylla* King) from Mexico's Yucatan Forests: Past, Present and Future', in R. Primack, D. Bray, H. A. Galleti, and I. Paciano (eds.), *Timber, Tourists, and Temples: Conservation and Development in the Maya Forest of Belize, Guatemala, and Mexico*. Washington DC, Island Press, 61–80.

Suazo, I. (1998). 'Aspectos ecologicos de la especie invasora *Pteridium aquilinum (L.)* Kuhn en una selva humeda de la region de Chajul, Chiapas, Mexico. M. S. Thesis'. Facultad de Biología Morelia: Michoacan, Universidad Michacana de San Nicolas de Hidalgo.

Swift, M. J., A. Russell-Smith, and T. J. Perfect (1981). 'Decomposition and Mineral Nutrient Dynamics of Plant Litter in a Regenerating Bush-Fallow in the Sub-Humid Tropics'. *Journal of Ecology* 69: 981–95.

Trejo, I., and R. Dirzo (2000). 'Deforestation of Seasonally Dry Tropical Forest: A National and Local Analysis in Mexico'. *Biological Conservation* 94: 133–42.

Turner, B. L., II (1974). 'Prehistoric Intensive Agriculture in the Mayan Lowlands'. *Science* 185: 118–24.

Turner, B. L., II, S. C. Villar, D. Foster, J. Geoghegan, E. Keys, P. Klepeis, D. Lawrence, P. M. Mendoza, S. Manson, Y. Ogneva-Himmelberger, D. P. Salicrup, R. Roy Chowdhury, V. Savitsky, L. Schneider, B. Schmook, and C. Vance (2001). 'Deforestation and Agricultural Change in the Southern Yucatan Peninsular Region: Integrative Land Change for Global Change Studies'. *Forestry, Ecosystems, and Management* 154: 353–70.

Uliassi, D. D., K. Huss-Danell, R. W. Ruess, and K. Doran (2000). 'Biomass Allocation and Nitrogenase Activity in *Alnus tenuifolia*: Responses to Successional Soil Type and Phosphorus Availability'. *Ecoscience* 7: 73–9.

Vitousek, P. (1984). 'Litterfall, Nutrient Cycling, and Nutrient Limitation in Tropical Forests'. *Ecology* 65: 285–98.

Whigham, D. F., P. Zugasty Towle, E. Cabrera Cano, J. O'Neill, and E. Ley (1990). 'The Effect of Annual Variation in Precipitation on Growth and Litter Production in a Tropical Dry Forest in the Yucatan of Mexico'. *Tropical Ecology* 31: 23–34.

Wieder, R. K., and S. J. Wright (1995). 'Tropical Forest Litter Dynamics and Dry Season Irrigation on Barro Colorado Island, Panama'. *Ecology* 76: 1971–79.

Xuluc-Tolosa, F. J., H. F. M. Vester, N. Ramirez-Marcial, J. Castellanos-Albores, and D. Lawrence (2003). 'Leaf Litter Decomposition of Tree Species in the Three Successional Phases of Tropical Dry Secondary Forest in Campeche Mexico'. *Forest Ecology and Management* 174: 401–412.

# 6

# Land Cover and Land Use: Classification and Change Analysis

*Rinku Roy Chowdhury and Laura C. Schneider*
*with Yelena Ogneva-Himmelberger,*
*Pedro Macario Mendoza, Sergio Cortina Villar,*
*Audrey Barker-Plotkin*

Despite its international designation as a hotspot of biodiversity and tropical deforestation (Achard *et al.* 1988), the micro-scale land-cover mapping of southern Yucatán peninsular region remains surprisingly incomplete, hindering various kinds of research, including that proposed in the SYPR project. This chapter details the methodology for the thematic classification and change detection of land use and cover in the tropical sub-humid environment of the region. A hybrid approach using principal components and texture analyses of Landsat TM data enabled the distinction of land-cover classes at the local scale, including mature and secondary forest, savannas, and cropland/pasture.

Results indicate that texture analysis increases the statistical separability of cover class signatures, the magnitude of improvement varying among pairs of land-cover classes. At a local level, the availability of exhaustive training site data over recent history (10–13 years) in a repository of highly detailed land-use sketch maps allows the distinction of greater numbers of land-cover classes, including three successional stages of vegetation. At the regional scale, finely detailed land-cover classes are aggregated for greater ability to generalize in a terrain wherein vegetation exhibits marked regional and seasonal variation in intra-class spectral properties. Post-classification change detection identifies the quantities and spatial pattern of major land-cover changes in a ten-year period in the region. Change analysis results indicate an average annual rate of deforestation of 0.4 per cent, with much regional variation and most change located at three subregional hotspots. Deforestation as well as successional regrowth is highest in a southern hotspot located in the newly colonized southern part of the region, an area where commercial chili production is large.

The objectives of this chapter are to describe and evaluate: (1) an experimental methodology that iteratively combines three suites of image-processing techniques (PCA, texture transformation, and NDVI); (2) the statistical separability of distinct land-cover signatures; and (3) a post-classification change detection for the region from 1987 to 1997 in order to derive regional deforestation rates, and identify the spatial pattern of deforestation and secondary forest succession. Specifically, a region encompassing 18,700km$^2$ (those land units completely within the defined region; Fig. 7.1) was mapped using a maximum likelihood supervised classification of lower-order principal components of Landsat TM imagery after tasseled-cap and texture transformations.

## 1. *Image Analysis for Tropical Forests*

Image classification and change analysis are increasingly important in monitoring and modeling global land-cover and land-use change, particularly in tropical environments where such change is largely concentrated. An understanding of human or environmental causes of this change requires knowledge of extant landscape patterns and ecosystems, monitoring extents and rates of change in these systems, and an analysis of the driving forces and consequences of change. Tropical deforestation has been a focus of international concern, and lately, high resolution satellite data are employed in monitoring deforestation as well as successional regrowth (Baltaxe 1987; Berta, Mausel, and Harrington 1990; Campbell and Browder 1992; Gilruth, Hutchinson, and Barry 1990; Green and Sussman 1990; Kummer 1992; Malingreau, Tucker, and LaPorte 1989; Mausel *et al.* 1993; Nelson and Holben 1986; Sader, Stone, and Joyce 1990; Woodwell *et al.* 1987). Satellite imagery is considered to be the most reliable source of quantitative information about deforestation, shifting or swidden cultivation, and other land-cover and land-use changes in the American tropics (Sader *et al.* 1994). Although much work has focused on the use of NOAA (National Oceanic and Atmospheric Administration) AVHRR (Advanced Very High Resolution Radiometer), and NASA (National

Aeronautics and Space Agency) Landsat Thematic Mapper (TM) data, the number of land-cover types typically classified from such imagery has been small, with few notable exceptions (e.g. Defries *et al.* 2000). While offering relatively lower spatial resolution at 1.1km, AVHRR data have proven to be rich data sources for the mapping of global land-cover at a high temporal resolution, particularly in drier eco-regions. Landsat TM, however, has been more effective at separating evergreen forest types in the humid tropics (DeFries and Townshend 1994; Hansen *et al.* 2000).

Until recently, remote sensing and GIS studies of forest-cover change focused on deforestation rates and patterns rather than those of successional regrowth. More recently, a better understanding of the extent and human and ecological dynamics of swidden cultivation (and its associated fallow successions) has emerged, as has the availability of higher temporal and spatial resolution imagery. This fortunate coincidence has refocused attention on using satellite imagery to monitor processes of successional regrowth and forest re-establishment, spurring efforts to map increasingly finer categories of tropical forest types, including various stages of secondary forest (Mausel *et al.* 1993; Sader 1995; Steininger 1996; Sohn, Moran, and Gurri 1999).

## 2. *Review of Main Preclassification Image Processing Techniques Employed*

Predefined classification schemes that assign each pixel of remotely sensed data to one among a set of discrete land-cover classes continue to be one prevailing tradition in image classification (DeFries *et al.* 1998, 2000; Loveland and Belward 1997). There are several disadvantages to such an approach, including: (1) the inability in post-classification change detection to identify within-class changes in land cover, such as progressive land degradation or continuously decreasing biomass levels, and (2) loss of information about within-class variation in vegetation characteristics (DeFries and Belward 2000; DeFries *et al.* 1995). Yet, when it comes to linking remote-sensing based land-cover classification with meso- and micro-spatial scale studies of the human dimensions of global change (Turner *et al.* 1995), it is crucial that land-cover classes relevant to land uses that prevail at those scales be discernible.

As is commonly accepted, most land-cover classification projects aim to explore methodologies that allow the distinction of the maximum achievable detail in land-cover categories (Saatchi *et al.* 2000). This classification effort in the SYPR project had the same objective. Image classification in phase-one research was initiated on the following principle: in order to link remote-sensing assessments of land-cover change effectively to social science land-use models utilizing socio-economic data about the agents of change, it is essential to distinguish with as much precision as possible the appropriate thematic categories of interest (Geoghegan *et al.* 1998). Ongoing image-processing research in the project aims to address the shortcomings of the thematic classification approach to land-cover mapping through techniques such as subpixel vegetation characterization, including fuzzy membership functions (Eastman 1999; Settle and Drake 1993; Wang 1990). An initial delineation of land-use and land-cover classes, however, is highly valuable at the local and regional scales for land-cover monitoring as well as spatially based econometric and agent-based models of land-use change (Chs. 12 and 13).

In the southern Yucatán, aside from forest-based economies involving timber and non-timber forest products, the land uses that extensively dominate are swidden and some market agriculture for which a particular forest type is preferred (medium to tall forest on well-drained but relatively level soils). Pasture is also important in certain regions, although secondarily so. Associated with agricultural activities are various stages of forest succession, involving diverse fallowing strategies that, contingent on data availability on individual decision-making linked to the fallow parcel, may be spatially modeled. In addition, the distinction of various classes of mature forest is critical for ecological studies of species diversity and ecosystem structure and function. Such distinction of forest types is also relevant to the study of forest economies, since different forest types vary in their economic potential. For instance, seasonally flooded forests are typically not managed for timber production due to the high costs and logistical impossibilities that inventories and extraction activities would involve in such terrain. Forestry activities focus instead on medium-tall forests located on well-drained upland soils.

## 2.1. *Principal Components Analysis*

A wide variety of image-processing techniques have been investigated better to map the diverse land-cover classes prevalent in tropical forest ecosystems. Principal components analysis (PCA) is one common technique long used to (1) reduce data dimensionality, (2) allow image enhancement and improve signal-to-noise ratios by removing random and systematic noise from satellite image bands, (3) perform digital change detection, and (4) characterize seasonal changes in cover types (Byrne, Crapper, and Mayo 1980; Eastman and Fulk 1993; Gillespie 1980; Townshend 1984). The use of PCA as a noise reduction tool is well documented ( Eastman 1999; Singh and Harrison 1985). In the absence of noise, pixels belonging to one land-cover class would demonstrate identical spectral signatures at different bands of the electromagnetic spectrum. In reality, the presence of class-dependent and class-independent noise results in the prevalence of variations in spectral signatures, even among pixels belonging to the same land-cover class (Sharma and Sarkar 1998). Class-dependent noise reflects natural variations in each land-cover class, while class-independent noise is caused by atmospheric scatter or absorption (random) and/or sensor calibration problems (systematic).

The principal components transformation is a multivariate statistical technique that produces uncorrelated linear combinations of the variables (pixel spectral response in electromagnetic bands), called principal components (Carr and Matanawi 1999; Gonzalez and Woods 1992; Jensen 1996). The technique produces each successive linear combination such that it has a smaller variance (smaller eigenvalues). Thus the first principal component is the linear transform that captures the most variation in the dataset, the second component captures a smaller variation not captured by the first component, and so on. The higher the linear intercorrelation among the original variables (bands), the larger the part of the total scene variance that is captured in the first few components (Singh and Harrison 1985).

PCA has been extensively used to reduce the number of variables in multispectral imagery, as higher-order components with smaller eigenvalues are assumed to contain relatively little information (ERDAS 1994; Lillesand and Kiefer 1994; Richards 1993). When used in discriminant or cluster analysis, higher-order principal components have been discarded to reduce the number of variables as well as required training data (Basili *et al.* 1994; Paradella *et al.* 1994). For instance, May (1986) used only the first two principal components of Landsat MSS imagery for discriminant mapping of vegetation classes. Other studies, however, caution that the use of PCA to reduce the number of variables for land-cover classification needs to be selective, owing to potential losses in the accuracy and power of land-cover discrimination as a result of discarding higher-order components that may contain important landscape information (Townshend 1984).

## 2.2. *Texture Analysis*

Textural transformations have been used to classify imagery for a variety of resource management applications, including sea ice determination using synthetic aperture radar (SAR) imagery, automated land-use mapping using black-and-white aerial photos, classification of urban land-use patterns, agricultural targets, and discrimination of vegetation types in humid tropical forests of the Amazon basin using JERS1-SAR data (Saatchi *et al.* 2000). Image texture means the spatial variation or distribution of (spectral) tones of pixels as a function of scale (Haralick 1979; Haralick, Shanmugam, and Dinstein 1973). Previous research in tropical forest environments has demonstrated the utility of texture information in the classification process, indicating that classification accuracy can be improved through the incorporation of spectral-spatial techniques in conjunction with maximum likelihood classification (Li *et al.* 1994). Spatial distributions of texture-tone have been modeled for use in image classification, with results indicating the utility of even lower-order texture measures in the process.

Local variance as a lower-order statistical measure of image texture can produce improvements in classification accuracy. For each band in an image, variance as a local texture measure is derived from the first-order histogram of a $3 \times 3$ window as follows (ERDAS 1994):

$$Variance(S_D^2) = \sum_{i=0}^{k-1} (i - S_M)^2 P(i)$$

where P(i) = N(i)/M, N(i) is the number of pixels of the same grey level in the window,
M is the number of classes,
i is the pixel grey level, and
k is the maximum possible grey level.

Essentially, variance measures heterogeneity within the pixel window such that the more different the grey level values are from their mean, the greater the variance. The pixel window mean is a first-order texture measure, and can be computed as

$$Mean(S_M) = \sum_{i=0}^{k-1} iP(i)$$

Third- and fourth-order texture measures such as skewness and kurtosis respectively measure degrees of asymmetry of pixel value distributions around the mean, and the relative peakedness or flatness of the distribution.

Texture analysis for land-cover mapping is challenged in environments that contain land-cover classes with similar spectral scattering properties. In such environments, texture analysis can none the less yield promising results if undertaken with rigorous ground-truthing and exhaustive training site datasets, as well as imagery processed to capture major axes of spectral variation, such as obtainable from principal components analysis. As a contextual technique incorporating neighborhood information, texture analysis can result in a reduction of classification error rates (Jhung and Swain 1996; Sharma and Sarkar 1998; Swain, Vardeman, and Tilton 1981; Welch and Salter 1971).

### 2.3. *Normalized Difference of Vegetation Index*

The Normalized Difference of Vegetation Index (NDVI) is a measure of relative biomass. It is calculated as

$$NDVI = \frac{IR - R}{IR + R}$$

where IR is the pixel value in the infrared band, and
R is its value in the visible red band.

NDVI may capture, depending on the season, variation among distinct vegetation types, and is useful in identifying vegetated land-cover patterns (Loveland *et al.* 2000). In fact, NDVI derived from the NOAA AVHRR platform is often considered the best dataset for land-cover classification outside the humid tropics, particularly in dry or strongly seasonal climates (DeFries and Townshend 1994; Stone *et al.* Townshend and Justice 1988; 1994; Tucker, Townshend, and Goff 1985). In general, vegetation indices are less useful in the humid tropics than in semi-arid regions. In regions where vegetation types display phenological differences that are strongly seasonal, however, vegetation indices may be useful tools in land-cover discrimination.

## 3.0. *Land-Cover Types and Target Classification Scheme*

This SYPR project study focuses on southwestern Quintana Roo and southeastern Campeche, Mexico (Fig. 1.1), a region dominated by seasonal tropical forests (Chs. 2, 4–5): semi-evergreen, well-drained upland forests (*selva mediana* or *alta*) classified elsewhere as medium to tall stature (but noted as mid-stature in Ch. 4 with crown heights ranging 15–20m) semi-deciduous forest (Miranda and Hernández-X 1963). Also extensive are short-statured (5–10m), less deciduous, seasonally inundated (*selva baja*), and to a much lesser extent, seasonally inundated savannas dominated by flood-tolerant forb and grasses (Flores and Espejel 1994).[1] Swidden or *milpa* cultivation is the predominant land use in the region (Fig. 6.1), along with pasture, forest management, and more recently, intensive chili cultivation and agroforestry. These uses generate the land covers evidenced in remotely sensed imagery capturing the region, along with secondary succession on abandoned or fallowed fields.

The target classification scheme was designed to be relevant to the socio-economic and ecological research components. From a land-use perspective, it was important to delineate swidden as well as pasture. Moreover, given the rise in jalapeño chili cultivation (Ch. 10), such cropping was an initial classification target. The preliminary ecological component of the project attempted to characterize the different natural vegetation types present in the area, notably upland and *bajo* forests. Research focused strongly on successional vegetation as well, significant to forest fallow swidden and the structure and function of secondary forests following this cultivation (Ch. 5). Informed by these needs and by local ecologists and social scientists familiar with the region, a target classification scheme was developed as detailed in Table 6.1.

## 4. *Image-Processing Methodology*

The overall image-processing methodology is detailed in Fig. 6.2. The image-processing steps involved geometric correction, noise removal using

[1] The forest nomenclature employed by different groups studying the regional landscape can be confusing. Some groups focus on the stature of the upland forest, recognizing *selva alta*, *mediana*, and *baja* (high, medium, and low, respectively) forests. Other distinguish upland (well-drained land) forest from *Bajo* (seasonally inundated depression, also known as *akalche*) forest (see Ch. 4). To avoid confisuion, the classification distinguishes upland forest, the majority of which is locally known as *mediana* and *bajo* forest, this latter term being less cumbersome than seasonal wetland forest. All potential subclasses of forest are aggregated into one of these two dominant forest types.

**Fig. 6.1.** Characteristic land covers.
*Photo*: P. Klepeis.

dehazing and principal components analysis, NDVI calculation, and texture analysis for neighborhood spatial context. The processed image was then used to develop training sites for target land-cover classes, which were then evaluated for separability and subsequently used in a maximum-likelihood supervised classification.

### 4.1. *Preprocessing: Haze Removal and NDVI*

Landsat TM images from several dates and three distinct WGS path/row sets (paths 19 and 20, row 47, and path 19, row 48) were used for this study. The individual images were first georeferenced to a UTM projection (UTM zones 15 and 16, Datum NAD27, Mexico) to under 0.5 pixel RMS error, then reprojected to geographic (latitude/longitude) coordinates. Following image registration, the TM scenes were dehazed using ERDAS Imagine's sensor-specific Tasseled Cap transformation designed for Landsat 4 TM and Landsat 5 TM imagery (Kauth and Thomas 1976; Lavreau 1991). The

**Table 6.1.** Target classification scheme

| Land-use/cover class | Comments |
|---|---|
| Water | Very little surface water present in region |
| Savanna type 1 | Locally designated '*tular*' |
| Savanna type 2 | Herbaceous wetland vegetation with green, leafy plants |
| Short stature, seasonally inundated forest (*bajo* forest) | Locally referred to as *bajos* or *selva baja* |
| Well-drained, medium-tall upland forest | Locally referred to as *selva mediana* or, on occasion, *montaña* |
| Cropland | Includes chili, which was not distinguishable from *milpa* fields in trial signature evaluations |
| Pasture | Mostly a few large private ranches, and some smaller-scale pastures, often ungrazed |
| Successional forests<br>herbaceous<br>shrub-dominated<br>arboreal | The three structural classes correspond loosely to three serial stages of 1–3-year, 4–9-year, and 10–15-year successional stages. Environmental and anthropogenic factors, however, cause many regional exceptions to this structure–age relationship |
| Successional *Pteridium* (bracken fern) | Invasive species whose emergence and prevalence is strongly tied to land-use history and site characteristics |
| Clouds | |
| Cloud shadows | |

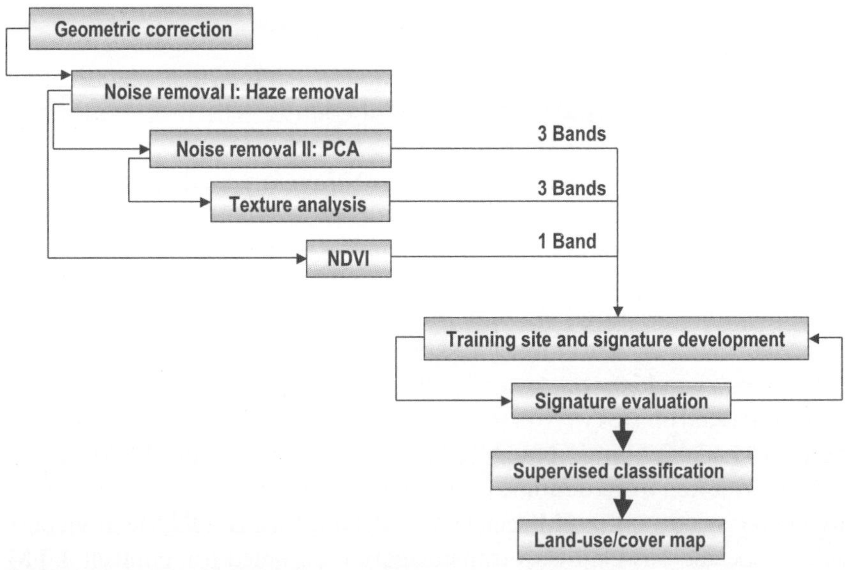


**Fig. 6.2.** Image-processing methodology.

transformation yields data components that are linear combinations of the TM bands, which correlate with brightness (albedo), greenness, wetness, and haze of the imagery. The brightness component captures, among other factors, principal variation in soil reflectance in a weighted sum of all bands. The greenness index is a contrast between the visible and near-infrared bands and correlates with amount of green vegetation in the scene. Wetness captures variation in soil and canopy moisture. For Landsat TM bands, an additional band correlated with haze is produced, and is removed from the image (ERDAS 1994).

As mentioned above, vegetation indices are used for semi-arid regions of the world rather than the humid tropics. In the study region, however, phenological characteristics (onset, peak, and seasonal duration of greenness) of vegetation associations combine with drainage patterns (seasonally flooded lowlands v. well-drained uplands) to produce a more diverse landscape than otherwise would be expected on a rolling, karstic topography with relatively little variation in soils, at least at the regional level (Martínez-Salas and Galindo-Leal n.d.). The prevalence of distinct vegetation types that vary in degrees of seasonality (deciduous character) and responses to anthropogenic use indicates that vegetation indices may be of benefit to classification. Therefore, an NDVI image was derived from the original red and infrared bands of the Landsat TM after radiometric correction for the removal of haze. NDVI images were produced for each image date for subsequent analysis, utilizing the red and infrared bands of each dehazed TM scene. The dehazed images were then processed using Principal Components Analysis to complete noise removal and reduce data redundancy.

## 4.2. *Principal Components Analysis (PCA)*

PCA was used here primarily for random and systematic noise removal and image enhancement for better visual interpretation, and secondarily to reduce data dimensionality. Random noise in an image scene is not uniformly distributed in multispectral space, but varies according to spectral bandwidth due to variable susceptibilities of band channels to atmospheric scattering. Thus, the bands of the visible spectrum display greater atmospheric scattering than do near- and mid-infrared channels.

Past studies argue that higher-order principal components may be discarded on account of their low variance, since the variance is a measure of information content and the information content remains unchanged under unitary transformation. However, other studies have shown that higher-order components of Landsat TM bands can contain important information for land-cover classification (Lark 1995; Townshend 1984).

To eliminate differentially distributed noise in band space, and simultaneously capture maximum spectral information, PCA was performed on two separate band sets: visible and infrared. The visible range bands (blue, green, and red, or bands 1, 2, and 3) of the Landsat TM images were transformed to produce three orthogonal components, the first component of which captured most of the visible band variance in the dataset (Fig. 6.3a). The second and third principal components of the visible bands captured most scattering/absorption related random atmospheric noise in the scenes, as well as striping (Fig. 6.3b). These higher-order visible band components were discarded from future analysis. In a separate step, PCA was run on the three infrared bands (bands 4, 5, and 7) of Landsat TM to produce again three principal components, the first two of which captured the major part of the scene variance and were used for subsequent analysis. The third component was discarded.

Rather than an approach that derives principal components transformations of all spectral variation jointly considered in 7-band spectral space, this two-tiered approach partitioned the principal components into two groups, later selecting the lower-order components from each group as representative of the most noise-free spectral information. Only the first principal component was selected from the visible group versus the first- and second-order components from the infrared group, owing to the higher

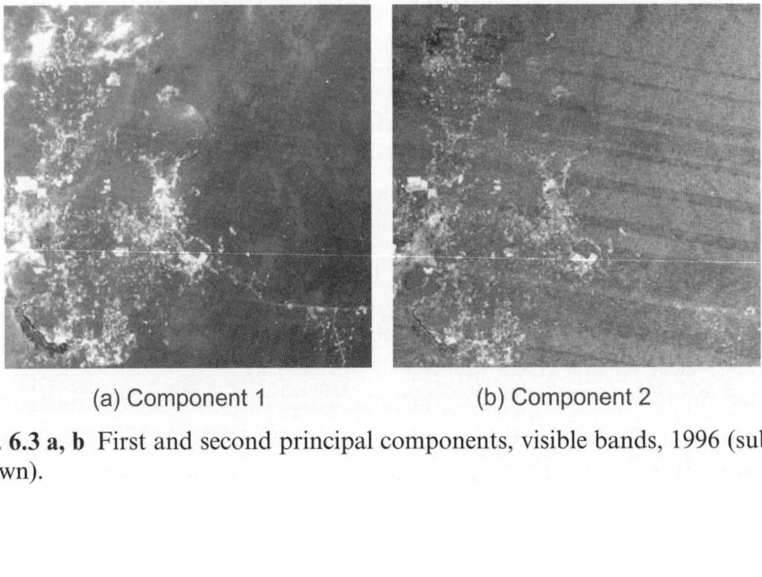

(a) Component 1 (b) Component 2

**Fig. 6.3 a, b** First and second principal components, visible bands, 1996 (subset shown).

atmospheric scattering and random noise in the first group and the relatively higher signal-to-noise ratios in the latter group. In this manner, the first and second principal components of the three infrared bands were stacked with the first principal component of the three visible bands to produce a 3-band PCA image that was used for all further analysis (Fig. 6.4a).

## 4.3. *Texture Analysis*

Next, indices of image texture were derived from the 3-band PCA image. By calculating spectral variance in spatial neighborhoods of pixels, texture analysis produced three additional bands of information based on the PCA image (Fig. 6.4b). The local texture measure was calculated from the first-order histogram of a $3 \times 3$ window using the formula for variance detailed in section 2.2. The variance characterizes the spatial heterogeneity of spectral reflectance around the central pixel of the chosen window. A small $3 \times 3$ window was chosen since terrain types were to be mapped on a pixel-per-pixel basis rather than as groups of pixels. Also, the small window size served to minimize edge effects observed in trial results of texture analyses using larger window sizes. The three texture bands were stacked with the three PCA bands to produce a 6-band image.

The NDVI image produced earlier from the originally dehazed red and infrared bands as detailed in section 4.1 was added to the PCA and texture

(a) 3 PCA bands (b) 3 texture bands

**Fig. 6.4 a, b** Three-band PCA image and 3-band texture image (subsets shown).

bands. Thus, a final 7-band image was generated for signature development and classification (Fig. 6.2).

### 4.4. *Training Sites and Ground Truth Data Collection*

Ground truth data was derived from a variety of sources: GPS-assisted field visits, topographic maps, vegetation and land-use maps, and detailed sketch maps of land-use history. Field studies for ground truth data were carried out in July 1997, January 1998, and March 1999. Trips were undertaken to identify areas representative of each land cover in the target classification scheme, and in later visits, to identify regions and causes of high classification uncertainty. The selection of areas for class-specific ground-truthing was strongly guided by project experts in local vegetation. As a result, sites were identified for seven cover classes, including water, two types of savannas, pasture, *milpa*, and *bajo* and upland forests. In addition, a diversity of successional forest patches ranging in age from 1 to 15 years were ground-truthed, as were areas of invasive bracken fern. Existing topographic and vegetation maps provided additional information on features such as water bodies, natural savannas and large, seasonally inundated *bajos*.

The ground-truthing benefited greatly from sketch maps derived in the field detailing smallholder land use in parcels over a period of 10–20 years. Sketch-mapping is a highly useful methodology in tropical forest environments to link better remote-sensing approaches to local-scale land change (Campbell and Browder 1995). Such maps were produced by project researchers during guided tours of agricultural plots belonging to interviewed farmers, and linked by GPS to the imagery (Fig. 6.5; also Ch. 9). In GPS-assisted visits, sketch-mapping documented the spatial pattern and history of land uses (Geoghegan *et al.* 2001; Turner *et al.* 2001). While map quality varied with farmer memory and surveyor accuracy, a ranked qualification by project personnel engaged in the survey aided in identifying the best maps for training site development. With the added ground truth data from the sketch maps, it was possible to increase the numbers of training sites for *milpa*, pasture, and chili, categories later coalesced into one agricultural cover class. Sketch maps also aided in the identification of additional training sites for successional forest classes. Since the maps were collected over a longer time period (one year) over a wider geographic area, they complemented well the more directed but shorter field visits for training site identification.

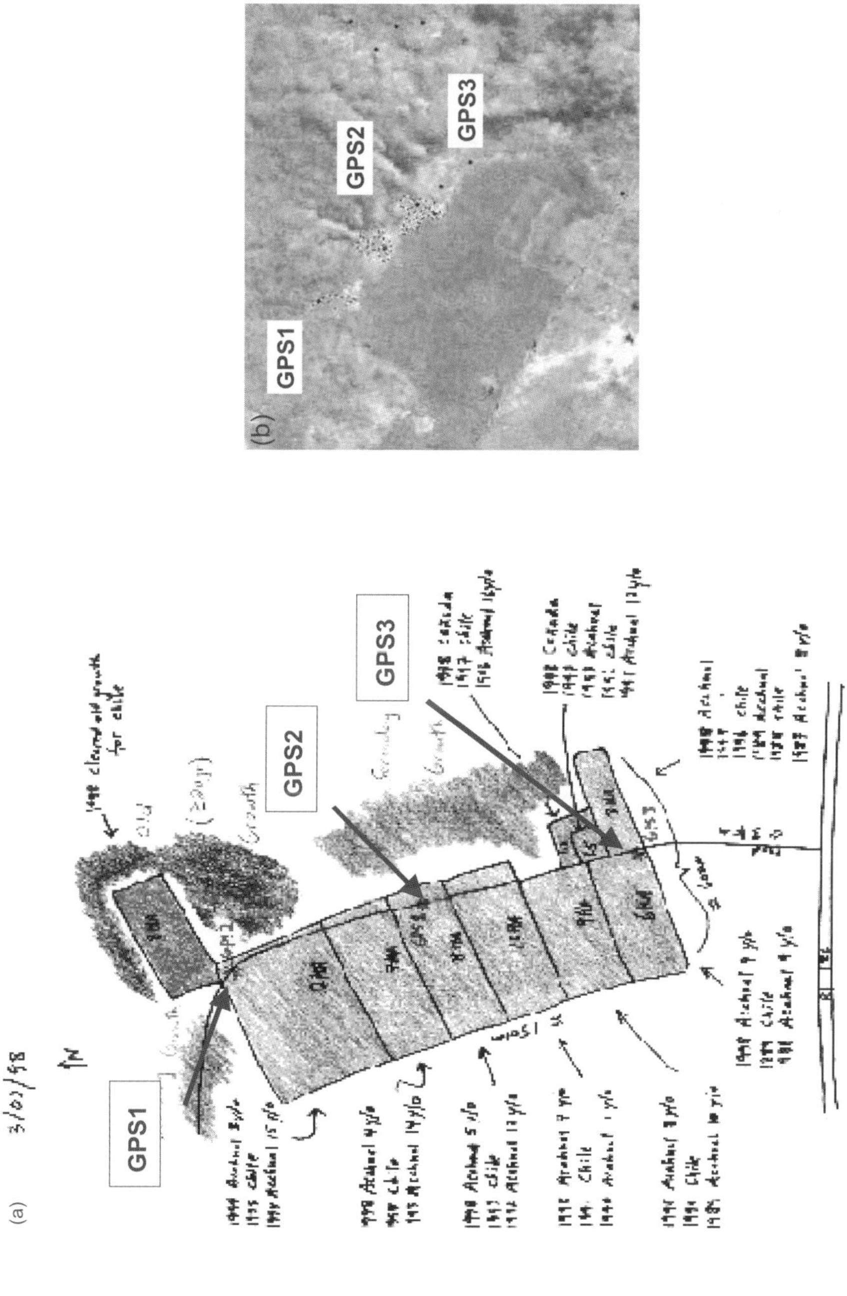


**Fig. 6.5.** Linking sketch maps to imagery.

## 4.5. *Training Site Evaluation*

Signature development involved the creation of training sites for land-cover classes of the target classification scheme. After the development of training sites, the corresponding land-cover signatures were evaluated using four accepted measures of separability: Euclidean distance, divergence, transformed divergence, and Jefferies-Matusita distance. The parameters of these measures provide the statistical separation for all pairs of land-cover classes in terms of their spectral response patterns. In general, the larger the magnitude of the separability index used, the greater the statistical distance between training sets, and the higher the probability of correct classification (Lillesand and Kiefer 1994). The indices were calculated using the transformations (Swain and Davis 1978):

For each formula,

$i$ and $j$ = two signature classes being compared
$C_i$ = covariance matrix of signature i
$|C_i|$ = determinant of $C_i$
$\mu_i$ = mean vector of signature i
$tr$ = trace function
T = transposition function

Jeffries-Matusita distance (ranges from 0 to 1,414)

$$\alpha = \tfrac{1}{8}(\mu_i - \mu_j)^T \left( \frac{(C_i + C_j)}{2} \right)^{-1} (\mu_i - \mu_j) + \tfrac{1}{2} ln \left( \frac{\left| \frac{(C_i + C_j)}{2} \right|}{\sqrt{|C_i| \times |C_j|}} \right)$$

$$JM_{ij} = \sqrt{2(1 - e^{-\alpha})}$$

Divergence

$$D = \tfrac{1}{2} tr\left((C_i - C_j)(C_i^{-1} - C_j^{-1})\right) + \tfrac{1}{2} tr\left((C_i^{-1} - C_j^{-1})(\mu_i - \mu_j)(\mu_i - \mu_j)^T\right)$$

and Transformed Divergence (ranges from 0 to 2,000)

$$TD_{ij} = 2\left(1 - \exp\left(\frac{-D_{ij}}{8}\right)\right)$$

## 4.6. *Supervised Classification*

The final signatures were then used in a maximum likelihood supervised classification to produce categorical land-use and land-cover maps for change detection and modeling. In the supervised classification approach, the spectral response vectors of each class are modeled to have multivariate normal distributions, and the parameters of the models are estimated from training samples (Sharma and Sarkar 1998). Several training samples are developed for each cover class in the target classification scheme. In the maximum likelihood technique, pixel assignments to cover classes are based on likelihood calculated at each pixel for its membership in each candidate cover class. In the absence of contextual information, a maximum likelihood classifier assigns each pixel to one among the set of candidate land-cover classes, depending solely on the spectral response of that particular pixel. In the approach developed in this study, however, pixel assignments were based not only on per-pixel noise-corrected spectral response, but also on its neighborhood characteristics as determined by the textural transformation, as well as the pixel NDVI. The Bayesian maximum likelihood classifier is (ERDAS 1994):

$$D = \ln(a_c) - [0.5\ln(|Cov_c|)] - [0.5(X - M_c)^T(Cov_c^{-1})(X - M_c)]$$

where

$D$ = weighted distance or likelihood
$c$ = land-cover class
$X$ = measurement vector of candidate pixel
$M_c$ = mean vector of the sample of land-cover class $c$
$a_c$ = probability that any candidate pixel is a member of class $c$ (in this study equal prior probabilities were assigned to each cover class)
$Cov_c$ = covariance matrix of pixels in sample class $c$
$|Cov_c|$ = determinant of $Cov_c$
$Cov_c^{-1}$ = inverse of $Cov_c$
ln = natural logarithm

$T$ = transposition

The pixel assignment is made to the class $c$ for which $D$ is lowest.

## 5. *Change Detection*

Assessments of land change for monitoring and projecting purposes require spatial specifications of where land change takes place, and are a critical component for land and forest management. Satellite remote sensing is a strong tool in the monitoring of forest clearing and land conversion over large areas such as the SYPR (Cihlar 2000; Lambin 1999).

### 5.1. *Image Mosaics*

Regional land-cover maps were created for the region to derive estimates of the major land changes in the past 30 years. The maps were created using the Landsat TM classified scenes as well as digitized aerial photographs. The study region is covered by portions of three Landsat scenes identified by the Landsat Worldwide Reference System as path 20, row 47; path 19, row 47; and path 19, row 48. Regional maps were created using TM images dating from 1984, 1985, 1987, 1988, 1994, 1995, 1996, and 1997. The regional land-cover map for 1969 was derived using aerial photographs that covered 63 per cent of the region. Most TM scenes pertained to the dry season (January to April) representing a relatively stable phenological status of the vegetation for these acquisition dates.

Two mosaics of the region were created using the TM scenes: one for the mid-1980s using scenes from 1984, 1985, 1987, and 1988, and the other for the mid-1990s using scene dates 1994, 1995, 1996, and 1997. The combination of scenes from different dates provides a more complete regional picture with less cloud coverage and solves some seasonal problems in class separability, such as those emerging from the deciduous behavior of both wetland and upland forest. As detailed in Table 6.1, eleven cover classes were initially identified, of which ten were land/vegetation based: water, wetland (*bajo*) forest, upland (*mediana*) forest, savanna (seasonally inundated), *tular* (herbaceous wetland), three stages of upland successional growth (herbaceous, shrub-dominated, arboreal), cropland, pasture, and one significant invasive species (bracken fern). The annual variability in precipitation throughout the region and the temporal variation in the imagery used to create the regional mosaics for the two periods of assessment, however, impeded such detail for the region at large. The ten land-cover classes were reduced to six for the regional analysis: wetland forest, upland forest, intermediate and late successional growth (4–15 years), *tular*-savanna ('natural' inundated and seasonally inundated grasslands, respectively), agriculture (cropland–pasture–early successional regrowth), and bracken fern.

Land-cover change estimates for the region were derived using the regional mosaics from TM and the 1969 map created from aerial photographs. Change detection was done through spatial cross-classification. Estimations of land-cover changes from 1969 to 1997 correspond only to 63 per cent of the total area (that covered by the 1969 aerial photographs), for three land-cover classes: forest, secondary growth (herbaceous and shrubby), and agriculture. Land-cover changes from 1987 to 1997 were estimated for the six classes. Rates of deforestation were estimated not only regionally but also for three subregional hotspots of change, to allow for quantitative comparisons across the region. The change hotspots included the Laguna Silvituc area in the west of the study region, the long-established *ejido* lands of Nicolas Bravo and Nuevo Becar in the east, and the recently opened southern area.

## 6. *Results and Discussion*

### 6.1. *Classification*

For each Landsat TM scene, the image used for classification included the 3-band output of the PCA, their three corresponding texture bands, and the NDVI band calculated before the PCA. Table 6.1 details the thirteen classes in the target classification scheme, and Table 6.2 the mean relative spectral response values of each class in each of the seven bands used. There was considerable regional variation in spectral reflectance within each land-cover class, inspiring trial classifications with merged and unmerged signature files, to test the effects of 'averaging' spectral signatures from various training sites for each land-cover class on the classification process. It was found that the merging of cover-class signatures leads to unsatisfactory classification results by strongly overclassifying particular land-cover classes such as wetland or upland forest. As a result, the final classification

**Table 6.2.** Mean pixel values for target land-cover classes

| Land cover | PC1 | PC2 | PC3 | Texture 1 | Texture 2 | Texture 3 | NDVI |
|---|---|---|---|---|---|---|---|
| Water | 36.77 | 5.77 | 4.52 | 2.61 | 1.94 | 3.23 | 36.79 |
| Savanna | 50.54 | 14.47 | 23.92 | 2.40 | 1.38 | 4.091 | 56.56 |
| *Bajo* forest | 47.85 | 40.10 | 58.81 | 2.53 | 1.51 | 2.91 | 102.70 |
| Upland forest | 46.40 | 48.05 | 61.88 | 2.58 | 2.89 | 3.81 | 107.80 |
| Cropland | 63.94 | 24.86 | 96.84 | 4.70 | 5.74 | 8.14 | 80.36 |
| Pasture | 62.52 | 33.79 | 103.02 | 4.25 | 4.71 | 8.77 | 89.36 |
| Herbaceous secondary | 57.25 | 43.06 | 87.55 | 3.45 | 4.54 | 5.51 | 100.28 |
| Shrubby secondary | 51.46 | 52.04 | 73.24 | 2.75 | 4.48 | 5.13 | 110.07 |
| Arboreal secondary | 48.80 | 59.77 | 72.96 | 2.53 | 3.77 | 4.03 | 115.67 |
| *Pteridium* (bracken fern) | 50.54 | 14.47 | 23.92 | 2.40 | 1.38 | 4.09 | 56.56 |
| Clouds | 58.62 | 40.72 | 81.91 | 3.70 | 1.91 | 4.58 | 83.29 |
| Cloud shadows | 37.16 | 23.86 | 22.56 | 2.52 | 1.62 | 3.17 | 90.74 |

was conducted with disaggregate signature files, akin to a family of signature replicates per land-cover class, to include fully the regional variation within each land-cover class in the target classification scheme. In the final result, ten signature families were employed with over ninety-eight individual signature (training site) files.

Figure 6.6a shows the complete spectral profiles of each target-over class in all the seven bands used for classification. Profiles are of the same order of magnitude in the PC and NDVI bands, but of a lower order in the texture bands. Figure 6.6b details separately the signature profiles in the principal components and NDVI bands, while Figs. 6.6c and 6.6d detail the characteristic texture or spatial variation of the signature profiles. Figure 6.6d specifically depicts, for each target cover class, the mean spatial variance of spectral reflectance in $3 \times 3$ pixel neighborhoods. For each class, the composite stacked bar graph depicts overall mean neighborhood variance in the three principal components bands. For example, average variance for water was 2.61 in texture band 1 (based on the first principal component of original visible TM bands), 1.94 in texture band 2 (first principal component of infrared TM bands), and 3.23 in texture band 3 (second principal component of infrared TM bands), for a total average variance of 7.78 over the three texture bands. Wetland forest has the lowest average variance overall, followed by the water and savanna classes. Most of this total variance for these three classes comes from the first and third texture bands, indicating that the signatures for these

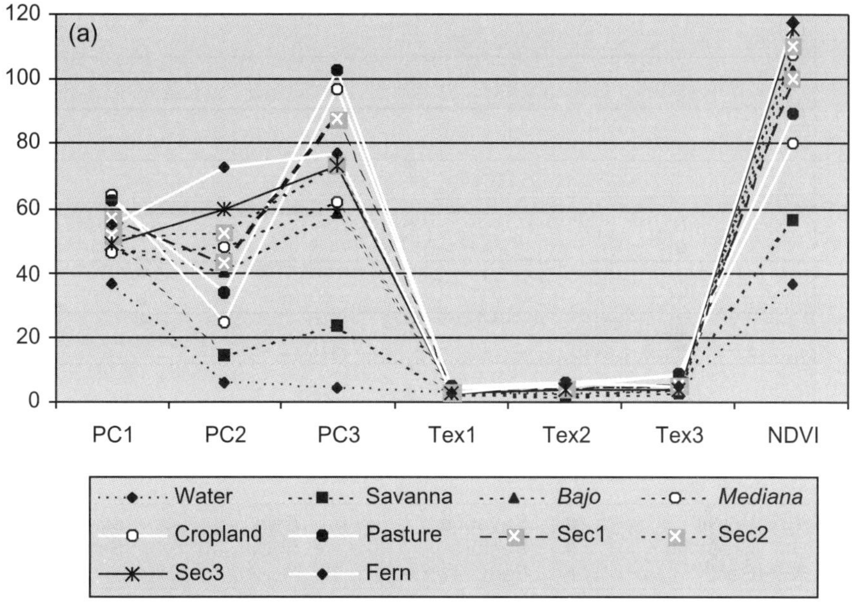


**Fig. 6.6 a** Signature profiles of major land-cover classes in all 7 bands in classifictaion.

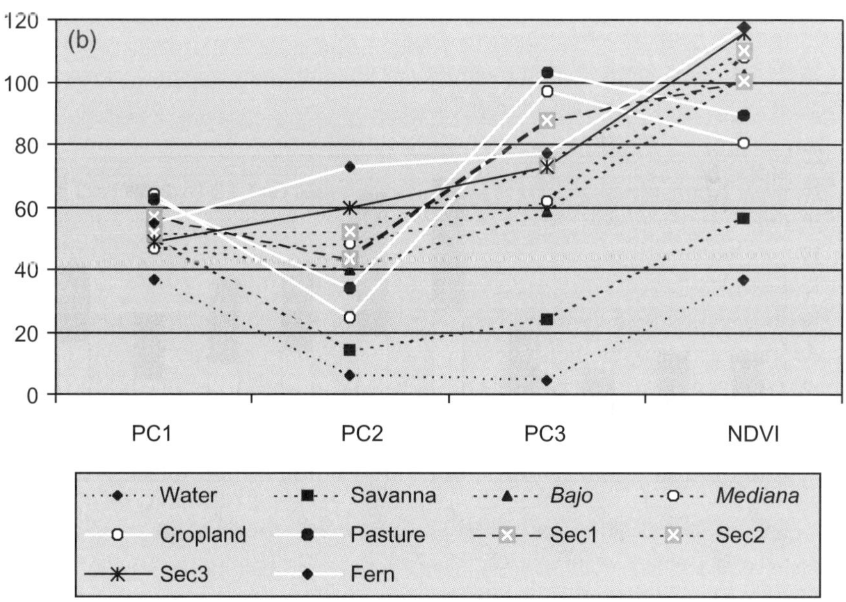


**Fig. 6.6 b** Signature profiles of major land-cover classes in PC and NDVI bands.

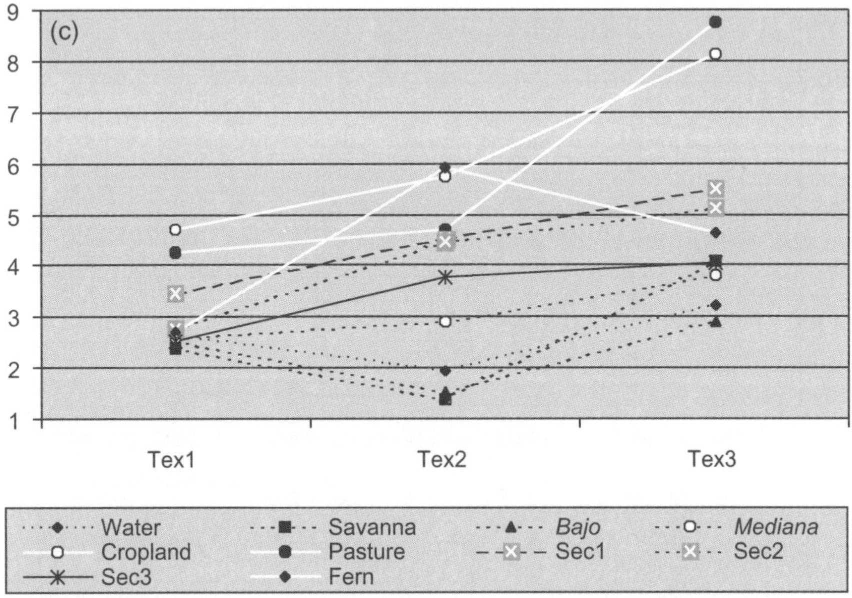


**Fig. 6.6 c** Signature profiles of major land-cover classes in texture bands.

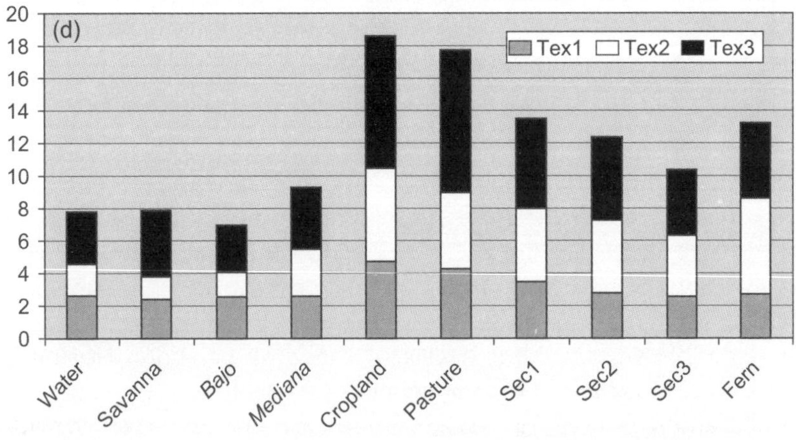


**Fig. 6.6 d** Mean texture (neighborhood variance) statistics of target land-cover classes.

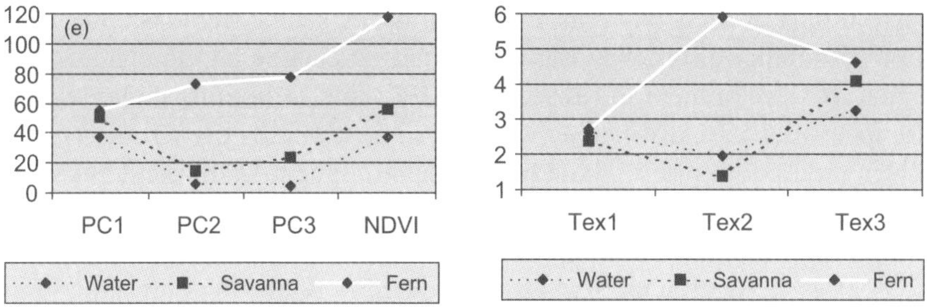


**Fig. 6.6 e** Signature profiles of spectrally distinct land-cover classes.

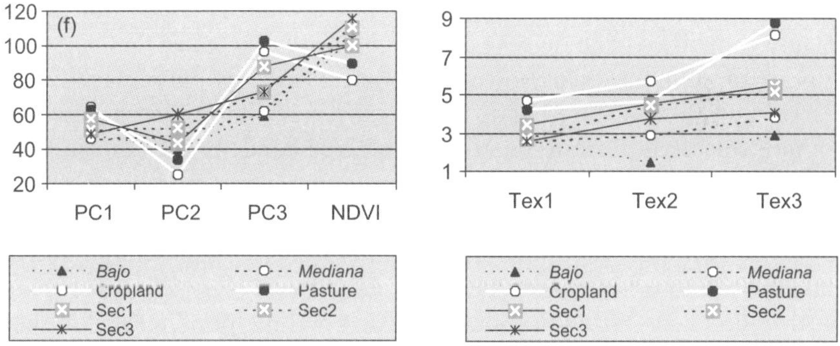


**Fig. 6.6 f** Signature profiles of spectrally similar land-cover classes.

classes exhibit relatively more spatial uniformity in the first principal component of the infrared bands.

Compared to water, savanna, and *bajo* forest, upland forest has higher total variance (9.28), to which all three texture bands contribute in equal degree (Fig. 6.6d; Table 6.2). Additionally, the per-texture band variance of upland forest is higher than that of *bajo* forest, most of the difference (1.51 in *bajo* v. 2.89 in *mediana*) originating in texture band 2, and in turn the first principal component of the infrared bands.

Some land-cover classes are distinct spectrally, and their signature profiles reflect their unique spectral properties (Fig. 6.6e). The invasive bracken fern (*Pteridium aquilinum* (L.) Kuhn.), largely prevalent in the eastern part of the study region (Ch. 4), is one such cover class. The high discriminability of bracken was confirmed for classification purposes and signature development in subsequent field visits. Other target classes that are spectrally quite distinct include water and naturally occurring, seasonally inundated savannas. These two classes are spectrally different but show similar textural properties.

Land-cover classes that have similar spectral signatures in the PC and NDVI bands are difficult to separate (Fig. 6.6f). A case in point is the spectral similarity among wetland and upland forest, as well as the three stages of successional vegetation. The signature profiles of these primary and secondary forest classes show areas of overlap in principal components band space. Particularly, upland and wetland forests are similar enough spectrally to cause some overlap in signature profiles. A closer look at the texture properties of these cover classes, however, indicates that variance appears to aid in the separation of the signature profiles, particularly in the second and third texture bands (based on the two lower-order principal components derived from the original infrared bands of the Landsat TM image). The spectral response of these two forest types may be similar, but the spatial pattern of their spectral variation differs. *Bajo* forest has the lowest spatial variance of spectral response, indicating it is a spectrally uniform class relative to other classes, such as upland forest (Fig. 6.6d). This may partly reflect actual structural characteristics of this type of forest. Another factor contributing to the low spatial variability of *bajo* forest signatures may be the fact that such forests are located on flat areas free from topographic effects such as hill shadows. The spectral signatures of upland forest, on the other hand, exhibit more spatial variation, accounted for by the local ecological variability of this thematic class, as well as slight topographic effects in regions of gently rolling hills of the study area. In preliminary studies, topographic normalization was attempted using an available DEM (digital elevation model). Due to low-slope relief and inadequate resolution of the DEM, the normalization algorithms did not achieve appreciable improvements. Another potential cause of confusion between *bajo* and upland forest is the presence of deciduous subformations, locally referred to as *subcaducifolia*, that may occur within either type of forest. This vegetation does not represent a forest type that is exclusive to the two main forest categories, and training sites were not separately developed for it. In future analyses, however, attempts are being made to discriminate spectrally this land-cover class within both the two forest types.

Secondary forest succession, pasture, and *milpa* are land-cover class pairs that present separability challenges. They represent either areas under a mix of different plant covers varying over short distances, or regions of changing vegetation, such as in a successional sequence, or both. The case of secondary successional classes is particularly problematic from a classification perspective, since sharp spectral thresholds do not exist among early, mid-, and late successional forest patches. Thus, early herbaceous successional vegetation blends spectrally and often spatially into mid-stage shrub-dominated succession. The exact spectral (or age) divide between shrubby mid-stage successional forest and late-stage, tree-dominated secondary forest is similarly difficult to identify. Rather than any linear or determinate process, succession is strongly influenced by factors such as land-use history, soils, slope and aspect, gap dynamics, and stand age.

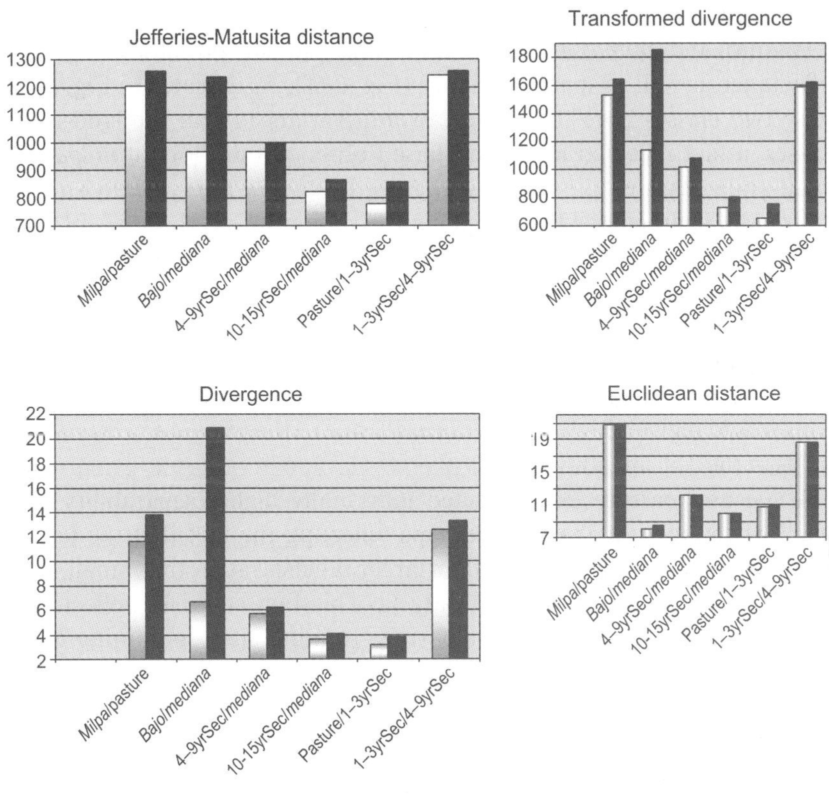


**Fig. 6.7.** Signature separability and texture measures.

These challenges are made all the more pressing by the fact that increasing numbers of research efforts identify the need to differentiate successional vegetation classes (Li *et al.* 1994; Mausel *et al.* 1993; Moran *et al.* 1994). These pairs are significantly improved with the incorporation of texture analysis, as demonstrated in Fig. 6.7. Most improvement attained through texture is between the cover pairs of early successional and crop/pasture, and mid- or late successional forest with mature upland forest.

Spectral signature separability varied greatly, depending on cover-class pairs. Figure 6.7 highlights the significant marginal improvements in signature separability obtained as a result of incorporating the three texture bands in the classification process. This improvement varies for different pairs of land-cover classes, and according to the separability index used (formulae in section 4.5). The highest marginal improvement in signature separability as a result of texture analysis was noted in distinguishing wetland and upland forest (Fig. 6.7). All separability indices used (Jeffries-Matusita distance, divergence, and transformed divergence) register the improvement.

Pasture and *milpa*, however, have very similar profiles in spectral reflectance as well as spatial variance (Fig. 6.6f). As depicted in Fig. 6.6d, these two classes have the highest overall variance of all the land-cover classes, a fact reflected in ground-based observation. Plots of *milpa* and small-scale pastures, particularly ungrazed pastures, are typically smaller in size and characterized by significant variation within the same plot, as well as variation among plots. This variation at a given moment (scene date) in existing vegetation at the scale of the individual parcels is due to a combination of site-specific environmental factors as well as differing land-use histories and strategies employed by the land manager. The smaller relative size of the plots also means that neighborhood variance calculations would register higher values overall, due to the presence of more edge pixels in the $3 \times 3$ windows used.

*Milpa* and pasture also reflected marginally higher separability as a result of the texture analysis, perhaps reflecting the fact that the largest areal extents of pasture are found on larger ranches rather than small, scattered plots on individual, *ejido* parcels. This fact could be reflected in the texture statistics, with most pasture training sites containing lower numbers of edge pixels in local pixel neighborhoods compared to *milpa* sites. In trial classifications, however, several areas of known pasture were misclassified as *milpa*, and vice versa. The classes of pasture and *milpa* were not satisfactorily separable at this stage of the classification process, and were merged into one agricultural cover class.

Although texture analysis aided in feature discrimination for some classes in the target scheme, useful information about some vegetation classes may have been lost as a result of the exclusion of the higher-order principal component bands, particularly in the visible bands. Future classification will revisit the PCA in order to manage better the differing signal-to-noise ratios in the TM bands. Also, future analysis might aim to identify the potential for developing texture functions using Fourier models, followed by testing the applicability of calibrated functions to other scenes within the same seasonal bounds. For the humid tropics, scene availability makes this a difficult issue.

The maximum likelihood Bayesian classification used relies heavily on a normal distribution of data in each input band, and tends to overclassify signatures with relatively large values in the covariance matrix. For classes such as upland forest, this is problematic since this class tends to have a large dispersion of pixels in the training samples and therefore large values in the covariance matrix for its signature. Additionally, the Bayesian classification was conducted with no preferential assignment of pixels to specific cover types. In other words, the prior probabilities for each land-cover class were assigned equal values. Future research will focus on the use of differential prior probabilities for certain land-cover classes that have relatively well-understood relationships to established GIS layers such as soils and topography.

## 6.2. *Change Detection of Land Use/Cover (1969–1997)*

The combination of aerial photographs and TM imagery analysis provides a spatially explicit picture of regional land cover and its change in the region (Fig. 6.8; color Pls. 1, 2; Tables 6.3–5). In 1969, just after the completion of Highway 186 but before its full paving, an area of 11,042km$^2$ of the forests in the photographed region (central SYPR) was intact (Table 6.3). By 1987, 6.2% (686 km$^2$) of this forest was lost, mostly within the older *ejidos* on the western and eastern edges of the study region. Over the next decade (1987–97), another 2.8% (288km$^2$) of this central area was deforested. At the full regional scale, 6.1% (970km$^2$) of the forest fell in the same decade (1987–97), reflecting the increased agricultural activity in the southeastern section not captured in the area covered by the 1969 aerial photographs (color Pl. 2; Tables 6.5–6). When successional regrowth (an area of 510 km$^2$) is included in the derivation of net deforestation rates, a 2.9% rate of deforestation results for the 10-year

**Table 6.3.** Land cover (km²) in 1969, 1987, and 1997

| Land-cover classes | 1969[a] | 1987[b] | 1997[b] |
|---|---|---|---|
| Forest (upland + *bajo*) | 11,042 | 10,356 | 10,068 |
| Secondary succession | 111 | 634 | 845 |
| Agriculture and pasture | 228 | 391 | 468 |

[a]Based on aerial photographs covering 63% of the study region or 11,318km².
[b]Based on TM Landsat imagery for same area as photographs.
*Source*: Turner *et al.* 2001.

**Table 6.4.** Land cover (km²) in the mid-1980s and mid-1990s

| Image mosaics | *Bajo* forest | Upland forest | Secondary (7–15 years) | Agriculture and pasture | Bracken fern | Semi-inundated savannas |
|---|---|---|---|---|---|---|
| 1984–8 | 3,286 | 12,502 | 784 | 524 | 39 | 48 |
| 1994–7 | 3,264 | 12,064 | 1,060 | 599 | 149 | 48 |

**Table 6.5.** Land-cover transitions from 1984–1988 to 1994–1997

| | | Land-cover classes in 1984–8 | | | | |
|---|---|---|---|---|---|---|
| | | *Bajo* forest | Upland forest | Secondary (7–15 years) | Agriculture and pasture | Bracken fern |
| Land-cover classes in 1994–7 | *Bajo* forest | 3,268 (99.2%) | 0.3 (0%) | 0 | 4 (0.8%) | 0 |
| | Upland forest | 0.3 (0%) | 11,591 (92.5%) | 324 (41.3%) | 182 (34.6%) | 0 |
| | Secondary (4–15 years) | 15 (0.5%) | 647 (5.2%) | 269 (34.2%) | 130 (24.8%) | 0 |
| | Cropland | 12 (0.4%) | 245 (2.0%) | 145 (18.4%) | 197 (37.6%) | 0.5 (1.4%) |
| | Bracken fern | 0 | 0 | 47 (6.0%) | 11 (2.1%) | 48 (98.6%) |

*Note*: Cover transitions are in km² with % change in parentheses.
*Source*: Turner *et al.* 2001.

period. Given the spatial incongruity between aerial photographs and TM images, precise deforestation rates in the overall region since the advent of the highway cannot be generated. Annual rates of deforestation derive to 0.32% (or 8.8% total) from 1969 to 1997, and 0.29% (or 2.9% total) from 1987 to 1997.

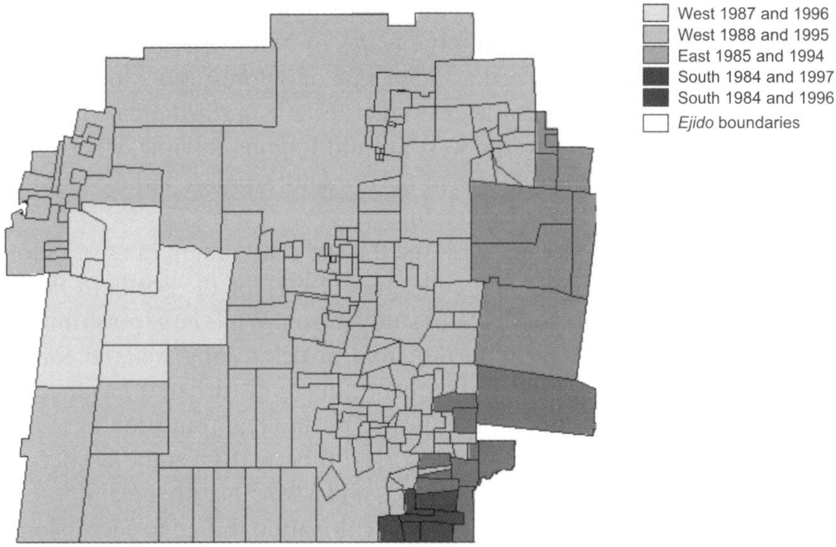


**Fig. 6.8.** Image Mosaics (each black-white tone represents a different date for TM scenes).

**Table 6.6.** Forest transitions (deforestation and succession) for three change hotspots (km$^2$)

| | Forest transitions from mid-1980s to mid-1990s (km$^2$) | | | |
|---|---|---|---|---|
| | Forest to forest | Non-forest to non-forest | Forest to non-forest | Non-forest to forest |
| Southern area | 2,547 | 139 | 462 | 261 |
| East-northeast area | 2,058 | 85 | 305 | 176 |
| Western area | 1,499 | 105 | 101 | 187 |
| Total SYPR | 15,064 | 664 | 1,125 | 992 |

There is considerable spatial variation in the average deforestation patterns described above. Significantly higher rates of deforestation are found in the southeastern *ejidos*, bordering the eastern edge of the Calakmul biosphere reserve. In this area of about 3,300km$^2$, a total of 15.8% of extant forest was removed between 1987 and 1997. Adjusting for secondary succession, the net decadal deforestation lowers to 11.1%, almost three times higher, nevertheless, than the deforestation rate for the total region

(Table 6.6). Another hotspot of change in the region is a 2,684km$^2$ area in the east and northeast, including the *ejidos* of Nuevo Becar and Nicolas Bravo. This region experienced a 12.9% rate of deforestation and 3.6% of forest regrowth over the decade examined. Finally a third area with high rates of change is located to the west around Laguna Silvituc, wherein an area of almost 1,900km$^2$ underwent 6.31% deforestation and 6.5% reforestation from 1987 to 1997 (Table 6.6).

These figures are consistent with the 0.4 per cent annual rates of deforestation estimated by Sader and colleagues (1994) for the northern Petén, Guatemala, immediately south of our study region. While large government-funded projects played an important role in deforesting wetland forests before the mid-1980s (Ch. 3), subsequent human disturbance has focused almost solely on upland forests, as indicated by the areas transitioning to and from the non-wetland land-cover categories (Table 6.5). Over the last decade examined, the amount of cultivated lands taken from mature upland forests seems to have decreased, and the focus of cultivation shifted to successional growth. This shift is illustrated by the amount of land cleared in the 1987 imagery and the amount fallowed (regrowth) in the 1997 imagery, especially along the southern roadway (Pl. 2). This shift may suggest a reduction in the *milpa* fallow cycle (less land taken from mature forest; more taken from early successional growth), a direction apparently indicated by the fourfold increase in area invaded by bracken fern (Table 6.5) and local claims of more intensive chili-swidden cultivation (Ch. 10). The area under bracken expanded from 39km$^2$ to 149km$^2$ in the past decade, much of the expansion occurring in the northeastern subregion centered around Nicolas Bravo (Fig. 6.9). This fern, the subject of future study, appears more prevalent in those areas entering second- or third-'generation' crop-fallow cycles and/or plots excessively depleted of their nutrients before fallowing (Ch. 5).

## 7. *Summary*

Landsat Thematic Mapper imagery was used to classify land cover in the subhumid tropics of the southern Yucatán peninsular region. Thematic information rather than within-class quantitative change was the primary goal of this classification effort, in order to link effectively to land-use change modeling studies. Texture analysis adds statistical dimensionality to spectral bands, incorporating information about the spatial context in pixel neighborhoods. The results of texture analysis were incorporated into

**Fig. 6.9.** Expansion of bracken fern in the east-central part of the region, 1987–97.

the classification process as additional bands in the multiple-band classification algorithm. The results highlight the utility of synergism in social research and remote sensing.

For phase one of the classification effort, these observations apply:

- Integrating parcel-level studies of land-use history with remote sensing improves the level of classification detail and the relevance of the target scheme to ecological and socio-economic modeling needs.
- PCA is effective at random and systematic noise removal and reducing data dimensionality, particularly in the two-tiered approach

employed. Some fine-scale variations in spectral reflectance of certain land-cover types, however, may have been lost as a result of discarding higher-order components that captured most striping-related noise.

- Statistical texture analysis improves signature separability between several pairs of land-cover classes, but results are limited to the datasets from which the texture measures are derived. Future analysis may develop texture functions using Fourier models.
- Topographic normalization may improve characterization of wetland and upland forests, but a finer-scale DEM is required.
- Subdeciduous vegetation in both wetland and upland forests causes confusion among these land-cover classes. This vegetation may be characterized by its structure, biomass, and moisture condition, and if correctly classified, could lead to better *bajo/mediana* separation.
- Secondary successional classes are separable at the local level, but at the regional level aggregation yields better generalizability. Trial experiments with tasseled cap transformations indicate that better separation may be achieved through this means among the different successional stages.
- Subpixel classification and fuzzy membership functions will yield richer, probabilistic land-cover information in the next stage of classification experiments.
- Accuracy assessment is underway to evaluate classification error and focus the development of additional training sites for more advanced classification efforts.
- Preclassification change detection can be combined with the results of transition maps derived in this stage to yield information on within-class changes, such as selective logging, land degradation, or decreasing biomass over the past decade.
- Annual deforestation rates are 0.32 per cent for 1967–97 and 0.29 per cent from 1987–97, most deforestation occurring from upland forests. There is significant regional variation in deforestation rates, rates of change being highest in the newly settled southern frontier.
- Large areas have undergone secondary succession, and the focus of cultivated lands appears to be shifting to successional forests, indicating a shortening of fallow cycles.
- Invasive species such as bracken fern have significantly expanded in areal extent, with potentially dramatic ecological and land-use consequences.

## REFERENCES


Achard, F., H. Eva, A. Glinni, P. Mayaux, T. Richards, and H. J. Stibig (1998). *Identification of Deforestation Hot Spot Areas in the Humid Tropics.* TREES Publications Series B. Research Report No. 4. Luxemburg, European Commission.

Baltaxe, P. A. (1987). 'The Application of Remote Sensing to Tropical Forest Cover Monitoring: A Review of Practices and Possibilities', in A. Cracknell, and L. Hayes (eds.), *Remote Sensing Yearbook*, 33–51.

Basili, P., P. Ciotti, G. D'Auria, F. S. Marzano, N. Pierdicca, and P. Quarto (1994). 'Assessment of Polarimetric Features to Discriminate Land Cover from the MAESTRO-1 Campaign'. *International Journal of Remote Sensing* 15: 2887–99.

Berta, M. S., P. W. Mausel, and J. A. Harrington (1990). 'Multidate Image Analysis of Forest Degradation in Equatorial Africa'. *Geocarto International* 5: 57–61.

Byrne, G. F., P. F. Crapper, and K. K. Mayo (1980). 'Monitoring Land-Cover Change by Principal Components Analysis of Multitemporal Landsat Data'. *Remote Sensing of Environment* 10: 175–84.

Campbell, J. B., and J. O. Browder (1992). *SPOT Survey of Agricultural Land Uses in the Brazilian Amazon.* ISPRS Commission VII, XXIX (Part B-7).

——(1995). 'Field Data Collection for Remote Sensing Analysis: SPOT Data, Rondonia, Brazil'. *International Journal of Remote Sensing* 16: 333–50.

Carr, J. R., and K. Matanawi (1999). 'Correspondence Analysis for Principal Components Transformation of Multispectral and Hyperspectral Digital Images'. *Photogrammetric Engineering and Remote Sensing* 65: 904–14.

Cihlar, J. (2000). 'Land Cover Mapping of Large Areas from Satellites: Status and Research Priorities'. *International Journal of Remote Sensing* 21: 1093–114.

DeFries, R. S. and A. S. Belward (2000). 'Global and Regional Land Cover Characterization from Satellite Data: An Introduction to the Special Issue'. *International Journal of Remote Sensing* 21: 1083–92.

DeFries, R. S., and J. R. G. Townshend (1994). 'NDVI-Derived Land Cover Classification at Global Scales'. *International Journal of Remote Sensing* 15: 3567–86.

DeFries, R., M. Hansen, J. R. G. Townshend, and R. Sohlberg (1998). 'Global Land Cover Classifications at 8 km Spatial Resolution: The Use of Training Data Derived from Landsat Imagery in Decision Tree Classifiers'. *International Journal of Remote Sensing* 19: 3141–68.

De Fries, R., M. Hansen, J. R. G. Townshend, A. Janetos, and T. Loveland (2000). 'A New Global 1-km Dataset of Percentage Tree Cover Derived from Remote Sensing'. *Global Change Biology* 6: 247–54.

DeFries, R. S., C. B. Field, F. Inez, C. O. Justice, P. A. Matson, M. Matthews, H. A. Mooney, C. S. Potter, K. Prentice, P. J. Sellers, J. Townshend, C. J. Tucker, S. L. Ustin, and P. M. Vitousek (1995). 'Mapping the Land Surface for Global Atmosphere-Biosphere Models: Towards Continuous Distributions of Vegetation's Functional Properties'. *Journal of Geophysical Research* 100: 20867–82.

Eastman, J. R. (1999). *IDRISI 32*. Worcester, Mass., Clark Labs.

Eastman, J. R., and M. Fulk (1993). 'Long Sequence Time Series Evaluation Using Standardized Principal Components'. *Photogrammetric Engineering and Remote Sensing* 59: 991–6.

ERDAS (1994). 'Field Guide' in *Field Guide*. Atlanta, Ga., ERDAS.

Flores, J. S., and C. Espejel (1994). 'Tipos de vegetación de la península de Yucatán'. *Etnoflora Yucatanense, Facículo* 3. Mérida, Universidad Autónoma de Yucatán.

Geoghegan, J. G., L. Pritchard, Y. Ogneva-Himmelberger, R. Roy Chowdhury, S. Sanderson, and B. L. Turner II (1998). '"Socializing the Pixel" and "Pixelizing the Social" in Land-Use/Cover Change', in D. Liverman, E. F. Moran, R. R. Rindfuss, and P. C. Stern (eds.), *People and Pixels: Linking Remote Sensing and Social Science*. Washington DC, National Academy Press, 51–69.

Geoghegan, J. G., S. Cortina Villar, P. Klepeis, P. Macario Mendoza, Y. Ogneva-Himmelberger, R. Roy Chowdhury, B. L. Turner II, and C. Vance (2001). 'Modeling Tropical Deforestation in the Southern Yucatán Peninsular Region: Comparing Survey and Satellite Data'. *Agriculture, Ecosystems and Environment* 85: 25–46.

Gillespie, A. R. (1980). 'Digital Techniques of Image Enhancement', in B. S. Siegal, and A. R. Gillespie (eds.), *Remote Sensing in Geology*. New York, John Wiley, 139–226.

Gilruth, P. T., C. F. Hutchinson, and B. Barry (1990). 'Assessing Deforestation in the Guinea Highlands of West Africa Using Remote Sensing'. *Photogrammetric Engineering and Remote Sensing* 56: 1375–82.

Gonzalez, R. C., and R. E. Woods (1992). *Digital Image Processing*. Reading, Mass., Addison-Wesley.

Green, G., and R. Sussman (1990). 'Deforestation History of the Eastern Rainforests of Madagascar from Satellite Images'. *Science* 248: 212–15.

Hansen, M. C., R. S. Defries, J. R. G. Townshend, and R. Sohlberg (2000). 'Global Land Cover Classification at 1 km Spatial Resolution Using a Classification Tree Approach'. *International Journal of Remote Sensing* 21: 1331–64.

Haralick, R. M. (1979). 'Statistical and Structural Approaches to Texture'. *Proceedings of the IEEE* 67: 786–804.

Haralick, R., K. Shanmugam, and I. Dinstein (1973). 'Textural Features for Image Classification'. *IEEE Transactions on Systems, Man and Cybernetics* 3: 610–21.

Jensen, J. R. (1996). *Introductory Digital Image Processing.* Englewood Cliffs, NJ, Prentice-Hall.

Jhung, Y., and P. H. Swain (1996). 'Bayesian Contextual Classification Based on Modified M-Estimates and Markov Random Fields'. *IEEE Transactions on Geoscience and Remote Sensing GE* 34: 67–75.

Kauth, R. J., and G. S. Thomas (1976). 'The Tasselled Cap—A Graphic Description of the Spectral Temporal Development of Agricultural Crops as Seen by LANDSAT'. *Proceedings of the 1976 Symposium on Machine Processing of Remote Sensed Data* 4B41–4B51. Purdue University, West Lafayette, Indiana.

Kummer, D. M. (1992). 'Remote Sensing and Tropical Deforestation: A Cautionary Note from the Philippines'. *Photogrammetric Engineering and Remote Sensing* 58: 1469–71.

Lambin, E. F. (1999). 'Monitoring Forest Degradation in Tropical Regions by Remote Sensing: Some Methodological Issues'. *Global Ecology and Biogeography* 8: 191–8.

Lark, R. M. (1995). 'Contributions of Principal Components to Discrimination of Classes of Land Cover in Multi-Spectral Imagery'. *International Journal of Remote Sensing* 16: 779–87.

Lavreau, J. (1991). 'De-Hazing Landsat Thematic Mapper Images'. *Photogrammetric Engineering and Remote Sensing* 57: 1297–302.

Li, Y., P. Mausel, J. Wu, and E. Brondizio (1994). 'Discrimination between Very Advanced Secondary Succession and Moist Mature Forest near Altamira, Brazil Using TM Data'. *ASPRS Proceedings*, 350–64.

Lillesand, T. M., and R. W. Kiefer (1994). *Remote Sensing and Image Interpretation*, 3rd edn. New York, John Wiley.

Loveland, T. R., and A. S. Belward (1997). 'The IGBP-DIS Global 1 km Land Cover Data Set, DISCover: First Results'. *International Journal of Remote Sensing* 18: 3289–95.

Loveland, T. R., B. C. Reed, J. F. Brown, D. O. Ohlen, Z. Zhu, L. Yang, and J. W. Merchant (2000). 'Development of a Global Land Cover Characteristics Database and IGBP DISCover from 1km AVHRR Data'. *International Journal of Remote Sensing* 21: 1303–30.

Malingreau, J. P., C. J. Tucker, and N. LaPorte (1989). 'AVHRR for Monitoring Global Tropical Deforestation'. *International Journal of Remote Sensing* 10: 855–67.

Martínez-Salas, E., and C. Galindo-Leal (n.d.). 'La vegetación de Calakmul, Campeche, México: clasificación, descripción y distribución'. Stanford University. Unpublished.

Mausel, P., Y. Wu, Y. Li, E. F. Moran, and E. S. Brondizio (1993). 'Spectral Identification of Successional Stages Following Deforestation in the Amazon'. *Geocarto International* 4: 61–71.

May, L. N. (1986). 'An Evaluation of Landsat MSS Digital Data for Updating Habitat Maps of the Louisiana Coastal Zone'. *Photogrammetric Engineering and Remote Sensing* 52: 1147–58.

Miranda, F., and E. Hernandez-X (1963). 'Tipos de vegetación de México y su clasificación'. *Boletín de la Sociedad Botánica de México* 28: 29–179.

Moran, E. F., E. Brondizio, P. Mausel, and Y. Wu (1994). 'Integrating Amazonian Vegetation, Land Use and Satellite Data'. *BioScience* 44: 329–38.

Nelson, R., and B. Holben (1986). 'Identifying Deforestation in Brazil Using Multiresolution Satellite Data'. *International Journal of Remote Sensing* 7: 429–48.

Paradella, W. R., M. F. F. Da Silva, A. D. N. Rosa, and C. A. Kushigbor (1994). 'A Geobotanical Approach to the Tropical Rainforest Environment of the Carajás Mineral Province (Amazon Region, Brazil), Based on Digital TM-Landsat and DEM Data'. *International Journal of Remote Sensing* 15: 1633–48.

Richards, J. A. (1993). *Remote Sensing Digital Image Analysis, an Introduction* (2). London, Springer-Verlag.

Saatchi, S. S., B. Nelson, B. Podest, and J. Holt (2000). 'Mapping Land Cover Types in the Amazon Basin Using 1 km JERS-1 Mosaic'. *International Journal of Remote Sensing* 21: 1201–34.

Sader, S. A. (1995). 'Spatial Characteristics of Forest Clearing and Vegetation Regrowth as Detected by Landsat Thematic Mapper Imagery'. *Photogrammetric Engineering and Remote Sensing* 61: 1145–51.

Sader, S. A., T. A. Stone, and A. T. Joyce (1990). 'Remote Sensing of Tropical Forests: An Overview of Research and Applications Using Non-Photographic Sensors.' *Photogrammetric Engineering and Remote Sensing* 55: 1343–51.

Sader, S. A., T. Sever, J. C. Smoot, and M. Richards (1994). 'Forest Change Estimates for the Northern Peten Region of Guatemala'. *Human Ecology* 22: 317–32.

Settle, J. J., and N. A. Drake (1993). 'Linear Mixing and the Estimation of Ground Cover Proportions'. *International Journal of Remote Sensing* 14: 1159–77.

Sharma, K. M. S., and A. Sarkar (1998). 'A Modified Contextual Classification Technique for Remote Sensing Data'. *Photogrammetric Engineering and Remote Sensing* 64: 273–80.

Singh, A., and A. Harrison (1985). 'Standardized Principal Components'. *International Journal of Remote Sensing* 6: 885–96.

Sohn, Y., E. Moran, and F. Gurri (1999). 'Deforestation in North Central Yucatan (1985–1995): Mapping Secondary Succession of Forest and Agricultural Land Use in Sotuta Using the Cosine of the Angle Concept'. *Photogrammetric Engineering and Remote Sensing* 65: 947–58.

Steininger, M. K. (1996). 'Tropical Secondary Forest Regrowth in the Amazon: Age, Area and Change Estimation with Thematic Mapper Data'. *International Journal of Remote Sensing* 17: 9–27.

Stone, T. A., R. Schlesinger, R. A. Houghton, and G. M. Woodell (1994). 'A Map of the Vegetation of South America Based on Satellite Imagery'. *Photogrammetric Engineering and Remote Sensing* 60: 541–51.

Swain, P. H., and S. M. Davis (1978). *Remote Sensing: The Quantitative Approach*. London, McGraw Hill.

Swain, P. H., S. B. Vardeman, and J. C. Tilton (1981). 'Contextual Classification of Multispectral Image Data'. *Pattern Recognition* 13: 429–41.

Townshend, J. R. G. (1984). 'Agricultural Land-Cover Discrimination Using Thematic Mapper Spectral Bands'. *International Journal of Remote Sensing* 5: 681–98.

Townshend, J. R. G., and C. O. Justice (1988). 'Selecting the Spatial Resolution of Satellite Sensors Required for Global Monitoring of Land Transformations'. *International Journal of Remote Sensing* 9: 187–236.

Tucker, C. J., J. R. G. Townshend, and T. E. Goff (1985). 'African Land Cover Classification Using Satellite Data'. *Science* 227: 369–75.

Turner, B. L., II, D. Skole, S. Sanderson, G. Fischer, L. Fresco, and R. Leemans (1995). *Land-Use and Land-Cover Change: Science/Research Plan*. IGBP Report No. 35 and HDP Report No. 7. Stockholm, SW, and Bonn, GR: International Geosphere-Biosphere Programme and International Human Dimensions Programme.

Turner, B. L., II., S. Cortina Villar, D. Foster, J. Geoghegan, E. Keys, P. Klepeis, D. Lawrence, P. Macario Mendoza, S. Manson, Y. Ogneva-Himmelberger, A. B. Plotkin, D. Pérez Salicrup, R. Roy Chowdhury, B. Savitsky, L. Schneider, B. Schmook, and C. Vance (2001). 'Deforestation in the Southern Yucatán Peninsular Region: An Integrative Approach'. *Forest Ecology and Management* 154: 343–70.

Wang, F. (1990). 'Fuzzy Supervised Classification of Remote Sensing Images'. *IEEE Transactions on Geoscience and Remote Sensing* 28: 194–201.

Welch, J. R., and K. G. Salter (1971). 'A Context Algorithm for Pattern Recognition and Image Interpretation'. *IEEE Transactions on Systems, Man and Cybernetics SMC* 1: 24–30.

Woodwell, G. M., R. A. Houghton, T. A. Stone, R. F. Nelson, and V. Kovalick (1987). 'Deforestation in the Tropics: New Measurements in the Amazon Basin Using Landsat and NOAA Advanced Very High Resolution Radiometer Imagery'. *Journal of Geophysical Research* 92: 2157–63.

# PART III

## *Agents and Institutions of Land Change: Household Economy and Land Change*

Land change in the region is undertaken largely by individual land managers, mostly *ejidatarios*. How these individuals manage lands, however, is strongly affected by institutions and policies that stake out the rules to land, capital, and other critical inputs as well as by various governmental and non-governmental organizations, each of which has some vested interest in the region and or its occupants. As demonstrated in Ch. 3, the character of the agents has changed over the past fifty years in terms of numbers (including landless individuals and households) as well as the conditions in which decisions are made. For the most part, open lands in the region have been drastically reduced, and most *ejidos* experience significantly increased land pressures, despite the generous 'average' size of holdings among households. Usufruct remains the dominant land tenure, but the basic orientation of government policy has shifted from state-sponsored large projects and subsidies to more local-input initiatives, and from agricultural expansion to attempts to localize production (i.e. intensification) in order to reduce pressures on older growth forests, the Calakmul Biosphere Reserve, and related programs that seek to preserve and conserve forests.

The *campesinos* of the region have not been passive agents in all these changes. They not only have become increasingly empowered through various initiatives, they have undertaken, independent of government and NGO programs, the largest ongoing land change in the southern Yucatán peninsular region, commercial jalapeño chili production. For the most part, this production is incorporated into the pre-existing swidden or *milpa* cultivation cycle and does not entail major shifts in agro-technological strategies, other than the use of chemicals to ward off pests and diseases on chili plants. In some cases, however, preferred lands are 'mechanized'—cleared mechanically with the intent of cultivating annually. Overall, chili production intensifies agriculture.

Despite these changes, *ejidatarios* remain semi-subsistence or semi-commercial farmers. The chili market is too volatile and the vagaries of nature too large to abandon subsistence cropping, and alternative land-use strategies remain in various stages of experimentation under conditions of thin markets. Should the farming population of the region continue to grow under current conditions, tensions will likely mount between the government's interest in the reserve and various external interests that have invested in El Mundo Maya and the Mesoamerican Biological Corridor.

The nature of the household, the *ejido*, and the myriad of organizations and institutions seeking to influence land and life in the region demonstrate that land-use dynamics are far more complex than perhaps imagined for a 'frontier'. This complexity, however, is critical for understanding the subsequent modeling exercises (Part IV), illustrating various gaps between the extant and required data for fine-tuned land-change models.

*The Editors*

# 7

# Institutions, Organizations, and Policy Affecting Land Change: Complexity Within and Beyond the *Ejido*

*Peter Klepeis and Rinku Roy Chowdhury*

Despite decades of colonization and development initiatives, the southern Yucatán peninsular region remains an economic frontier. The term 'frontier', however, hides a complex political economy of social, political, and economic structures in which land managers operate. Presently, multiple interest groups vie for influence, increasingly positioning themselves around sustainability concerns, and attempting to reconcile the competing goals of economic development and environmental preservation. The major political institutions and organizations promoting conservation and development in the region fit into five categories: federally decreed land management regimes, federal and state secretariats, local community-based groups and institutions, national non-governmental organizations (NGOs), and international accords.

These institutions and organizations aim to influence land-use decisions in the dominant land access unit, the *ejido*. The relationships among *ejidos,* social movements, NGOs, government policy, and international activity in the region are examined here, highlighting how even within a frontier economy, conservation and development visions increasingly influence resource use.

We thank Colin Vance for his help in crafting the structure and content of this chapter, Eric Keys and Birgit Schmook for valuable comments in earlier drafts, and various individuals in Mexico for their insights.

## 1. *The* Ejido *as an Institution*

Before the Mexican revolution of 1910–17, 96 per cent of Mexico's rural people were landless (Sinha 1984). These rural poor supported the revolution, in large part, to break up grand *haciendas* (estates) and to allow *campesinos* (peasants) access to agricultural land. *Ejidos*, one of four land-tenure types federally mandated, were designed to provide *campesinos* access to land that could not be transferred easily and thereby taken from them. Based on interpretations of pre-Hispanic land tenure, Article 27 of the Constitution established *ejido* land to be communal, ruled by an *ejido* assembly (consisting of all members with land rights in the *ejido*, or *ejidatarios*), and used in *ejido*-defined usufruct. Prior to 1992, when the law was reformed, *ejidatarios* were prevented from selling their land, renting it, or using it as collateral, and from negotiating deals with private investors.[1]

Perhaps more important than these official guidelines, however, are the perceptions of *ejidos* by state officials. Established, in part, to protect 'indigenous' people and not open to privatization, the *ejido* was stigmatized as ill-suited for modernization (Oasa and Jennings 1982). A bimodal Mexican agrarian policy followed (de Janvry 1981; Tomich, Kilby, and Johnston 1995) in which the potential productive role of *ejidatarios* was largely ignored (Oasa and Jennings 1982; Sonnenfeld 1992; Tomich, Kilby, and Johnston 1995). This bimodality was exacerbated in the study region given its sparse occupation, and when agrarian attention was given, it focused first on government-led, large-scale projects (Ch. 3).

## 2. *Changing Views of* Ejido *Agriculture and the Colonization of the Region*

A prerequisite to the beginning of the 1970s boom of colonization in the region was a change in governmental attitudes about the *ejido* as a vehicle for commercial agriculture (Szekely and Restrepo 1988). In the face of basic food-production shortfalls nationally, the federal government sought to implement technological changes in agriculture beyond the 'green revolution'

[1] Article 27 dictated that an *ejidatario* lost access to land rights if two consecutive years passed without cultivation, and that land rights could not be sold. In everyday practice, however, such regulations have been violated when necessary or convenient to *ejidatarios*, often with the blessing of the *ejido* assembly.

systems in northern Mexico. Colonization of the sparsely populated frontiers of southern Mexico became a means of resolving problems of land shortage elsewhere in the country, low agricultural production nationally, and political tensions, especially in central Mexico (Bassols Batalla *et al.* 1976). The new *ejidos* were referred to as NCPEs (Nuevo Centros de Población Ejidal—New Ejidal Population Centers), selected in areas of low occupation and with sufficient natural resources to support settlers (Fuentes Aguilar 1980).

With respect to the tropical forests of southeastern Mexico, the NCPE program had two principal goals: to provide workers currently active in forestry with opportunities in agriculture and livestock rearing, and to reduce deforestation that officials perceived to be associated with swidden agriculture, deemed a maladaptive and destructive cultivation strategy (ibid.). Secondarily, colonization and the NCPE initiative peopled Mexico's lands adjacent to British Honduras (now Belize) and Guatemala, as well as the territory of Quintana Roo, which needed a minimum of 80,000 people to become a state.

The Comisión Intersecretarial para Nuevos Centros de Población Ejidal (Inter-Secretarial Commission for New Population Centers), or COINCE, became law in 1971, designed to boost basic food production by coordinating the directed settlement of underdeveloped lands (Gates 1993). COINCE incorporated the Yucatán Peninsula in a program called *Cuenca del Sureste* (Basin of the Southeast), believing the southern Yucatán Peninsula possessed significant agricultural potential (Szekely and Restrepo 1988). The land grants of the 1970s that were associated with these development programs were a generous 100ha/*ejidatario* (Dachary and Arnaiz 1983), plus 25ha for *ejido* centers.

The paving of the Chetumal–Escárcega road (Highway 186) by 1972, enabled the subsequent dramatic settlement and development of the region (Revel-Mouroz 1980; Vogeler 1974). As word spread about available land, settlers spontaneously migrated to the southern Yucatán Peninsula. The vast majority of respondents from the 1997–8 SYPR survey (Ch. 8) identified land availability as the reason for their move (Table 8.2). To a much lesser degree, environmental (e.g. drought and decreasing soil fertility) and political factors pushed people to the study region.

### 2.1. *Expansion of* Ejido *Lands*

The acceleration of *ejido* land grants in the southern Yucatán peninsular region after 1970 was dramatic. Most of the *ejidos* were small in size and established along the north–south improved but unpaved road. The number of *ejido* land grants and the population of the region went up significantly in the 1980s (Fig. 7.1), triggered by the eruption of the volcano Chichonal in 1982 (4,000 Chol Maya from Chiapas) and political refugees from Guatemala (Cortez Ruiz 1998; Stedman-Edwards 1997). Today, many *ejidos* no longer accept new members, and instead, attract *pobladores* (settlers with no land rights who rent and borrow land and sell their labor) and squatters on government lands (Stedman-Edwards 1997).

## 2.2. *Predominant* Ejido *Land Tenure and Uses*

As mandated by Article 27, *ejido* members could not own land, but were given usufruct rights. These rights are administered differently in different *ejidos*, however. For the most part, land rights in the southern Yucatán peninsular region involve usufruct over assigned plots of land; that is, a designated area or number of parcels are assigned (delimited but rarely demarcated) to each *ejidatario*. As long as the rules of the *ejido* (established in the

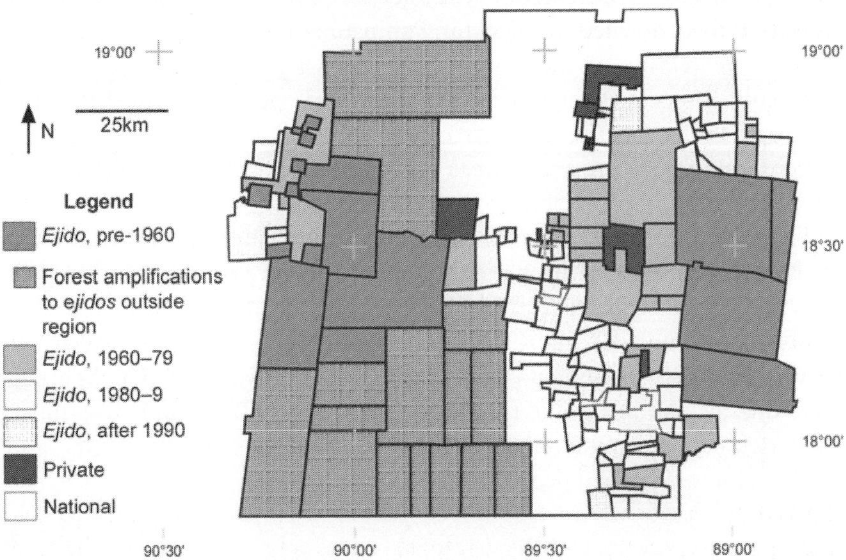


**Fig. 7.1.** Land tenure.
*Sources*: Field sketch maps; INEGI 1985–7; Landsat TM imagery; RAN 1940; SRA 1985; and various maps from the archive, Archivo General del Estado de Campeche.

*ejido* assembly) are followed, the farmer is assured that the assigned plots are part of the farm household's portfolio, including inheritance to offspring. A few *ejidos*, including one of the eleven in the SYPR survey (Ch. 8), follow less well-defined parcel use, at least in terms of repeated use by the same household. No *ejidos* in the region have moved to a private property system, an option made available after the reform of Article 27 in 1992.

Most *ejidatarios* in the region are semi-subsistent, using agricultural systems that simultaneously represent two distinct forms of organizing production: the profit-oriented family farm and the subsistence-oriented household (Ch. 11). In the absence of fully developed markets and insurance or subsidies to mitigate risk, most farmers follow hybrid consumption-commercial production, which translates into three basic land uses: swidden or *milpa* (Ch. 9); pasture; and jalapeño chili (Ch. 10). *Milpa* traditionally involves the intercropping of maize, squash, and nitrogen-fixing legumes as part of a rotation cycle (short to medium fallow) based on slash-and-burn technologies with minimal reliance on purchased inputs. Jalapeño chili is the predominant commercial crop, requiring significant capital inputs (human and physical). Pasture, much of which is maintained absent a significant number of livestock, is the third major land use. Perhaps, it should best be seen as investment that bets on future, expected programs fostering livestock.

## 3. *Forging a More Complex Politics of Land Use*

Development and agrarian policy shifts, national forest policy, social movements (coalitions of *ejidatarios* in various organizational forms), and the increasing involvement of a broad range of international actors in local use decision-making have complicated the institutional landscape. The context in which land-use decisions are made is, therefore, more complicated than that of farmer and *ejido*.

### 3.1. *Agrarian Policy Intervention*

This complexity is anchored in a shift in Mexican policy toward *ejidos* from state paternalism to neoliberal reform, beginning in earnest in the mid-1980s by the Salinas administration. For *ejidos* and Mexico at large, this shift resulted in the deregulation and privatization of large sectors of the economy, and social reinvestment programs to aid the transition away from government subsidies to more marginalized zones of the nation (Galletti 1998).[2] Perhaps the most important nation-wide policy initiatives linked to these reforms and with the potential for affecting land change on *ejidos* in the study region is land-tenure reform and PROCAMPO (Programa de Apoyo Directo al Campo—Direct Rural Support Program), although others have also been significant.

### 3.1.1. *Neoliberal Reforms*

More than a mechanism for distributing land, *ejidos* have served as the main conduit through which the state channeled resources to the region's smallholder households. Various government agencies disbursed a package of entitlements to agricultural inputs, credits, and other expenditures (e.g. for roads, schools, clinics) to *ejidatarios* to hasten the adoption of modernized agricultural techniques (Gates 1988). In addition, the parastatal marketing agency CONASUPO (Compania Nacional de Subsistencias Populares—National Company for Popular Subsistence) guaranteed prices for basic staples, such as maize, as an incentive for market participation. These support mechanisms were often conditioned on a series of restrictions on factor use intended to give the government political and economic leverage over *ejido* members.[3] Moreover, the government's credit extension program was commonly tied to the choice of specific crops and technologies, and output could often only be marketed through parastatal agencies.

A radical change in public support of agriculture occurred by the late 1980s. As part of its broader agenda to foster economic liberalization, the Salinas administration torpedoed the model of heavy state intervention that guided the Mexican government's rural development policies for most of the twentieth century. In 1986 Mexico entered into the General Agreement on Tariffs and Trade (GATT), the impact of which reached the agricultural sector by 1990, when tariffs on most products were dropped or drastically lowered, subsidies on inputs were withdrawn or sharply reduced, and the guaranteed price was eliminated for all crops but maize

[2] PRONASOL (Programa Nacional de Solidaridad—The National Solidarity Program) and SEDESOL (Secretaría de Desarollo Social—The Secretariat for Social Development) were the two main institutions established to implement the social reinvestment programs (Galletti 1998).

[3] Restrictions included the prohibition of employing wage labor on *ejidal* property and the renting or selling of land, limitations on plot size, and restrictions on patterns of land use (e.g. bans on intercropped maize) (de Janvry *et al.* 1996).

and beans (Foley 1995: 62). The continuation of these reforms was secured under the terms of the North American Free Trade Agreement (NAFTA), effective in 1994, obligating Mexico to liberalize fully its agriculture, including maize and beans, over a fifteen-year period.

### 3.1.2. *Land-Tenure Reform*

On the legal front, Article 27 of the constitution, which had served as the embodiment of the government's commitment to the rural poor in post-revolutionary Mexico was amended in 1992 to terminate the continued extension of *ejido* land grants and to permit existing *ejidatarios* the right to enter into business arrangements with outside investors and obtain private title to their lands, all of which were prohibited under the original terms of Article 27. The original intent of *ejidos* to prevent smallholder loss of land and land concentration among largeholders was now seen as an impediment to both the modernization and commercialization of the sector. Paraphrasing President Salinas, speaking in 1992, Article 27 anticipated both the capitalization of the countryside and the opening of productive options by establishing a legal framework guaranteeing property ownership.[4]

The effect of tenure reform on land use is difficult to isolate in the study region, however. No *ejido* has voted to privatize their land, and few indicators signal the adoption of the suite of other options opened to *ejidos* in the revised Article 27, including engaging in business ventures with off-farm investors. In short, the option of land-tenure reform has not generated a move to private holdings and integration into agricultural markets (Klepeis and Vance 2003).[5]

### 3.1.3. *PROCAMPO*

Land reform has been complemented by the introduction of alternative income support mechanisms that specifically seek to delink the extension

[4] The comments of President Carlos Salinas de Gortari, speaking in 1992, were cited in Foley (1995: 62).

[5] While small-scale business transactions have taken place between individual *ejidatarios* and outside investors both prior to the 1992 reform (albeit illegally) and afterwards, the SYPR survey (Ch. 8) identified only one case in which an *ejido* negotiated access to private credit. In this case, a private bank gave *ejidatarios* an early disbursement of their PROCAMPO payment to be repaid with interest following the harvest. In another *ejido*, a small group of *ejidatarios* collectively were able to solicit permission from their *ejido* council to obtain private legal title to a land parcel for independent agricultural enterprise within the designated *ejidal* common use zone.

of aid from the direct command of production decisions. One of the most pre-eminent of such programs is PROCAMPO, which extends agricultural support via payments for the continued cultivation of a fixed area of land over fifteen years, a period corresponding to the time in which agrarian price supports are to be phased-out under NAFTA. Introduced in 1993, with payments first awarded in 1994, the PROCAMPO program extends a per hectare direct payment on the production of a wide range of crops, including maize, beans, wheat, soybeans, sorghum, rice, and cotton.[6] The number of hectares receiving payment is based on the area that was cultivated in any of the aforementioned crops in 1994 (SARH 1993) and *ejidatarios* not registering with the program at that time could not subsequently join. Because aid is not tied to the cultivation of a specific crop, the terms of PROCAMPO are intended to be consistent with the increased *ejidatario* autonomy conferred by the legal changes to Article 27. It was the hope of SAGARPA (Secretaria de Agricultura, Ganadería, Desarollo Rural, Pesca, y Alimentacíon/Secretariat of Agriculture, Livestock, Rural Development, Fishing, and Food), the agency now administering the subsidy, that PROCAMPO would decrease environmental degradation by reducing distortions in the supply of agricultural commodities. Since the area and location of payment is fixed over the life of the program, the assumption by government officials has been that the funds are to be used to intensify production and thereby decrease pressure on remaining forest.

Analysis of results from the SYPR survey indicates that PROCAMPO support is a significant determinant of deforestation, contrary to its intent, a result that may be due to inherent contradictions in the program's terms and objectives (Klepeis and Vance 2003). PROCAMPO's policy of payment by lands cultivated in 1994 implies the intent of sedentarizing or intensifying cultivation—thus reducing the expansion of cultivated lands. The policy has had a statistically significant, positive effect on chili cultivation, consistent with PROCAMPO's aims, but has had even greater impact on the hectarage in pasture. Farmers are more likely to invest PROCAMPO money in clearing lands for pasture, even absent livestock for its use, given that pasture is a relatively inexpensive use to initiate and maintain compared to chili. In this regard, the program's relatively insubstantial payments of N$484/ha ($US64/ha)—fixed in real terms—appears to provide

[6] In 1994 the payment was N$350/ha and in 1995 rose to N$440/ha. In 1996, PROCAMPO payments were fixed in real terms (i.e. adjusted for inflation) at N$484/ha ($US64) until the end of the program.

insufficient capital to produce the kinds of land improvements intended. Such results elsewhere in Mexico lead some authors to regard this support as essentially welfare (de Janvry, Gordillo, and Sadoulet 1997; Myhre 1998). For example, Sadoulet, de Janvry, and Davis (2001) found that 70 per cent of households in a nationwide panel survey (taken in 1994) used PROCAMPO funds to purchase agricultural inputs, although respondents claimed the subsidy was too small to stimulate change in cropping strategies or in area planted. This characterization fits well with how the program is perceived in the study region where it is cynically referred to as *PRI-*CAMPO, in reference to PRI (Partido de la Revolución Institucional—Institutional Revolutionary Party), Mexico's dominant political party throughout most of the past century and author of PROCAMPO.

### 3.2. *National Forest Law: Exclusion and Participation*

Mexico is distinctive in that nearly 80 per cent of its forested lands are communally owned (Castilleja 1993; Klooster 1997; World Bank 1995: 22). Although forest legislation in Mexico dates back to 1884, forest law has historically been subordinate to agrarian development (World Bank 1995). In the absence of effective federal regulation, timber concessionaires of the late 1800s and early 1900s systematically logged significant volumes of precious hardwoods (Ch. 3). After the 1910 revolution and the initiation of agrarian reform, Mexican forests were gradually transferred to *ejidos* over six decades of land grants. Most concessions continued until the 1980s, however, exploiting timber at the expense of local communities that controlled the forests but lacked the capital to exploit them (Guerrero 1988; Klepeis 2000; Klooster 1997).[7] Based on its 1926 precursor, the 1992 Forest Law mandates that timber extraction from forested areas requires permission from the environment ministry (SARH 1992: Articles 34 and 35, Fracción II). Inventories of timber in each *ejido* form the basis for annual allowable cutting volumes, and the signatories to the contract assume responsibility for reforestation under the national forest law.

[7] Timber production in Campeche remains beset by various problems: the absence of an organizational structure that unifies the timber-producing forest *ejidos* around their common resource and sectoral interests; a lack of financial and technical expertise; and an inability to market tree species other than the traditional mainstay of forestry in the region, mahogany.

The National Reforestation Program (Programa Nacional de Reforestación), PRONARE, was formally created in 1995 to address the degradation of forest resources through ecological restoration and the establishment of forest plantations. The provision of employment and income opportunities is one of PRONARE's mandates (Galletti 1998). PRONARE projects in the study region were initiated through a strong local inter-*ejido* alliance (see CRASX below), and focused on providing technical assistance for timber extraction in selected forest *ejidos*.

The National Forest Development Program (Programa de Desarollo Forestal), PRODEFOR, facilitates community access to technical expertise required to improve forest management (SEMARNAP 1997). This program focuses on native forest management and production rather than plantations, and funds technical assistance and forestry training for *ejido* members, as well as non-timber forest product industries.

Along with these programs, a series of revolving funds for small businesses is available to local producers of forest-based enterprises. Rural development funds such as Alianza para el Campo (Rural Alliance) and FONAES (Fondo Nacional de Apoyo a Empresas Sociales—National Fund for Assistance to Social Enterprises) finance extractive activities such as apiculture, chicle production, or reforestation programs. Federal resources from diverse sources not only fund forestry initiatives but can also fund up to 45 per cent of costs of technology improvements in agricultural and livestock production, as well as apiculture (Ford 1998).[8]

## 3.3. *Social Movements and Smallholder Agency*

The demise of logging and the large government-directed agricultural projects in the late 1970s and early 1980s (Ch. 3) opened up opportunities for increasing local agency in land management. In 1979, a union of five *ejidos* (UEPMC—Unión de Ejidos Pueblo Maya de Campeche or the inter-*ejido* Union of Maya Peoples of Campeche) was formed—despite resistance by the state, which viewed peasant organizations as a threat to state authority—in order to reclaim the rights of *ejidatarios* to the control and use of

[8] In the end, these 'forest' programs affect the land managers' livelihood options, and thus decisions about agriculture. In future work the SYPR project will account for this influence in various modeling ventures. Data limitations prevent the project from considering these factors in detail now. Understanding them, however, guides our interpretations of the results.

their forest resources, which up to that point had been largely under the control of the Mexican state and forest concessions.

Through political mobilization by *ejidatarios* (e.g. UEPMC), state, and NGO sponsorship of *ejido* development initiatives have increased. Two important examples of the interaction of the state, NGOs, and local social movements are reviewed here, one from Quintana Roo and the other from Campeche: Plan Piloto Forestal (Pilot Forestry Plan); and the Consejo Regional Agrosilvopecuario y de Servicios de Xpujil (CRASX), or the Regional Council of Solidarity and Agriculture, Livestock, and Forestry Services of Xpujil (Campeche).

### 3.3.1. *Plan Piloto Forestal (PPF)*

The end of forest concessions to parastatal companies gave *ejidatarios* the right to extract logs, produce their own processed lumber, and sell it (Snook 1998). Outside interests concerned with conservation and the social welfare of *ejidatarios* arrived quickly to offer aid and guidance to the *ejidos*. Federal, state, and NGO involvement led to the creation of PPF, which is now world-renowned in conservation circles. Bray (2000) refers to the 'PPF process' because, rather than a single coherent program of action, it represented a process of adaptive management by local *ejidatarios* to changing social and ecological conditions that was stimulated by state, federal, and international actors. This process started in 1983 to empower *ejidatarios* to gain a livelihood from forest extraction in sustainable ways, complemented by national and international conservationist aims to preserve the region's biodiversity and forest habitat in the face of rapidly expanding cropland. In addition, state governors sought to create a landscape consistent with public images of the El Mundo Maya (The Maya World) tourism scheme (see below; Bray 2000).

These broad goals were accomplished by providing *ejidatarios* within forestry lands (i.e. *ejidos* with large land grants that were designated originally for forest extraction; Ch. 3) with the technical and management skills necessary to manage their own forest resources. The PPF approach grew out of an agreement between SARH (Secretaría de Agricultura y Recursos Hidráulicos—Secretary of Agriculture and Hydraulic Resources) and the GTZ (Gesellschaft für Technische Zusammenarbeit—German Association of Technical Cooperation), referred to as the Mexican-German Accord (Galletti 1998).[9] The goal of the German group was to encourage the

[9] PPF also had the support of the governor of Quintana Roo, and organizations such as the British Overseas Development Agency (ODA), the John D. and Catherine T. MacArthur Foundation, and Conservation International (Galletti 1998).

establishment of local institutions that would continue to manage the forest sustainably even after the various NGOs and outside technical advisors left the area (Martens 1999). The goals of more sustainable forest extraction and institution building were furthered by numerous approaches at the *ejido* level: permanent forest extractive reserves were established; forest inventories helped to formulate sound management practices; civil organizations were created to represent *ejidatarios*, provide technical assistance, and explore marketing strategies; and *ejidatarios* were trained in all aspects of forest planning and management (Bray 2000).

The PPF ended in 1998, and the degree to which it met its goal of sustainable forest management is questionable. Perhaps its primary legacy lies less in forest management *per se* but in institution building and the enhancement of the political skills of *ejidatarios* who gained valuable organizational, political, and technical capabilities (ibid.). The program served to engage *ejidatarios* in land management beyond agriculture and ranching, and brought international conservation and development attention to the study region.

### 3.3.2. *Consejo Regional Agrosilvopecuario y de Servicios de Xpujil (CRASX)*

The Salinas administration of the late 1980s reasserted its political reach into the study region through the institution of the federal 'solidarity brigades' of PRONASOL (Programa Nacional de Solidaridad—National Solidarity Program) and the INI (Instituto Nacional Indigenista—National Institute for the Indigenous) (Galletti 1998). The projects implemented by the INI's Fondo Regional (revolving funds for regional development) program were populist and participatory. The Xpujil Regional Solidarity Council was formed by *ejidos* in the Calakmul region in order to negotiate the terms of projects in the local communities funded by Fondo Regional. The initial Solidaridad funds were directed towards subsidies and extension programs for agriculture, focused primarily on chili cultivation, cattle ranching, and apiculture.

The Solidaridad metamorphosed in the CRASX (Consejo regional Agrosilvopecuario y de Servicios de Xpujil—Regional Council of Solidarity and Agriculture, Livestock, and Forestry Services of Xpujil), representing over half of southeastern Campeche's *ejido* communities, and serving as the vehicle for almost all development and conservation projects targeting the region. Primary concerns of the CRASX were managing water, marketing of cash crops, improving roads, and facilitating the implementation of numerous NGO and government-sponsored conservation, education, and development programs, most notably ecotourism in recent years (Haenn 2000; Haenn 2003). Initial activities centered on development rather than conservation.

In the ensuing years, CRASX went on to claim 3,000 members in fifty-eight *ejidos*, and oversaw the establishment of five nurseries in the region, the reforestation of 1,500 ha, establishment of agroforestry plots in 150ha in seven *ejidos*, the establishment of 75,200ha of permanent forest areas in seventeen *ejidos*, and the promotion of organic agriculture and apiculture in several *ejidos* (CRASX 2000). Due to organizational problems, regional tensions within geographically distinct *ejido* communities, corruption, financial mismanagement, and perhaps most significantly, the creation of the municipality of Calakmul in 1997, however, CRASX is no longer the major political entity it once was (Haenn 2000; Haenn 2003).

Like the PPF, the most important legacy of the CRASX is its impact on the regional profile of the southern Yucatán peninsula. For much of the 1990s CRASX was *the* channel for political organization and lobby for international and national government attention in southeastern Campeche, as well as the sole conduit for all federal, state, and non-governmental funding. Both CRASX and PPF can be seen as paving the way for subsequent large-scale initiatives, most notably the establishment of the Calakmul Biosphere Reserve and municipality of Calakmul, Bosque Modelo (Model Forest project), the El Mundo Maya tourism scheme, and the Mesoamerican Biological Corridor (MBC) initiative.

## 3.4. *Calakmul and Increasing Conservation Concerns*

The establishment of the Calakmul Biosphere Reserve (CBR) was a watershed event for the study region. The 1989 federal decree that established the reserve attracted the attention of researchers and environmental NGOs and set the tone for federal, state, and local politics predicated on the conservation-development dialectic.

### 3.4.1. *The Calakmul Biosphere Reserve*

In 1993 Calakmul was designated a UNESCO (United Nations Educational, Scientific, and Cultural Organization) Man and the Biosphere site, explicitly drawing human activities into the reserve's scope of management, responsibility, and vision, leading the reserve to seek the active participation of CRASX in its management decisions.[10] The reserve's technical advisory committee represents a plethora of federal, state, and local organizations (e.g. *ejido* authorities), NGOs, and academic institutions.[11] Goals for regional development include both economic development and environmental conservation: environmental education, organic agriculture, composting, waste management, home gardens, and tourism programs focused on archaeological sites, biodiversity, and Maya culture (SEMARNAP 2000). The reserve's larger agenda includes controlling fire, deforestation, and both settlement and cropland expansion into the reserve, resolving tenure and boundary conflicts, and promoting social development programs.

The reserve did not officially function until 1995, when its first administrative and management committee was founded. Reserve boundaries, however, were defined in the first official site decree in 1989, although without appropriate biological and land-tenure information. Some well-preserved areas remain unprotected, therefore, located outside the reserve's legal boundaries, and other areas that are within reserve limits were cultivated for a long time prior to reserve declaration. These planning errors highlight inconsistencies about the relative importance of regional development versus conservation goals, and have sparked considerations about re-demarcating the reserve better to ensure long-term security of the reserve's ecosystems.[12]

### 3.4.2. *Bosque Modelo de Calakmul (Calakmul Model Forest)*

During the early 1990s the Canadian Model Forest network, an international NGO, negotiated with the Mexican government the establishment of a network of 'model' forested areas of better management practices funded by the GEF (Global Environmental Fund, UN). Bosque Modelo

[10] The reserve is funded federally by SEMARNAT (Secretaría del Medio Ambiente y Recursos Naturales—Environment and Natural Resources Secretariat), and internationally by the GEF (Global Environmental Fund, United Nations) and the World Bank. The Nature Conservancy, through the Parks in Peril program, oversees and evaluates annual work plans of the reserve in a joint partnership with SEMARNAT and the World Bank. The reserve receives additional support from the World Wildlife Fund, the Ford Foundation, and the US Fish and Wildlife Service in an assortment of conservation and research activities.

[11] J. Rodríguez de la Gala, Director, Calakmul Biosphere Reserve, personal comment.

[12] J. Miranda, Fire Prevention and Control Program Coordinator, Calakmul Biosphere Reserve, personal comment.

(Calakmul) was instituted in 1993, operating in an area of 380,000ha lying between the Quintana Roo border and the biosphere reserve, 300,000ha of which is comprised of forested *ejido* land (Ericson 1996). Bosque Modelo incorporated local (CRASX), national (SARH), and international interests (e.g. International Development Research Centre).

Defining its mandate as the promotion of 'productive ecology', Bosque Modelo focused its efforts on the sustainable management of its forested lands and sustainable development of the communities within its boundaries. Its primary goals were to decrease swidden agriculture by promoting agroforestry, to encourage the management of wildlife, reforestation, and apiculture, to facilitate the administration and management of water, and to promote the restoration of ruins and ecotourism (International Model Forest Network 1994). Bosque Model worked to decrease human disturbance of the forest by promoting what it assumed were sustainable uses while at the same time improving the quality of life for local people. Subsequently, Bosque Modelo has refocused its attention to the original Canadian mandate of promoting sound forest management practices and strengthening community forestry in Calakmul's forest *ejidos*. The swing back to forest conservation rather than community development projects follows evaluations of the project's ineffectiveness at promoting a regional vision of better forest management owing to its myriad development projects.[13]

### 3.4.3. *Other NGOs*

A diverse group of NGOs now shares two basic common goals for the study region, its actions often channeled through the biosphere reserve, Bosque Modelo, or CRASX: (1) the promotion of sustainable natural resource management and conservation, including better forest management practices and extractive industries, and (2) agricultural 'sedentarization' intended to reduce cultivated land expansion into extant forest.

The most prominent NGO still active in the region is ProNatura Peninsula de Yucatán (ProNatura Yucatán Peninsula), founded in 1988. ProNatura is the Mexican partner of The Nature Conservancy in the funding and evaluation of the Calakmul Biosphere Reserve's annual management plans. ProNatura began its activities in the region focusing on flora and fauna research and environmental education in schools. The organization later

[13] L. Poot, General Coordinator, Bosque Modelo para Calakmul, personal comment.

evolved to include communities in conservation and development projects. In this regard, ProNatura is focused on the sedentarization of agriculture and generating income from sustainable development projects. Projects in multiple *ejidos* (totaling over 1,000ha) promote apiculture, environmental education, and organic agriculture using green fertilizers and reduced chemical inputs and burning.

The International Center for Research in Agroforestry (ICRAF) is active to the east of the biosphere reserve. Between 1994 and 1999, approximately 700ha of agroforestry plots were established in forty-two different *ejidos* in the buffer zone of the reserve (Snook and Zapata 1998). The plots consist of alternating rows of timber trees, mahogany, cedar, and orange trees as well as maize and other crops associated with *milpa* (Ch. 9). In addition, NGOs such as Naturaleza Compartida (Nature Shared) and Tropico-Rural Latinoamericana (Tropical Rural Latin America) recognize the problems stemming from inadequate commercialization of sustainable forest products, and are exploring initiatives, such as product certification, to boost forest management practices and production (Boege and Fuentes 2000).[14] As demonstrated in Quintana Roo, initiatives such as forest certification can affect not only forest ecology but also local livelihoods and social relations by providing new income opportunities and broadening the base of participating actors (Maynard and Robinson 1998).

### 3.5. *Pluralization of Institutions*

As the southern Yucatán peninsular region moves away from its long-held status as an economic frontier, the diversity of land uses has increased following the synergy of numerous local-to-international agents and organizations and the governing institutions they promote. New or expanded land uses/associations include apiculturalists, chili cultivators, timber producers, allspice gatherers, *chicleros*, and ranchers, to name a few. The first four uses/associations were among the most important producer groups to believe that their needs and demands were not addressed by the organizational structure of CRASX in the early–mid-1990s. These groups, assisted by INI in some cases (apiculture), went on to form their own unions, as did

[14] R. Delgadillo, Coordinator, Apiculture ProNatura and NaCom and A. Villaseñor, Forest Technician/Consultant, Tropico-Rural Latinoamericana and Naturaleza Compartida, personal comments.

other sector-based groups, such as those focusing on organic agriculture, timber, and chicle production.[15]

Local unions to international programs give rise to three distinct 'institutional' areas within the study region—the Quintana Roo portion that is to the east of the CBR, the centrally located CBR-Bosque Modelo area, and the region west of the reserve (Fig. 1.1). Among members of CRASX, the *ejidos* of the southern frontier with Guatemala were relatively isolated and marginalized, as were a group of *ejidos* west of Xpujil in the region of Laguna Silvituc. They split off to form the Consejo Regional Indigena y Popular de Xpujil (CRIPX—Regional Indigenous and Popular Council of Xpujil), representing the southern portion of the municipality of Calakmul, and the Consejo Arteecoloagropecuario (Art-Ecological-Agropastoral Council), representing the western subregion of the municipality. These new unions, aligned with Calakmul, focus not only on specific agricultural products but also small business enterprises in their communities.

A spatial picture of relative marginalization and government investment in *ejidos* of the Calakmul region is detailed in Fig. 7.2. This figure is based on a partial reconstruction of highly diverse development projects and conservation initiatives over the period 1991–2000 in the approximately eighty *ejidos* and communities in the greater Calakmul region, on the Campeche side of the study region. As noted in this chapter, numerous programs funded these projects. The variation is cumulative over the decade indicated.

## 4. *Beyond the CBR*

From an institutions perspective, the southern Yucatán peninsular region is no longer a frontier. Numerous local social movements interact with the vision of external organizations for conservation and development in the region. Increasing settlement growth and establishment of the CBR spurred organizations and movements at various sociopolitical scales,

[15] Among those exploring market-specialized production is the allspice producers' union. It is shifting its strategy from promoting allspice as a non-timber forest product to be gathered from forested *ejido* lands, to a product that can be cultivated in forest plantations and agroforestry parcels under cooperative arrangements among groups of *ejidatarios*, thereby profiting from economies of scale (L. Guzman, President, Bosque Modelo de Calakmul and Allspice Union (Sociedad en Solidaridad Social de Pimienteros de Xpujil, SPR), personal comment).

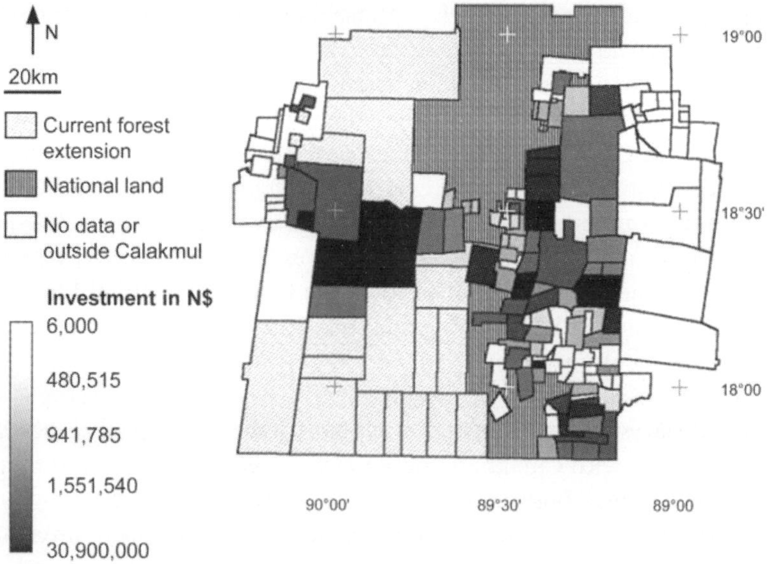


**Fig. 7.2.** Relative marginalization and government investment, 1991–2000.

affecting lands throughout the region, regardless of their formal incorporation into the reserve. Nevertheless, the CBR and the complementary visions of El Mundo Maya and the Mesoamerican Biological Corridor increasingly define the character of the region. The government redrew existing municipal boundaries of Hopelchen and Champotón (in Campeche) to create Mexico's first 'ecological' *municipio*, Calakmul, covering 1,383,911ha of which 723,185ha are located within the CBR (Ericson, Freudenberger, and Boege 1999) and includes sixty-one *ejidos* and twenty-one unrecognized communities. These entities as well as the fourteen *ejidos* on the Quintana Roo side of the study region have begun to feel the impacts of El Mundo Maya and the Mesoamerican Biological Corridor.

### 4.1. *El Mundo Maya (The Maya World)*

Begun in 1993, El Mundo Maya is perhaps the most important federal/state-instituted policy to affect the study region in recent decades, with the potential to affect dramatically its landscape in the future. The program constitutes a common goal for Belize, El Salvador, Guatemala, Honduras, and Mexico to develop their Maya lowland region economically via tourism (Garrett 1989). Building off the experience in Cancún and the eastern coast of the Yucatán Peninsula, the aim is to use both Maya ruins, replete throughout the study region, and the forest itself as commodities to capture a portion of the rapidly expanding international market for ecological and archaeological tourism (ecoarqueoturismo). A few *ejidos* containing reconstructed ruins have financed 'rustic cabins' for tourists, but tourist dollars circulate largely within the Cancún-based economy of prepackaged tours, with accommodations typically arranged at the very few and expensive hotels in the region (e.g. Ecovilla to the west of Xpujil). Private spin-offs from the industry include new motels and restaurants in selected locales, such as Xpujil. In addition, various NGOs (e.g. ProNatura, Bosque Modelo) and government agencies have jointly organized training workshops for local ecotourist guides. For the most part, however, the economic return and the ability to maintain the tourist sector in the study region remains unproven, and the effect El Mundo Maya will have on smallholder productive options and land resources remains uncertain.

### 4.2. *The Mesoamerican Biological Corridor*

The Mesoamerican Biological Corridor, formalized in 1997, is an ambitious international project that aims to combat habitat fragmentation and improve the connectivity of regional ecosystems in southern Mexico and Central America (Fig. 7.3; Kaiser 2001; Miller, Chang, and Johnson 2001). The broad objectives of the project are to (1) facilitate genetic flow among established forest reserves, and (2) integrate regional policy, land use, and conservation for sustainable development.[16] Many different stakeholders are involved in the development of corridor(s) (e.g. producers' associations, research institutions, and planning committees), composed of 'nodes' and

[16] The Mesoamerican Biological Corridor is led by the respective country governments, and supported by the GEF, World Bank, United Nations Development Program, GTZ, and many other national and international organizations. The steering committee of MBC-Mexico is comprised of federal actors from eight government ministries, including SEMARNAT, SEDESOL, SAGARPA (Secretaría de Agricultura, Ganadería, Desarollo Rural, Pesca y Alimentación—Secretariat for Agriculture, Livestock, Rural Development, Fishing and Nutrition), SEP (Secretaría de Educación Pública—Secratariat for Public Education) and SCT (Secretaría de Comunicaciones y Transportes—Secretariat for Communications and Transportation), as well as the National Ecology Institute. The MBC-Mexico project is budgeted at an estimated $US79.9 million, and includes the states of Campeche, Quintana Roo, Yucatán, Chiapas, and Tabasco, with additional linkages to protected areas in Oaxaca and Veracruz (GEF 1999).

'connectors' (Galletti 1999). Nodes are pre-existing, legally protected conservation areas such as the Calakmul Biosphere Reserve. Connectors are the areas between nodes where various conservation-based management strategies will aim to conserve forest cover. The southern Yucatán peninsular region is a corridor between the CBR and Sian Ka'an reserve (node) to the northeast and reserves (nodes) in Guatemala to the south. If implemented, this corridor would try to protect 600,000ha of forest in the study region deemed important for genetic flow and ecosystem health. It portends significant land-use consequences because MBC aims to create or preserve corridors by considering biophysical processes and ecosystem functions in a particular place in relation to the various human use systems employed.

The corridor's strategy in the Calakmul region is to focus on (1) *ejidos* and private lands located along Highway 186, extending from Veinte de Noviembre located east of Xpujil to Chichonal located in the west, (2) smaller agricultural *ejidos* located along the northeastern border of the

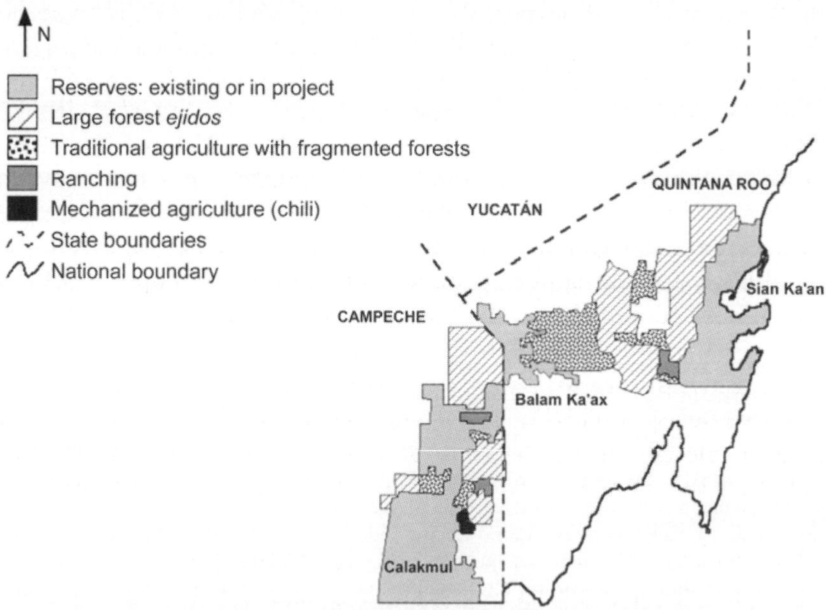


**Fig. 7.3.** Proposed biological corridor between Calakmul and Sian Ka'an
*Source*: figure adapted from Boege *et al.* 2000.

Calakmul Biosphere Reserve near La Montaña, (3) the chili-growing region around the *ejido* Polo Norte, and (4) large forest *ejidos* (Alvaro Obregon, Nuevo Becal, Veinte de Noviembre, and Conhuas). The Zoh Laguna–Xpujil and Las Montañas focal areas lie in the center and north of the Calakmul region respectively, and are particularly critical to the corridor's regional development and conservation initiatives. The Zoh Laguna focal area is also critical to establishing landscape connectivity between the northern and southern blocks of the CBR dissected by the trans-peninsular Highway 186. To achieve these aims, participatory workshops have maintained an overarching emphasis on increasing 'sustainable' agricultural productivity to reduce the amount of land required in cultivation and to expand markets for forest products, thus increasing the resource value of forests for local occupants (Boege and Fuentes 2000; CBM 2000).

## 5. *Summary*

The increasing institutional complexity of the southern Yucatán peninsular region makes uncovering the factors leading to land-use and land-cover change all the more difficult. The abrupt shifts in regional development initiatives over the past half century moved land change in different directions. The current broad goals of increasing conservation-tourism and the economic productivity of the land, no matter how laudable, remain unproven. These goals, placed in context of neoliberal reforms and a proliferation of agencies and organizations seeking to stake out land and resource institutions in the study region, mirror the local-to-global dynamics that are a part of globalization processes everywhere. For the SYPR project at large, accounting for the direct and indirect land impacts of these institutions and organizations is not easy. At this stage in our work, grappling with the cacophony of programs and projects helps us to understand variation in land dynamics across subregions and to assess better the results of our modeling work (Chs. 12–13). In subsequent phases of research we hope to entertain spatial data on 'protected' forests and to incorporate more fully those policies and institutions that affect farming decisions in our model design.

For phase one of the project, the following lessons apply.

- The *ejido* is the region's main land-access unit.
- Over the last two decades, conservation and development initiatives have complicated the context in which land managers make decisions.

- Social forestry and environmental conservation initiatives have designated a significant amount of land (*mediana*) off-limits to forest clearing.
- Social movements and NGO activities that empower smallholders and more liberal agrarian policies have created more diversified land uses region-wide.
- Important changes in regional political economy, such as those linked to the Calakmul Biosphere Reserve and El Mundo Maya, hold important implications for land change and attempts to project future dynamics.
- This last development accentuates the need for refined understanding of the human–environment dynamics in the region as they play out in terms of land-use and land-cover change—the base rationale for the SYPR project.

## REFERENCES

Bassols Batalla, A., M. Aceves Garcia, C. Bustamente Lemus, and A. Ortiz Wadgymar (1976). *Estudio Geográfico y Socio-Económico de Estado de Quintana Roo*. México DF, Sociedad Mexicana de Geografía y Estadística.

Boege, E., and T. Fuentes (2000). 'Documentación y sistematización de 10 años de experiencia de ecodesarollo y conservación de ProNatura península de Yucatán A.C. en Calakmul, Campeche, México'. Report funded by the USAID (Biodiversity Support Program) and the Nature Conservancy. Unpublished.

Boege, E., H. A. Galletti, and N. Poot (2000). *Evaluación social del corredor biológico Mesoamericano Sian Ka'an—Calakmul: Área focal no. 1 Xpujil—Zoh Laguna y area focal 2 'La Montaña', Campeche, Mexico*. Report to the Mesoamerican Biological Corridor Project.

Bray, D. B. (2000). 'Adaptive Management, Organizations and Common Property Management: Perspectives from the Community Forests of Quintana Roo, México', in Eighth Biennial Conference of the International Association for the Study of Common Property. Bloomington, Ind. Unpublished.

Castilleja, G. (1993). 'Changing Trends of Forest Policy in Latin America: Chile, Nicaragua and Mexico'. *Unasylva* 44: 29–35.

CBM (Corredor Biologico Mesoamericano) (2000). 'Memoria Primer Taller: Información y Organización'. Report on the first workshop of the CBM. Zoh Laguna, Campeche (27 March). Unpublished.

Cortez Ruiz, C. (1998). 'Hay futuro para la poblacion de la selva lacandona? (colonizacion y desarrollo en Marques de Comillas)'. Paper presented at the Annual Meeting of the Latin American Studies Association. Chicago, Ill. September. Unpublished.

CRASX (Consejo Regional Agrosilvopecuario y de Servicios de Xpujil) (2000). Unpublished internal document of CRASX, prepared by Eulalio Lopez, former president.

Dachary, A., and S. Arnaiz (1983). *Estudios socioeconomicos preliminares de Quintana Roo: sector agropecuario y forestal (1902–1980)*. Puerto Morelos, CIQRO (Centro de Investigaciones de Quintana Roo).

de Janvry, A. (1981). *The Agrarian Question and Reformism in Latin America*. Baltimore, Johns Hopkins University Press.

de Janvry, A., G. Gordillo, and E. Sadoulet (1997). *Mexico's Second Agrarian Reform: Household and Community Responses, 1990–1994*. San Diego, Calif., Center for U.S.-Mexican Studies, University of California.

de Janvry, A., E. Sadoulet, B. Davis, and G. Gordillo de Anda (1996). 'Ejido Sector Reforms: From Land Reform to Rural Development', in L. Randall (ed.), *Reforming Mexico's Agrarian Reform*. Armonk, NY, M. E. Sharpe, 71–116.

Ericson, J. (1996). 'Conservation and Development on the Border of the Calakmul Biosphere Reserve'. M. S. Thesis, Humboldt State University.

Ericson, J., M. S. Freudenberger, and E. Boege (1999). 'Population Dynamics, Migration, and the Future of the Calakmul Biosphere Reserve'. *Occasional Paper No. 1, Program on Population and Sustainable Development*. Washington DC, American Association for the Advancement of Science.

Foley, M. W. (1995). 'Privatizing the Countryside: The Mexican Peasant Movement and Neoliberal Reform'. *Latin American Perspectives* 22: 59–76.

Ford, L. (1998). *Policies Affecting the Adoption of Agroforestry Systems in the States of Campeche and Quintana Roo, Mexico*. Chetumal, ICRAF (International Center for Research in AgroForestry).

Fuentes Aguilar, L. (1980). 'Desmontes y Colonizacion'. Paper presented at the Symposium, *Memorias de Simposio Quintana Roo-Problematica y Perspectivas*. Cancún, Quintana Roo. October. Unpublished.

Galletti, H. (1998). 'The Maya Forest of Quintana Roo: Thirteen Years of Conservation and Community Development', in R. B. Primack, D. Bray, H. A. Galletti, and I. Ponciano (eds.), *Timber, Tourists, and Temples: Conservation and Development in the Maya Forest of Belize, Guatemala, and Mexico*. Washington DC, Island Press, 33–46.

——(1999). 'La situación forestal en el área del corredor biológico Calakmul—Sian Ka'an'. Report to the Mesoamerican Biological Corridor Project. Unpublished.

Garrett, W. E. (1989). 'La Ruta Maya'. *National Geographic* 176: 424–79.

Gates, M. (1988). 'Institutionalizing Dependency: The Impact of Two Decades of Planned Agricultural Modernization on Peasants in the Mexican State of Campeche'. *The Journal of Developing Areas* 22: 293–320.

——(1993). *In Default: Peasants, the Debt Crisis, and Agricultural Challenges in Mexico*. Boulder, Westview Press.

GEF (Global Environmental Fund) 1999. United Nations. <http://www.conabio.gob.mx/cbm-m> accessed 2 December 2001,

Guerrero, G. (1988). 'Estado, madera y capital'. *Ecología Política/Cultura* 2: 5–13.

Haenn, N. (2000). 'Biodiversity is Diversity in Use: Community-Based Conservation in the Calakmul Biosphere Reserve'. *The Nature Conservancy's Working Paper Series América Verde* 7.

——(2003). 'Risking Environmental Justice: Culture, Conservation, and Governance at Calakmul, Mexico', in S. Eckstein, and T. Wickham-Crawley (eds.), *Social Injustice in Latin America*. New York, Routledge, 81–101.

INEGI (Instituto Nacional de Estadística Geográfica e Informática) (1985–7). *Cartas topográficas*. Aguascalientes, INEGI.

International Model Forest Network (1994). *Calakmul Model Forest: Productive Ecology Proposal, Canada-SARH*. Quebec, Ministry of Supply and Services.

Kaiser, J. (2001). 'Bold Corridor Project Confronts Political Reality'. *Science* 293: 2196–9.

Klepeis, P. (2000). 'Deforesting the Once Deforested: Land Transformation in Southeastern México'. Ph.D. Dissertation, Clark University.

Klepeis, P. and C. Vance (2003). 'Neoliberal Policy and Deforestation in Southeastern Mexico: An Assessment of the PROCAMPO Program'. *Economic Geography*, *79* (3), 221–240.

Klooster, D. J. (1997). 'Conflict in the Commons: Commercial Forestry and Conservation in Mexican Indigenous Communities'. Ph.D. Dissertation, University of California, Los Angeles.

Martens, R. (1999). 'Participación y Capacitación en el Mundo Maya'. <http://www2.planeta.com/mader/ecotravel/mexico/yucatan/ronmartens/1.html> accessed 7 December 2001.

Maynard, B., and D. Robinson (1998). *Ethical Trade and Sustainable Rural Livelihoods—Case Studies: Quintana Roo Forest Study Certification Case*. <http://www.nri.org/NRET/nret.htm> accessed 7 December 2001.

Miller, K., E. Chang, and N. Johnson (2001). *Defining Common Ground for The Mesoamerican Biological Corridor*. Washington DC, World Resources Institute.

Myhre, D. (1998). 'The Achille's Heel of the Reforms: The Rural Finance System', in W. A. Cornelius, and D. Myhr (eds.), *The Transformation of Rural Mexico: Reforming the Ejido Sector*. San Diego, Calif., Center for U.S.-Mexican Studies, University of California San Diego, 39–65.

Oasa, E. K., and B. H. Jennings (1982). 'Science and Authority in International Agricultural Research'. *Bulletin of Concerned Asian Scholars* 14: 30–44.

RAN (Registro Agrario Nacional) (1940). *Carpeta Básica de Xbonil*. México DF, RAN.

Revel-Mouroz, J. (1980). *Aprovechamiento y Colinización de Trópico Húmedo Mexicano: La Vertiente del Golfo y del Caribe*. México DF, Fondo de Cultura Economica.

Sadoulet, E., A de Janvry, and B. Davis (2001). 'Cash Transfer Programs with Income Multipliers: PROCAMPO in Mexico'. *World Development* 29(6): 1043–56.

SARH (Secretaria de Agricultura y Recursos Hidráulicos) (1992). *La Nueva Ley Forestal. Articles 34 and 35, Fracción II*. México DF, SARH.

——(1993). *PROCAMPO: Vamos al grano para progresar (PROCAMPO: Progress Through Getting Back to Basics)*. México DF, SARH.

SEMARNAP (Secretaria de Medio Ambiente, Recursos Naturales y Pesca) (1997). *Anteproyecto de reformas y adiciones a la Ley Forestal. Con el Comité Nacional Forestal, Subcomité de Legislación y Normatividad, la Comisión de Silvicultura y Recursos Hidráulicos del Senado, y la Comisión de Bosques y Selvas de la Legislatura. Procuraduría Federal de Protección al Ambiente*. México DF, SEMARNAP.

——(2000). *Programa de manejo de la reserva de la biosfera de Calakmul, México*. México DF, SEMARNAP.

Sinha, R. (1984). *Landlessness: A Growing Problem*. Rome, Food and Agriculture Organization, United Nations.

Snook, L. K. (1998). 'Sustaining Harvests of Mahogany (*Swietenia macrophylla* King) from Mexico's Yucatán Forests: Past, Present, and Future', in R. B. Primack, D. Bray, H. A. Galletti, and I. Ponciano (eds.), *Timber, Tourists, and Temples: Conservation and Development in the Maya Forest of Belize, Guatemala, and Mexico*. Washington DC, Island Press. 61–80.

Snook, A., and G. Zapata (1998). 'Tree Cultivation in Calakmul, Mexico: Alternatives for Reforestation'. *Agroforestry Today* 10: 15–18.

Sonnenfeld, D. A. (1992). 'Mexico's "Green Revolution", 1940–1980: Towards an Environmental History'. *Environmental History Review* 16: 29–52.

SRA (Scretaria de Reforma Aagrara) (1985). 'Estado de Campeche, 1:500,000'. Unpublished.

Stedman-Edwards, P. (1997). 'Socioeconomic Root Causes of Biodiversity Loss: The Case of Calakmul, Mexico'. México DF, World Wildlife Fund. Unpublished.

Szekely, M., and I. Restrepo (1988). *Frontera Agrícola y Colonización*. México DF, Centro de Desarrollo.

Tomich, T. P., P. Kilby, and B. F. Johnston (1995). *Transforming Agrarian Economies: Opportunities Seized, Opportunities Missed.*: Ithaca, Cornell University Press.

Vogeler, I. K. (1974). 'The Cultural Ecological Setting of Southeastern Campeche', R. E. W. Adams (compiler), *Preliminary Reports on Archaeological Investigations in the Rio Bec Area, Campeche, Mexico*. New Orleans, Middle American Research Institute, No. 31, Tulane University, 110–12.

World Bank (1995). *México: Estudio de Revisión del Sector Forestal y Conservación de Recursos*. Regional Office: Latin American and the Caribbean. Report No. 13114-ME.

# 8

# The *Ejido* Household: The Current Agent of Change

*Colin Vance, Peter Klepeis, Birgit Schmook, Eric Keys*

The *ejidatario* is the primary agent of land change in the region today, recognizing that this decision-maker is influenced by various policies and programs, both federal and NGO in origin (Ch. 7). Although colonization of the region, especially during the past forty years, has produced a diverse mix of ethnicities, cultural attributes, and economic characteristics, it is still appropriate to speak of a typical *ejido* household. The characterization of these households detailed here is based in large part on a SYPR project survey conducted in 1997 and 1998, designed to represent the entire study region. Subsequent surveys have been directed to specific issues, such as chili (Ch. 10) and the role of institutions; information from these studies is specified when used here.[1]

Most production within the *ejido* sector of the study region is organized around the semi-subsistence farm (or dual) household (Ch. 11). This unit of production is both a family and an enterprise (Ellis 1993), with the implication that decisions concerning what, how, and how much to produce are made in response to both market signals and the biological and cultural imperatives of the family unit. This chapter explores the socio-economic and

The views expressed in this chapter are those of C. Vance and the other authors and do not necessarily represent the official views of the US Environmental Protection Agency.

[1] The 1997–8 household survey was undertaken largely by Peter Klepeis and Colin Vance and involved portions of their doctoral dissertation work in geography and economics, respectively. Birgit Schmook and Eric Keys assisted them in some cases. These individuals and the project thanks Pedro Marcario Mendoza, ECOSUR senior researcher and SYPR project member, himself a former *ejidatario*, for insights.

farm characteristics of the sampled *ejidatario* households and presents relevant descriptive statistics. The data demonstrate that the *ejido* household has been exceptionally dynamic since the late 1960s, moving from a primarily maize-production orientation to a more diversified household economy.

## 1. *SYPR Survey, 1997–1998*

A standardized questionnaire with formal and open-ended components was used to elicit socio-economic and land-use data from *ejidatario* heads of household. In order to ensure a representative sample, data collection proceeded according to a stratified, two-stage cluster sample, with *ejidos* as the first stage unit and *ejidatarios* as the second stage unit (Deaton 1997; Warwick and Luinger 1975). The strata were delineated such that *ejidos* from across the region were represented in the sample, thereby capturing variability in both environmental conditions and in the influence of proximity to roads and markets. This strategy resulted in the random selection of eleven *ejidos* followed by the random selection of 199 *ejidatario* households (see Vance 2000 for details), which produced 188 completed surveys.

In the formal component of the survey, a panel of data was collected on household demographics and living conditions, land-use systems, labor and sources of income, and involvement in government programs (Table 8.1). The informal, open-ended component explored respondent perceptions of land use, government policy, and land-tenure reform. Completion of this component involved a guided tour of the respondent's agricultural plot. Using a geographic positioning system, the interviewer created a geo-referenced sketch map detailing the distinct fields within the plot as the farmer-respondent provided an interpretation of the use of these fields (Ch. 9). The process of visiting the parcel and drawing the map took anywhere from two to six hours, depending on the location of the parcel and the willingness of the respondent to engage in conversation. The sketch map not only documented the spatial configuration of contemporary land uses, including forested areas, but also the land-use transition histories of the principal areas of activity, emphasizing the period between 1986 and 1997 (see Klepeis and Turner 2001). This information is used in modeling efforts (Chs. 11 and 12) to calculate various indices of land-use change, such as the area planted in different crop types since acquisition of the plot, as well as to inform the interpretation of available Thematic Mapper (TM) satellite imagery (Ch. 6).

**Table 8.1.** Key results of the SYPR household survey, 1997 and 1998

| | Mean | Stand. dev. | Min | Max |
|---|---|---|---|---|
| *Household demographics* | | | | |
| Household size | 6.43 | 3.60 | 1 | 31 |
| No. of children of household head | 5.30 | 3.79 | 0 | 27 |
| Age of male head | 46.28 | 15.10 | 18 | 86 |
| Age of female head | 40.60 | 12.90 | 16 | 80 |
| Years of education of male head | 2.80 | 3.16 | 0 | 14 |
| Years of education of female head | 2.93 | 3.05 | 0 | 14 |
| Years of education of oldest son if older than 17 | 6.61 | 4.12 | 0 | 16 |
| Years of education of oldest daughter if older than 17 | 6.26 | 3.48 | 0 | 14 |
| *Household characteristics (%)* | | | | |
| Electricity | 71 | 45 | | |
| Cement walls | 11 | 32 | | |
| Cement floor | 53 | 50 | | |
| Potable water | 66 | 48 | | |
| Radio | 27 | 44 | | |
| Television | 37 | 48 | | |
| *First language of household head (%)* | | | | |
| Spanish | 70 | 46 | | |
| Maya | 24 | 43 | | |
| Totonaco | 6 | 25 | | |
| *Farm capital and inputs* | | | | |
| Access to tractor (%) | 11 | 32 | | |
| Access to chainsaw (%) | 37 | 48 | | |
| Owning vehicle (%) | 16 | 37 | | |
| Applying fertilizer | 46 | 50 | | |
| Applying insecticide (%) | 51 | 50 | | |
| Applying herbicide (%) | 43 | 50 | | |
| Total expenditure for hired labour (N$) | 4,173 | 12,353 | 0 | 120,000 |
| *Incidence and area of principal land uses* | | | | |
| Land endowment (ha) | 88.23 | 50.45 | 10 | 390.5 |
| Households with forest cover on some parcels (%) | 100 | 0 | | |
| Forest (ha) | 51.58 | 52.25 | 1 | 297 |
| Households planting swidden (%) | 98 | 13 | | |
| Swidden (ha) | 4.51 | 4.04 | 0.5 | 30 |
| Households planting chili (%) | 52 | 50 | | |
| Ha of chili among planters | 1.33 | 1.36 | 0.06 | 9 |
| Households planting pasture and owning grazing animals (%) | 26 | 44 | | |
| Pasture among planters with grazing animals (ha) | 25.36 | 28.43 | 0.5 | 144.5 |
| Households planting pasture and not owning grazing animals (%) | 23 | 42 | | |
| Pasture among planters without grazing animals (ha) | 11.03 | 17.17 | 0.4 | 95 |

**Table 8.1.** (*cont.*)

| | Mean | Stand. dev. | Min | Max |
|---|---|---|---|---|
| *Indicators of access to farm plot and to market* | | | | |
| Distance from *ejido* to nearest primary market (km) | 71.35 | 35.57 | 17.33 | 125.38 |
| Distance from household to plot (km) | 6.59 | 6.16 | 0 | 28.60 |
| Access to plot on foot (%) | 42 | 50 | | |
| Access to plot by bike (%) | 19 | 39 | | |
| Access to plot by vehicle (%) | 11 | 31 | | |
| Access to plot by horse (%) | 22 | 41 | | |
| Access to plot by hitch-hiking (%) | 6 | 24 | | |
| *Number of domestic animals* | | | | |
| Cows | 3.38 | 11.12 | 0 | 91 |
| Turkeys | 3.57 | 5.91 | 0 | 30 |
| Chickens | 14.89 | 16.76 | 0 | 100 |
| Pigs | 3.56 | 5.36 | 0 | 32 |
| Horses | 0.87 | 1.33 | 0 | 8 |
| Sheep | 1.22 | 5.17 | 0 | 45 |
| *Non-farm sources of income* | | | | |
| Apiculture (%) | 25 | 43 | | |
| Extraction of chicle (%) | 8 | 27 | | |
| Wage income (N$) | 10,328 | 22,313 | 0 | 200,000 |
| *Income from government credit/subsidy* | | | | |
| Credit (N$) | 2,328 | 2,236 | 0 | 14,850 |
| PROCAMPO payment (N$) | 2,050 | 1,968 | 0 | 13,076 |

## 2. *Basic Household Characteristics*

Despite being located in the heart of the Maya lowlands (Ch. 2), Maya-speaking people do not dominate *ejidos* in the study region because the majority of colonizers entered the region after 1970 from non-Maya parts of Mexico, especially from the Gulf Coast states. These colonists bring with them agricultural experiences that differ from that prevalent among the Maya peoples to the north and south of the region. For example, small-scale cattle production gained importance in the 1980s and 1990s in large part because many colonists came from cattle-producing regions, such as Tabasco and Veracruz. Likewise, jalapeño chili, an increasingly prevalent commercial crop in the region, is linked to Gulf Coast peoples who marketed the crop in their homelands. Even those of Maya ancestry, originating from the state of Yucatán, tend to embrace such activities as beekeeping to supplement the *milpa* of the local land.

The average length of occupancy in the region is 19.4 years. Twenty-four per cent of the sample originates from Veracruz, where both jalapeño and pasture are found in abundance. The remainder of the sample originates primarily from other neighboring states or from within the region itself: 13% from Chiapas; 17% from Tabasco, 4% from Michoacan; and 40% from one of the three states of the Yucatán Peninsula (Table 8.2). Virtually the entire sample engaged in farming activities in their places of origin, with 40% having access to their own land and the remaining 60% working as farmhands or in some other—generally unskilled—employment. Whether the respondents owned land or not, the most commonly stated reason for migrating was to acquire more land; 33% stated other reasons, such as family disputes or adverse ecological conditions in their place of origin (Table 8.2).

Substantial disparities exist across households with regard to basic indicators of material well-being. These disparities, largely a function of the *ejido*'s proximity to a major road and political clout in leveraging various services from the government, are most apparent at the *ejido* level. Eight of

**Table 8.2.** Regions of origin of *ejidatarios* in the SYPR survey and reason for relocation

| | Before 1970 | 1970–9 | 1980–9 | 1990–8 | TOTAL |
|---|---|---|---|---|---|
| *State of origin* | | | | | |
| Campeche | 11 | 11 | 9 | 14 | 45 |
| Chiapas | 0 | 2 | 17 | 7 | 26 |
| Michoacan | 0 | 0 | 5 | 2 | 7 |
| Quintana Roo | 5 | 1 | 0 | 2 | 8 |
| Tabasco | 0 | 19 | 10 | 5 | 34 |
| Veracruz | 3 | 9 | 28 | 8 | 48 |
| Yucatán | 11 | 3 | 0 | 1 | 15 |
| Native to study region | 1 | 3 | 5 | 2 | 11 |
| Other | 0 | 0 | 3 | 0 | 3 |
| *Reason for move*[a] | | | | | |
| Land (pull) | 9 | 29 | 61 | 31 | 130 |
| Work (pull) | 12 | 10 | 5 | 2 | 29 |
| Environment[b] (push) | 5 | 2 | 6 | 1 | 14 |
| Conflict[c] (push) | 2 | 2 | 4 | 0 | 8 |
| Other | 2 | 6 | 3 | 3 | 15 |
| TOTAL | 36 | 47 | 76 | 40 | 199 |

[a] In some cases, two factors for an individual respondent are listed if the individual had two equally important reasons for deciding to move (e.g. the need for land and to escape political turmoil); thus the column total and the total may not be the same.

[b] The category 'environment' refers to factors such as drought, declining soil fertility, high incidence of pests, volcanic eruptions, which respondents claim were reasons to leave their region of origin.

[c] 'Conflict' refers to various forms of social upheaval, such as armed conflict or familial conflict.

the eleven sampled *ejidos* are supplied by electricity, while only two are serviced by a continually staffed health clinic. Among sampled respondents, 11% live in cement homes (the material most preferred by inhabitants), 71% have electricity, and 30% have regular access to health services. These data do not fully capture stark intra-*ejido* inequities, however. Within two of the *ejidos* sampled, for example, it is common to observe decrepit huts with dirt floors and thatch roofs alongside cement homes with satellite dishes.

Five languages are represented in the sample, the most common of which is Spanish, the basic language of 70% of household heads surveyed. Another 24% of households speak one of three Maya languages. The remaining 6% originate from the state of Veracruz and speak Totonaco. Not surprisingly, the incidences of these languages, except for Spanish, are unevenly distributed across the region and tend to be clustered in village communities. The respondents speaking Totonaco, for example, all live in the community of Nuevo Becar, located in the northern section of the Quintana Roo portion of the study region. With two exceptions, all households speaking an indigenous language were also able to communicate effectively in Spanish.

The combination of migrants and high fertility rates leads to high rates of population growth, presently about 4.0% p.a. or a doubling time of slightly less than 20 years (Ericson, Freudenberger, and Boege 1999). Household sizes are well above the national average, the sample average being 6.4 members, compared to the 4.6 members nationally (INEGI 1996*a*,*b*). This figure can partly be explained by a relatively high incidence of infant mortality—24% of women lost a child within the first year of life—coupled with poor access to family planning services. Differences in household size by ethnicity are notable. The mean household size of non-native Spanish speakers is 7.4 members, substantially higher than the corresponding figure of 5.9 members among native Spanish speakers. Likewise, the average number of children of the former group, 5.75 children, is higher than the average 5.1 children borne by native Spanish-speaking household heads.

Increasing population notwithstanding, the rate of out-migration among *ejido* household children appears to be considerable. Of the 33% of households with children who had migrated, 75% settled in a city compared with 25% who settled in another rural locale. Among those migrating, slightly less than half (42%) sent home remittances.

Formal education is minimal among household heads: on average, males completed 2.8 years of formal schooling while females completed 2.9 years.

Moreover, 36% of household heads received no schooling whatsoever, and only 24% completed primary school. First generation residents of the region have received slightly more exposure to formal education. The average duration of schooling for males over 17 years of age in this generation is 6.6 years, and for females, 6.3 years.[2]

## 3. *Division of Labor and Sources of Income*

### 3.1. *On-Farm Labor*

Both household size and composition affect the organization of production within semi-subsistence households (Ellis 1993). Male members of the household throughout the region, beginning at around age 12, generally carry out field labor, while women primarily engage in domestic activities, including the raising of small animals, such as pigs and chickens, and the maintenance of house gardens. Nevertheless, in 48% of the households surveyed, women also reported to occasionally help in agricultural activities, notably planting, weeding, fumigating, and harvesting.

*Ejidatarios* earn cash income from a broad spectrum of land-based outputs, including citrus fruits, squash, and grasses used to make thatch roofs for the tourist industry along the coastline of Quintana Roo. The two most common sources of crop income, however, are through the sale of maize and jalapeño. Fifty-two per cent of the sample produced chili for sale in 1997–8, and 41% of all respondents reported to sell maize (Table 8.3). While many farmers are unable or unwilling to engage in what is widely perceived as the risky undertaking of chili cultivation, others depend on that crop's production for a significant portion of their annual income. The mean gross revenue from chili sales among growers was N$17,143 per hectare (approximately $US1,714 in 1997 and 1998), dramatically higher than the corresponding figure for maize of N$494 pesos. It should be noted, however, that farmers selling maize generally consume a substantial portion—roughly 60%—of their output. Moreover, the input costs of chili production are as much as five times those of maize, depending on the mix of labor and chemicals employed (Ch. 10).

[2] Surprisingly little difference existed in education levels among households speaking Spanish as their first language, rather than an indigenous language. The former (70% of sample) had an average schooling, for first-generation males over 17 years of age, of 6.9 years and for females, 6.5 years. The latter (30% of the sample) revealed 6.1 and 5.8 years, respectively.

**Table 8.3.** Gross per hectare income from maize and chili, 1998–1998 (pesos)

| Crop | % of sample selling | Mean income among sellers (N$) | Standard deviation income (N$) | Median income (N$) |
|---|---|---|---|---|
| Chili | 52 | 17,143 | 17,065 | 12,506 |
| Maize | 41 | 494 | 393 | 347 |

Over the past two decades, honey production has increased among peasant producers seeking to diversify income sources without impinging on the production calendar of agriculture (Ewell and Merrill-Sands 1987). About 25% of households sampled engaged in apiculture. Chicle tapping, of historic importance in the region (Ch. 3), is now undertaken by only 8% of the households.

### 3.2. *Off-Farm Labor and Engagement of the Market*

Households in the region are not of the autarkic type characterized in the classic Chaynovian model, allocating family labor solely to on-farm tasks with no reliance on hired labor (Chayanov 1966). To the contrary, an overwhelming majority of households in the sample participate in the market as sellers and/or buyers of labor, with only 6% of households completely self-sufficient and slightly over half both selling and hiring labor at particular times in the year (Table 8.4). On average, households earned N$1,032 from off-farm employment for the year of questioning and expended N$4,173 on hired labor to aid in agricultural production. The most important regional development in the selling and hiring of labor is chili farming, which provides significant amounts of employment for people in the region in tasks such as harvesting, weeding, and fumigating (Ch. 10). Chili farmers with mechanized land, for example, use an average of 246 days of labor per growing season on their chili plots, with an average of 89 days paid to non-family members.

Among the 80% of households selling labor off the farm, the majority participated as day laborers on the farms of neighbors in the village. It is also not uncommon, however, for family members to leave the village temporarily to take advantage of employment opportunities afforded by the government (e.g. restoration work in the local ruins or in road and other infrastructure projects). In less than 10% of the sample did households have members working full-time off the farm throughout the year.

**Table 8.4.** Participation in labor markets

| Households | Selling labor (%) | Not selling labor (%) |
|---|---|---|
| Hiring labor | 56 | 14 |
| Not hiring labor | 24 | 6 |

In response to hurricane events and drought, as well as in the search for extra income to maintain the farming enterprise (Kearney 1996), many smallholders seek temporary wage labor in urban locations, especially the cities of Campeche, Chetumal, and Cancún. In 1993 and 1994, for example, drought affected Tomas Garrido, causing *ejidatarios* to abandon their fields and to find work in Cancún and other coastal areas for up to four months. After hurricane Roxanne in 1995, workers in La Lucha, Campeche, also went to the Quintana Roo coast to find work. Since El Mundo Maya was initiated in 1993 (Chs. 3 and 7), opportunities for off-farm labor increased in the study region proper due to the large number of projects to improve infrastructure and services.

Categorical data on historical practices gives some evidence that the high degree of participation in labor and output markets may be part of an overall trend toward a greater reliance on markets for goods and services. Between 1986 and 1997 various indicators show an increasing percentage of households from the sample engaged in some market-based activity (Table 8.5). While the first indicator, the percentage of *ejidatarios* producing maize for sale, changes minimally, the percentage of *ejidatarios* planting pasture, cultivating chili, and using chemical inputs (such as herbicides, pesticides, and fertilizers) has increased steadily. That the percent selling maize has remained fairly stable may reflect government price supports for certain staples that have shielded producers from international market fluctuations through 1992. The slight dip in market participation in 1993 may be partly attributed to an initial response to Mexico's constitutional reforms of the previous year as well as the first reductions in price supports.

### 3.3. *Sources of Credit*

Until the changes to the Constitution in 1992, *ejidatario* farmers were prohibited from receiving credit from financial institutions. While this prohibition is no longer the case, the governing assemblies of *ejidos* in the region have not extended to their members titles of private property that could

**Table 8.5.** Indicators of market orientation

| Year | Selling maize (%) | Planting pasture (%) | Cultivating chili (%) | Using chemical inputs (%) | Hiring labor (%) | Selling labor (%) | n |
|---|---|---|---|---|---|---|---|
| 1986 | 44.9 | 8.2 | 19.0 | 24.5 | 52.0 | 54.5 | 89 |
| 1990 | 41.3 | 14.9 | 31.5 | 38.4 | 54.7 | 58.4 | 114 |
| 1993 | 34.3 | 24.1 | 41.7 | 46.6 | 60.0 | 61.7 | 134 |
| 1997 | 41.9 | 49.0 | 52.6 | 70.4 | 71.1 | 80.0 | 186 |

serve as collateral for loans. In rare cases, individuals have been able to receive loans from private lending institutions for the development of chili cultivation and its subsequent transportation to market centers. Most *ejidatarios*, therefore, remain dependent on the government for financial support, as has been the case since *ejidos* were established. Most of this support presently comes from two forms. The first of these, PROCAMPO (Programa de Apoyo Directo al Campo/Direct Rural Support Program), is a fixed, per hectare direct payment on the production of a wide range of products that was introduced in 1994 to ease the adjustment to NAFTA. By replacing the price supports on basic staples that obtained before 1992, PROCAMPO was intended to reduce distortions in the supply of agricultural commodities and promote more efficient use of land (OECD 1997). PROCAMPO payments are based on the acreage planted in staple crops in 1994 and are fixed in real terms since 1996. In the southern Yucatán peninsular region, PROCAMPO support is widespread and is received by 88.3% of the sample. The area covered ranges from 1 to 27ha, with a mean of 4.8ha. While two-thirds of the households surveyed use the payment for household goods, education, or health care, as opposed to investment in agricultural inputs, indications are that the payment is tied to expansion of hectares in pasture (Klepeis and Vance 2003). Unusual and innovative uses of PROCAMPO funds exist, however. In Narciso Mendoza, for example, farmers grouped together to leverage their future PROCAMPO earnings against half the price of a tractor.

The other major source of government support, PRONASOL (Programa Nacional de Solidaridad/National Solidarity Program), was introduced in 1990. PRONASOL's implementation coincided with a cutback in the budget of a precursory program, BANRURAL (Banco Nacional de Crédito Rural/National Rural Credit Bank), which since 1975 had served as the main official rural credit institution. While BANRURAL credits were extended past 1990, the organization retargeted its client base to exclude those farmers without access to high-quality rainfed or irrigated lands (Myhre 1998; OECD 1997). PRONASOL was intended to fill the resulting credit vacuum for the rural poor, although, as noted by Myhre (1998), the meager payments of the program have done little beyond subsidizing subsistence cultivation. From 1990 to 1994, these payments were fixed at 300 pesos per ha for a maximum of 3ha, to be planted with a crop of the farmer's choice. Beginning in 1995, the payment was raised to N$400 per hectare. Farmers are not required to provide collateral for the loans. Failure to repay carries no penalties other than exclusion from the program in subsequent years. Of the sample, 27.6% had received some support from the PRONASOL program since its inception, with the mean total amount received being N$2,926.

Throughout the region, many other sources of subsidy exist from governmental and non-governmental sources. While not universally received, sources of money and technical aid further bolster farmers' portfolio approaches to livelihood in the region. Over 50% of *ejidatarios* in the southern region of Calakmul Municipality (Campeche State), for example, report receiving subsidies and assistance from more than four sources.

## 4. *Land Assets and Capital, Parcel Access, and the Land Market*

### 4.1. *Land Assets and Farm Capital*

Parcel sizes in the sample range from 10 to 390ha, and the majority of households additionally maintain a house garden. Forests and successional growth within the total parcels of a household were of various ages. On average, what respondents refer to as *montaña* (forest of at least 15 years in age) covered 43% of their parcels, with 11% under the cultivation of perennial crops. The remaining 46% was covered by a combination of *acahual*, a term for secondary growth ranging from 1 to 15 years old, and pasture, planted by 49% of the sample. The incidence of mechanically cleared or cultivated land in the region is relatively rare. While 19% of respondents reported an average of 11.3ha of mechanized land, these were often tracts that were not regularly tilled (see Ch. 10). In this regard, only 11% of farmers reported continual access to a tractor.

The use of other capital inputs in production depends largely on the crop grown and on the household's engagement of the market. The *milpa* or swidden relies primarily on human labor and land; farm capital is often limited to an ax, machete, and digging stick for planting seeds, with generally little reliance on purchased inputs. Less than half the sample, 39%, reported to apply chemical inputs in maize cultivation, primarily herbicides. Jalapeño chili production, by contrast, necessitates repeated applications of chemical inputs throughout the growing season. Of the sample, 87% reported to apply some combination of fertilizer, insecticide, and herbicide to their jalapeño crop over three intervals from mid-summer to early fall. Moreover, data obtained from the creation of the sketch maps indicates that mature forest is more commonly cleared for jalapeño production than for the *milpa*, making accessibility to a chainsaw an important determinant in the production of this crop. In 27% of the parcels where chili was observed, mature forest (> 15 years) had been directly converted, with the corresponding figure for maize at 15.3%.

### 4.2. *Parcel Access*

Contrasting with the now well-known fishbone pattern of settlement identified in parts of the Brazilian Amazon (Tucker, Holden, and Goff 1984; Woodwell *et al.* 1987), where the coterminous relation of households and parcels creates a pattern of narrow swaths extending from the roadside into the forest, *ejidatario* households in the southern Yucatán peninsular region are clustered in villages away from their plots. Most farmers thus commute—sometimes considerable distances—with the result that plot accessibility is an important factor in determining both the area cultivated and crop mix. Travel from the village house to the plots takes anywhere from five minutes to several hours, depending on the condition of the path or road, the mode of transportation, and the proximity of the plot to the roadside. In a small percentage of cases, farmers must remain at their plot for days at a time due to the distance separating it from their home in the *ejido* center. Those farmers with plots close to a paved road often commute by bicycle or may take a bus or hitch a ride. Many farmers, however—some 42%—access their plots by foot, often over muddy trails not passable by vehicle. These conditions, in addition to an average on-road distance to the nearest primary market center of 71 kilometers, makes the transaction costs of market participation for the majority of farmers extremely high.

### 4.3. *Borrowing, Renting, and Selling Land*

Despite laws prohibiting land exchange before 1992 (Ch. 7) *ejidatarios* have long been renting and selling as well as borrowing land. Thirty-seven per cent of those surveyed reported to be borrowers, while 22% reported to be renters. In the majority of cases, these land exchanges, which average 6.6ha, are among *ejidatarios* and are based on informal barter arrangements, such as land for labor swaps. Among the roughly 17% engaged in monetary exchanges, the average amount paid per hectare is N$356.9. There are also rare cases in which *ejidatarios* report that they have bought or sold land; 14% reported to be buyers and 5% sellers, with the average amount exchanged being 8.73ha at a rate of N$718 per ha.

The increasing incidence of pasture in the region may also be indicative of a latent land market. Initially buoyed by government programs, some *ejidatarios* have planted pasture in the hope that other programs might support livestock purchases. Between 1986 and 1997, the percentage of sampled farmers maintaining some amount of pasture increased from 8% to 49%. Of these, only slightly more than half owned grazing animals (cattle or sheep) as of 1997, with the non-owners invariably stating their hopes of acquiring livestock at some point in the future (Fig. 8.1). Moreover, among the livestock owners, the ratio of cattle-head to hectares in pasture is 0.47 to 1, far less than the density of approximately one head per hectare that locals believe the land can support. Pasture can thus be regarded as a relatively land extensive use, the economic motivations of which may not be entirely connected to animal husbandry. In this regard, a common explanation for the maintenance of pasture is that it increases the value of the land should the *ejidatario* decide to rent or sell in the future.

## 5. *Ranchos and Large Pasture*

### 5.1. *Ranchos*

In fact, a land market external to the *ejido* sector has existed in the region since the late 1960s when various individuals, almost all absentee landlords, were able to purchase private lands. Almost all these lands, covering some 200km$^2$ (less than 3%) of the region on roughly sixty-seven different ranches, have been made into small ranchos, eleven of which were included in the field survey. While these ranchers generally had access to relatively larger land endowments, with an average plot size of 328ha compared with 88ha for *ejidatarios*, their land-use portfolios did not distinguish themselves markedly. All the ranch managers surveyed planted maize and many also planted chili. It bears noting that along a number of qualitative criteria, the ranch managers as a group were no different than the *ejidatarios*. Because of their small representation in the sample, statistical evidence for this assertion relies on the Pearson chi-square exact significance tests.[3] Table 8.6 shows the results of this test, whereby the traits of chili cultivators, sellers of maize, and users of chemicals in production are all independent of whether the household is a ranch. Ranchers do, however, distinguish themselves from *ejidatarios* with respect to the increased likelihood that they own cattle and maintain pasture. This makes obvious sense as all original private land was purchased with cattle production in mind.

### 5.2. *Large Pasture*

While orchestrated by the state government of Quintana Roo and not representative of land rental region-wide, the case of Laguna Om (Nicolas Bravo) bears noting due to the large area of pasture present there. After the failure of government-sponsored rice projects in the early 1980s (Ch. 3), the land was maintained clear of regrowth and used for either cattle or small patches of rice production until the early 1990s. The *ejidatarios* were unable to make rice economically viable, however, despite the project's remaining infrastructure, proximity to a processing plant and market, and price sub-

**Table 8.6.** Pearson chi-square exact tests for differences between ranchers and *ejidatarios* with respect to qualitative criteria

| | Exact significance (p-values of two-sided test) |
|---|---|
| Plant chili | 0.124 |
| Sell maize | 0.371 |
| Apply chemicals | 0.999 |
| Own cattle | 0.040 |
| Maintain pasture | 0.010 |

[3] Exact significant tests are appropriate when there are concerns about the size, distribution, or balance of the data; they yield accurate results when the assumptions of the asymptotic method cannot be met.

sidies.[4] Eliminating most of the remaining rice production, Laguna Om converted large tracts of rice fields into pasture populated with cattle for milk and beef production.[5]

## 6. *The* Ejido *Household in Review*

Despite the few *ranchos* and pasture schemes, the *ejido* household remains the dominant force in regional agricultural production and in land change. These households display overwhelming uniformity in that each engages in swidden production in which maize is the principal staple of choice, and each draws on several basic government programs (e.g. PROCAMPO) to supply modest credits and subsidies for various agricultural activities (Ch. 7), and increasingly they take on commercial chili cultivation (Ch. 10). The technology of production and the rationale for production is more or less uniform throughout the region, therefore. Beyond these characteristics, *ejido* households display a considerable range in the diversity of 'landed' activities in which they engage and means by which they engage in them. As noted in Ch. 7, a considerable portion of this variation is determined by the *ejido* occupied, inasmuch as the *ejido* determines the precise tenure rules and which government programs will be engaged in, if they require *ejido*-level approval. The remainder of the variation, however, involves the character of the individual household and the decisions that follow from it. Critical elements of the household, beyond its wealth and access to land, include the past agricultural experiences of the household, usually elsewhere, the connectedness of the household to critical agents and the programs they promote, and the degree to which households engage the market. These qualities lead to different household land and capital portfolios, as they fit the general rules of land uses imposed by the government (Ch. 7).

[4] In 1994 the government negotiated an agreement with the *ejidatarios* of Laguna Om to rent a portion of prepared rice land to Mennonites. Rent is collected in a unique arrangement by which the government pays *ejidatarios* the PROCAMPO subsidy for the hectares worked by the Mennonites. In essence, the Mennonites work the land on behalf of the *ejidatarios*. The Mennonites are also given access to the rice processing plant. They supply all the labor and equipment to work the fields in what has apparently been a profitable endeavor. As of 1998, the government was in negotiation with the Mennonites over the sale of the rice plant to them.

[5] A similar pattern of conversion occurred in the *bajos* surrounding Laguna Silvituc, where large tracts of land formerly developed for rice production were planted with grasses for cattle, some of which is reportedly exported for sale in the US prior to slaughter.

The critical lessons for understanding land-use decisions taken from the *ejido* household are several:

- The *ejidatario* is the principal land manager and agent of land change in the region.
- The average land controlled by individual *ejido* households is 88ha, ranging from 10–390ha—the *ejidatario* is land rich in terms of hectares controlled.
- Virtually every household produces maize for direct consumption, and a majority of households sell a portion of their maize crop or engage in commercial chili production.
- About 50% of the households have converted lands to pasture (about 10–30% of individual total household lands), although only one-quarter of them own any livestock, indicating the power of government subsidies to influence land-use decisions.
- All markets in the region are thin, and household access to capital is low.
- These conditions place the *ejido* household within the economic frame of smallholders elsewhere in the world, variously described as peasants, dual producers, or 'penny capitalists' (Tax 1953).

## REFERENCES

Chayanov, A. V. (1966). 'Peasant Farm Organization', in D. Thorner, B. Kerblay, and R. E. F. Smith (eds.), *A. V. Chayanov in the Theory of the Peasant Economy*. Homewood, Ill., Richard D. Irwin, 29–69.

Deaton, A. (1997). *The Analysis of Household Surveys: A Microeconomic Approach to Development Policy*. Baltimore, The Johns Hopkins University Press.

Ellis, F. (1993). *Peasant Economics: Farm Households and Agrarian Development*, 2nd edn. Cambridge, Cambridge University Press.

Ericson, J., M. S. Freudenberger, and E. Boege (1999). *Population Dynamics, Migration, and the Future of the Calakmul Biosphere Reserve*. Washington DC, Program on Population and Sustainable Development. American Association for the Advancement of Science.

Ewell, P., and D. Merrill-Sands (1987). 'Milpa in Yucatan: A Long-Fallow Maize System and its Alternatives in the Maya Peasant Economy', in B. L. Turner II and S. B. Brush (eds.), *Comparative Farming Systems*. New York, Guilford Press, 95–129.

INEGI (Institutio Nacional de Estadística, Geografica e Informatica) (1996*a*). *Conteo de Población y Vivienda 95: Resultados Definitivos Tabulados Basicos. Estado de Campeche*. Aguascalientes, INEGI.

——(1996*b*). *Conteo de Población y Vivienda 95: Resultados Definitivos Tabulados Basicos*. Estado de Quintana Roo. Aguascalientes, INEGI.

Kearney, M. (1996). *Reconceptualizing the Peasantry: Anthropology in Global Perspective*. Boulder, Westview Press.

Klepeis, P., and B. L. Turner II (2001). 'Integrated Land History and Global Change Science: The Example of the Southern Yucatán Peninsular Region Project'. *Land Use Policy* 18: 27–39.

Klepeis, P., and C. Vance (2003). Neoliberal Policy and Deforestation in Southeastern Mexico: An Assessment of the PROCAMPO Program, *Economic Geography*, 79 (3), 221–240.

Myhre, D. (1998). 'The Achille's Heel of the Reforms: The Rural Finance System', in W. A. Cornelius, and D. Myhr (eds.), *The Transformation of Rural Mexico: Reforming the Ejido Sector*. San Diego, Center for U.S.-Mexican Studies, University of California San Diego, 39–65.

OECD (Organisation for Economic Co-Operation and Development) (1997). *Review of Agricultural Policies in Mexico.* Washington DC, OECD.

Tax, S. (1953). *Penny Capitalism*. Washington DC, US Government Printing Office.

Tucker, C. J., B. N. Holden, and T. E. Goff (1984). 'Intensive Forest Clearings in Rondonia, Brazil, as Detected by Remote Sensing'. *Remote Sensing of Environment* 15: 255–61.

Vance, C. (2000). 'Land-Use Change, Deforestation, and Peasant Farm Systems: A Case Study of Mexico's Southern Yucatán Peninsular Region'. Ph.D. Dissertation, Clark University.

Warwick, D. P., and C. A. Luinger (1975). *The Sample Survey: Theory and Practice.* New York, McGraw Hill.

Woodwell, G. M., R. A. Houghton, T. A. Stone, R. F. Nelson, and W. Kovalick (1987). 'Deforestation in the Tropics, New Measurements in the Amazon Basin Using Landsat and NOAA Advanced Very High Resolution Radiometer Imagery'. *Journal of Geophysical Research* 92: 2157–83.

# 9

# Subsistence Sustained: Swidden or *Milpa* Cultivation

*Peter Klepeis, Colin Vance, Eric Keys, Pedro Macario Mendoza, and B. L. Turner II*

From the modern settlement of the southern Yucatán peninsular region to the present, smallholder farmers have followed a system of cultivation variously labeled swidden, slash-and-burn, or shifting within agricultural typologies (Watters 1971; but see Denevan 1992), and known as *milpa* in the Yucatán and Maya lowlands.[1] *Milpa* cultivation has been so pervasive historically and geographically throughout the peninsula and the subject of such an extensive literature, that its description is only briefly reviewed here.[2] Understanding the character and dynamics of this system of cultivation, including its long-term prognosis for continued use within the development

[1] The term *milpa* has its origins in Nahuatl (language of central Mexico) and literally means 'maize field'. Given that these fields in the Maya lowlands have historically been cultivated in some sort of swidden system, the literature fuses the crop and cropping procedure such that *milpa* means swidden cultivation in which maize is the dominant crop. Swidden itself connotes burning, but the term is applied consistently in typologies of agriculture to rotational, slash-and-burn practices (Turner and Brush 1987).

[2] For various swidden or *milpa* systems in the larger region examined at different times see: Bernstein and Herdt 1977; Chamberlain 1948; Cook 1921; Emerson and Kempton 1935; Emerson 1953; Hernández Xolocotzi, Bello Baltazar, and Levy Tacher 1994–5; Lundell 1933; Nations and Nigh 1980; Reina 1967; Reina and Hill 1980; Rojas Rabiela 1985, 1991; Roys 1943; Steggerda 1941; Turner 1983; Wagner 1968; Wilhelmy 1981; Wilken 1971, 1987. Most of these systems exist(ed) within the peninsular region (between the north coast of Yucatán and the central lakes of Petén, Guatemala). That by Nations and Nigh (1980) is in the Lacandón rainforest of Chiapas, where environmental conditions and cropping systems differ from the peninsular region. It is included because of its Maya (Lacandón) heritage.

of the region, is essential for understanding land changes and modeling them, although recent changes in cropping strategies portend the emergence of a 'new' kind of *milpa*.

Swidden cultivation in the region, as elsewhere in Middle America, is invariably undertaken as an outfield activity—located at some distance from the farmstead—and is accompanied by small but complex house-gardens or *solares* adjacent to and surrounding the house (e.g. Killion 1992; but see Gómez-Pompa 1987). The house-garden not only provides shade for the abode, it provides fruits, nuts, medicinal and ornamental plants, and a place for cropping experiments (Keys 1999). The extent and elaboration of house-gardens in the southern Yucatán peninsular region appears to be tied to the length of residency and, perhaps, ethnicity of the resident. Maya people, for example, tend to maintain large and elaborate *solares* (Fig. 9.1). House-gardens tend to be smaller, even unrecognizable to the untrained eye, in the few densely settled communities in the region (e.g. Xpujil). As these gardens are not yet a central element of the broader dynamic of land change in the region, they are not given further attention here.

**Fig. 9.1.** *Solar* or housegarden.
*Photo*: P. Klepeis.

## 1. *Maize Swidden*

The outfield or swidden supplies maize (*Zea mays* L.), planted in several varieties and serving as the consumption staple. Depending on the household, some portion of the maize crop may be sold. This 'dual' production function has been part of swidden in the region at least since the opening of Highway 186, reflecting government policy promoting commercial maize production (Ch. 7) and the abundant land awarded to individual *ejidatarios* at that time. As early as 1969, Vogeler (1974, 1976) reported that 41 per cent of the initial smallholders in Xpujil (Fig. 1.1), situated in the center of the study region, sold about 49 per cent of their maize crop grown on swiddens. Today, almost all households attempt to be self-sufficient in maize, although the SYPR-project household survey (Ch. 8) indicated only 33 per cent claimed to be self-sufficient. Despite controlled prices, some portion of maize (an average of 40% of the selling household's production) taken from swiddens is still sold by 41 per cent of the households in the region.

### 1.1. *Swidden Practice*

The basic steps in swidden cultivation are consistent throughout the Maya realm: plot selection, cutting-drying, burning, planting, weeding and doubling or folding, and harvesting (see n. 2). In January (winter dry season), smallholders select which 'new' parcels will be cut from older growth upland forest (*mediano* or *montaña*) or younger successional growth (*acahual*), as well as which of the previous year's cultivated parcels will be resown. In the late 1960s and 1970s smallholders selected upland forest sites, especially those possessing ancient Maya housesites and rockwall features, perhaps reflecting that lands once considered prime for cultivation remain so today. In Xpujil, for example, parcels to the east of the village were favored (Vogeler 1970), coinciding with ancient Maya farmsteads and hillside terraces abundant there (Turner 1974, 1983).[3] Today, the selection of new parcels tends to be restricted by usufruct access (Ch. 8), involves, more often than not, successional growth from previous swiddens, and tends to include parcel rotation emphasizing the proximity of recultivated and newly cultivated plots. The cutting of older growth takes place first

[3] Given that the means of parcel delegation for Xpujil in 1969 is not known, we cannot be sure that this eastern location was by 'choice' of the farmer or reflected some other decision factor. Because these parcels were the first in Xpujil, we suspect it reflected the opinion of the initial *ejidatarios* of which lands were most suitable for *milpa*.

because it must dry longer to burn properly, in some cases from late January to April or May (Fig. 9.2). Recultivated plots are cut as early as six weeks before the anticipation of the onset of rains. Currently, some parcels are cleared with a chainsaw and turned by tractor and disk. These mechanized lands are intended for 'intensive' commercial chili cultivation, and in 2000–01, the older of these parcels had been cultivated for 5–7 years continuously. For these reasons, mechanized cultivation is not considered to be swidden and is addressed in Ch. 10.

In April or early May the parcels are ready for burning. Conventional wisdom, as captured in much of the literature, indicates that parcels are set to fire in the afternoon when strong east–west winds develop to push the burn across the cut biomass. More often than not, however, parcels are burned in the morning as part of a routine 'fieldwork day' (06.00 to 13.00) of the *milpero*. Regardless, the goal is to match the peak dryness of the cut vegetation with strong winds to create a complete burn (Fig. 9.3). A successful burn supplies abundant ash, a much-needed source of phosphorus (Ch. 5), across the parcel; an incomplete burn leaves much intact, dead

**Fig. 9.2.** Cut swidden.
*Photo*: P. Klepeis.

**Fig. 9.3.** Burned swidden.
*Photo*: P. Klepeis.

biomass, which must be planted around, and insufficient ash.[4] Early rains can also reduce the quality of the burn. Vogeler (1970) reported few out-of-control burns, the fire being 'absorbed' by the forest edge. Fire outbreaks are reported, however, especially in the eastern sections of the region that have been cultivated longer and whose forests are more prone to wind damage from hurricanes and tropical storms. In 1998 and 1999, numerous swidden burns raged out of control in the central part of the region.

Planting tends to begin about 15 May (ibid.), but coincides with the annual flux in onset of the rains. The timing between the burn and planting is critical to ensure plenty of ash is washed into the soil. Using a digging stick, the rain-softened soils are seeded with maize kernels stored from the previous harvest. Holes are punctured in soils about 1 meter apart in which seeds of maize, beans (*Phaseolus* spp.), and squash (*Cucurbita* spp.) are

[4] Substantial research demonstrates that burning mineralizes nutrients and 'fertilizes' the burned plot (Ewel *et al.* 1981; Juo and Manu 1996; Kauffman *et al.* 1993). Lightning that accompanies the early rains may also help to fix nitrogen, but this input is only a fraction of that likely to be fixed by soil microbes during this early wet period.

inserted. The parcel is subsequently weeded by machete, involving little more than hacking the higher standing weeds (Fig. 9.4). In late August or early September, mature maize cobs are folded (*doblado*) to dry; the upper half of the stalk is bent downwards to protect the cob from rain and birds. Maize harvesting begins as early as October in non-chili swiddens, and beans even earlier. Maize is placed in bags and stored on site, historically in *trojas* (elevated maize cribs; Fig. 9.5). The bags are taken to the house by truck, horse or mule, or bike. Squash is left on site and collected as needed.

While the basic steps in swidden or *milpa* cultivation in the region still strongly resemble that described by Vogeler (1970) for 1969, several agricultural strategies and technologies are changing (Fig. 9.5). For example, the long-standing Mesoamerican intercropped 'triad' of maize-beans-squash is farmed less than in previous years. Increasingly, beans are planted to the side of maize fields, farmers citing the tangle of the vines as an impediment to maize care. Experiments are underway with *nescafe* (*terciopelo* or velvet bean, *Mucuna pruiens* (L.) DC), planted after maize as a groundcover and nitrogen-fixing crop. Squash is still mixed with maize in the same

**Fig. 9.4.** Weeding by machete in a traditional *milpa*. Note the maize bent downwards (*doblado*)
*Photo*: P. Klepeis.

**Fig. 9.5.** New or hybrid swidden. Increasingly slash-and-burn techniques used to clear the forest are mixed with disking by tractor and planting maize in rows without intercropping, giving rise to a 'new' *milpa*.
*Photo*: P. Klepeis

plot, but increasingly serves as an important source of income, especially in years when chili prices fall. Today as well, farmers experiment with small patches of various crops in their swidden, including sugar cane, bananas, and pineapples.

To clear land, many farmers employ chainsaws or pay chainsaw owners to complete that service. In addition, a minor but potentially increasing trend employs the application of agricultural chemicals in maize farming, a capital input not followed in earlier years. Complementing the direct application of fertilizers, herbicides, and pesticides, farmers may crop maize on a former chili field, perhaps taking advantage of the inputs used previously (Ch. 10). This strategy may be considered a new version of swidden, one which lengthens the cropping portion of the cycle but still employs fallow.

## 1.2. *Swidden Size, Yields, and Rotation*

Theoretically and simplistically, the amount of swidden land cleared annually and the rotational cycle employed reflect household consumption needs, the basic crops in question, and the quality of environment used (Turner and Brush 1987). Consumption needs and expected yields determine the total amount of land cultivated each year, and the regenerative capacity of the land and amount of land to which the smallholder has access determines the crop–fallow cycle. Conforming to this 'rule', the amount of land used or cut for non-chili swidden appears not to be affected by access to chainsaws. Such access does correlate with area planted in chili, however, indicating that responses to commercial production is affected, even if indirectly, by the techniques of cutting forest.[5] Regardless, modern occupation of the southern Yucatán peninsular region involves mixed subsistence-market producers, affecting the decisions in question (Ch. 11), and the more recent emergence of commercial chili cultivation plays an important role in parcel selection and crop–fallow cycles for some farmers (Ch. 10).

### 1.2.1. *Swidden Size*

Vogeler (1970, 1974, 1976) informs us that average size of a swidden in Xpujil in 1969 was about 4.3ha, located on average about 4.0km from the smallholder's house, and about 4.0ha in Tomas Aznar Barbachano (La Moza), to the east of Xpujil. The range in size, however, was substantial, about 0.02ha to 19.2ha! No explanation was given about the large range in the sizes of the swiddens at that time. Today, the region maintains *ejidos* of various sizes and established with different aims in mind, thus possessing very different smallholder-to-land access ratios (10 to 390ha/household; mean 88.2ha; also Ch. 7). Despite this variation, the amount of land

[5] The correlation coefficient between chainsaws and area planted in chili is 0.27 and is significant at the 1% level. A Heckman model of chili cultivation further corroborates that access to a chainsaw is a significant determinant of area planted in this commercial crop (see Ch. 11).

cultivated annually per household throughout the region may reflect the level of engagement with chili production more so than land access (Ch. 10). Among 188 households surveyed in eleven *ejidos* in 1997–8, the average amount of swidden cultivated annually was 4.6ha, surprisingly close to that reported by Vogeler (1974) for 1969—and an additional 1.3ha was added to those households growing chili by way of swidden (Table 9.1). Subsequent surveys in five additional *ejidos* (Ch. 10) that focus on chili production support these findings. Non-chili swiddens in these *ejidos* averaged about 5.4ha in size in 1999, with 2.5ha on average dedicated to maize production.

### 1.2.2. *Swidden Yields*

Variability in swidden yields is large and linked to the basic crop and quality of environment. Evidence from the Maya area focuses almost exclusively on maize production and largely ignores the other crops raised in the plot. Vogeler (1974, 1976) recorded that maize yielded on average 1,700kg/ha from first- and second-year plots in 1969, a figure above the State of Campeche average (1,050kg/ha; no age of plot reported) at the time, but less than the higher yielding swiddens elsewhere in the greater Maya area (Turner 1983: 24). Vogeler (1974, 1976) also reported that yields increased by 400kg/ha in second-year plots and that high yields were sustained in years three and four of cultivation absent fallow.[6] In contrast, controlled swidden experiments in Uaxactún (north of the study region)

**Table 9.1.** Total area planted in five communities in Southern Calakmul Municipio

| 1999 | All farmers (n = 152) | Non-chili farmers (24) | Swidden chili farmers (81) | Mechanized chili farmers (47) |
|---|---|---|---|---|
| ha swidden | 3.2–3.5 (range 1–20) | 2.5–3.0 (range 1–7) | 2.9–3.2 (range 1–8) | 4 (range 1–20) |
| ha maize | 3.2 | 2.5–3.0 | 2.9 | 3.77 |

[6] Popenoe (1957) long ago reported that the cutting and burning of *mediano* leads to the following soil impacts: bulk density increases, acidity decreases and percent base staturation increases in the top 2 inches, nitrogen and carbon decrease through the soil profile, potassium increases throughout the profile, especially in the top 2 inches, and calcium and magnesium decreased in the top 2 inches. After cultivation, however, almost all these factors reverse: bulk density decreases, acidity increases, and percent base staturation decreases in the top 2 inches, nitrogen and carbon increase in the top 2 inches but carbon decreases below that, potassium decreases through the profile, and calcium continues to be translocated to the lower horizons.

indicate that yields of maize decline in excess of 1,000kg/ha after two years of same-plot cultivation (Urrutia 1967: 91). Vogeler was quick to note, however, that only twelve farmers in his survey reported plots cultivated more than two years, suggesting that the claims about sustained yields should be taken cautiously.

Today yields are much lower than those reported for the 'new' lands cultivated in 1969. The households surveyed by the SYPR project revealed that average maize yields were only 790.8kg/ha (s.d. = 50.45), across eleven different *ejidos*. This difference may reflect the qualities of Xpujil lands compared to the region at large (doubtful), a 'tired' landscape entering its third or fourth swidden crop–fallow cycle (Ch. 5; Fig. 5.8), or less attention given to maize cultivation today, given the alternative activities in which farmers and their households are engaged (Chs. 8 and 10). Possible indicators of 'tired' land are increasing, especially the swidden land taken by invasive species, such as tajon (*Viguiera dentata* [Cav.] Spreng), cecropia (*Cecropia peltata* L.; locally *garumbo*), especially in the southeastern part of the region, and bracken fern (*Pteridium aquilinum* L.; locally, *helecho*) [Cav.] Spreng) (Ch. 6; Turner *et al.* 2001). Bracken is particularly difficult to rid from the plot where it takes over large areas, and tends to force smallholders to abandon the lands invaded. Influenced by the activities of Plan Piloto Forestal (largely east and north of the region; Acopa and Boege 1998), Bosque Modelo, and other agencies and programs (e.g. Haggar 1999), various experiments are underway to enhance soil nutrients, increase yields, and potentially lower fallow times without increasing the risk of invasive species. Variations on slash and mulch systems are promoted today, especially the use of *nescafe* (above), a crop and method employed throughout Middle America (Garcia Espinoza, Quiroga Madrigal, and Granados Alvarez 1994). To date, more claims are made about *nescafe*'s achievement than rigorous evidence supports.

### 1.2.3. *Swidden Rotation*

Early settlers in Xpujil claimed that they could approximate a 1:1 crop–fallow cycle, although no one in the region had experienced this cycle. By 1973, most farmers employed some plot rotation that converted to a 1:3–4 cycle (e.g. 3 years crop, 9–12 years fallow) (Turner 1983: 25). This last cycle is consistent with the cropping experiments by Urrutia (1967) further north, indicating that after two years of cultivation, ten years were needed to regenerate the soils to produce acceptable yields. Today, swidden rotations appear to be more variable than in the past, depending on land endowments by *ejido* and household, as well as the role of chili and various introduced cropping strategies (above). Farmers use their different plots in a temporal mosaic, but simplifying, crop–fallow rotations range from 1:3 to 1:6, cycles usually predicated on 3 to 4 years of cultivation and 9 to 24 years of fallow. The cycle shortens where chili is incorporated into the swidden among households that average 40ha of land or less; it lengthens when chili is not cultivated and where households average 70ha of land or more (usually in older *ejidos*).

Overall, the region has increasingly grown more complex in its swidden practices and cycles, as smallholders experiment with new cropping strategies and land uses not part of the swidden cycle, such as mechanized chili production (Ch. 10) and pasture. Information obtained from GPS-referenced sketch maps of each *ejidatario*'s parcel, created during the SYPR-project field survey (Ch. 8), illustrates trends in land use that have evolved over the last twenty years. One of the more prevalent of these trends is the increasing incidence of pasture, seen in the sketch map and corresponding window from the TM imagery (Fig. 9.6a). Farmers often intersperse exotic grass seeds with maize in the first year of cultivation following the clearance of primary forest. By the second year the grasses have taken hold, and in subsequent years the fields are either burned or weeded to maintain the pasture free of weeds and woody vegetation. Unanswered questions arise as to how long these pastures may maintain productivity and avoid problems with pests and weeds. Figure 9.6b illustrates typical land-use sequences associated with chili cultivation, another crop increasingly featured in the *ejidatario*'s land-use portfolio. Chili is commonly planted in the first-year plot (cut from forest), followed by a year or two of maize cultivation before leaving the land in fallow. An alternative strategy, also illustrated in the figure, is to reverse the sequence, planting maize in the first year, followed by a year of chili. Both strategies aim to maximize soil nutrient content during the year of chili cultivation; the 'chili second' strategy relies on root decomposition for nutrients.

Despite diversification, *milpas* as maize swiddens remain dominant across the cultivated landscape, and fallow periods are relied on to replenish once cultivated land, more so than chemical and other inputs. Through the years, the average size of the swidden plots planted annually has remained remarkably constant, slightly above 4ha, implying that 16 to 24ha or more of each *milpero*'s land is in some stage of fallow or successional growth. This observation is supported by the condition of the plots selected for cultivation each year, ranging from those plots cultivated the

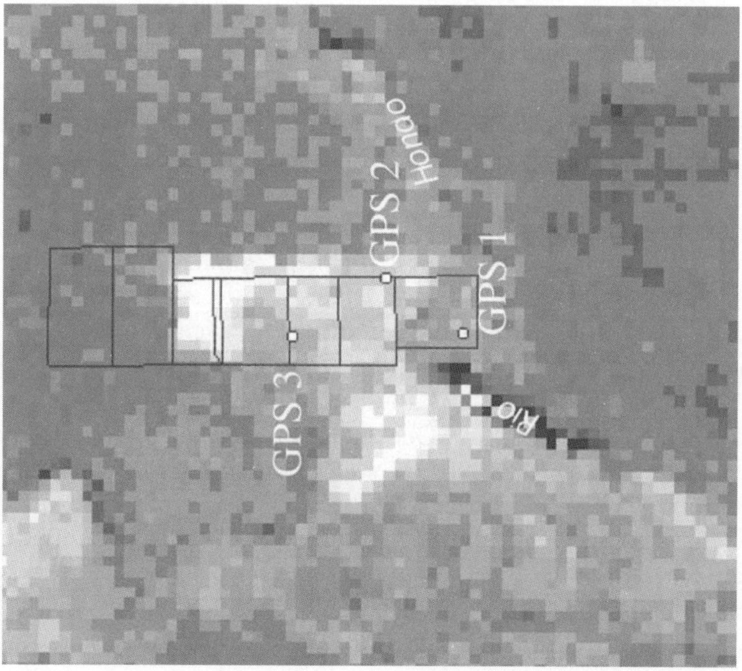


Parcels linked to imagery by GPS

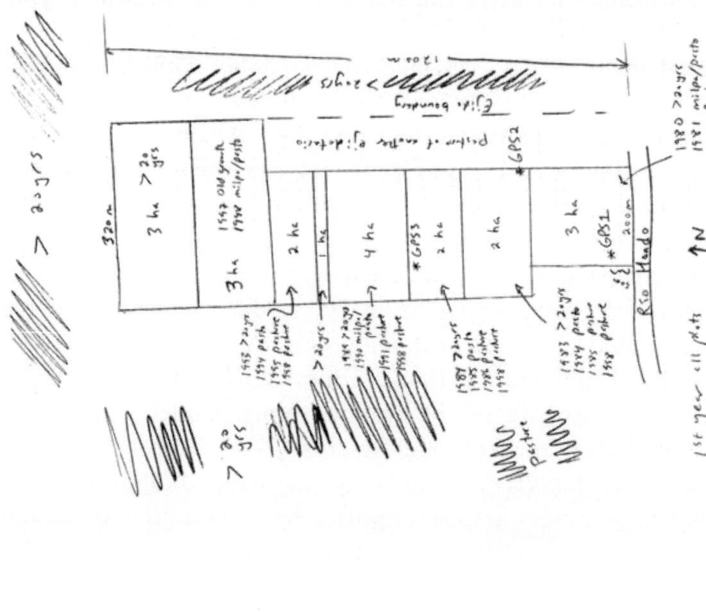

Land-use history (last 20 years) via sketch map

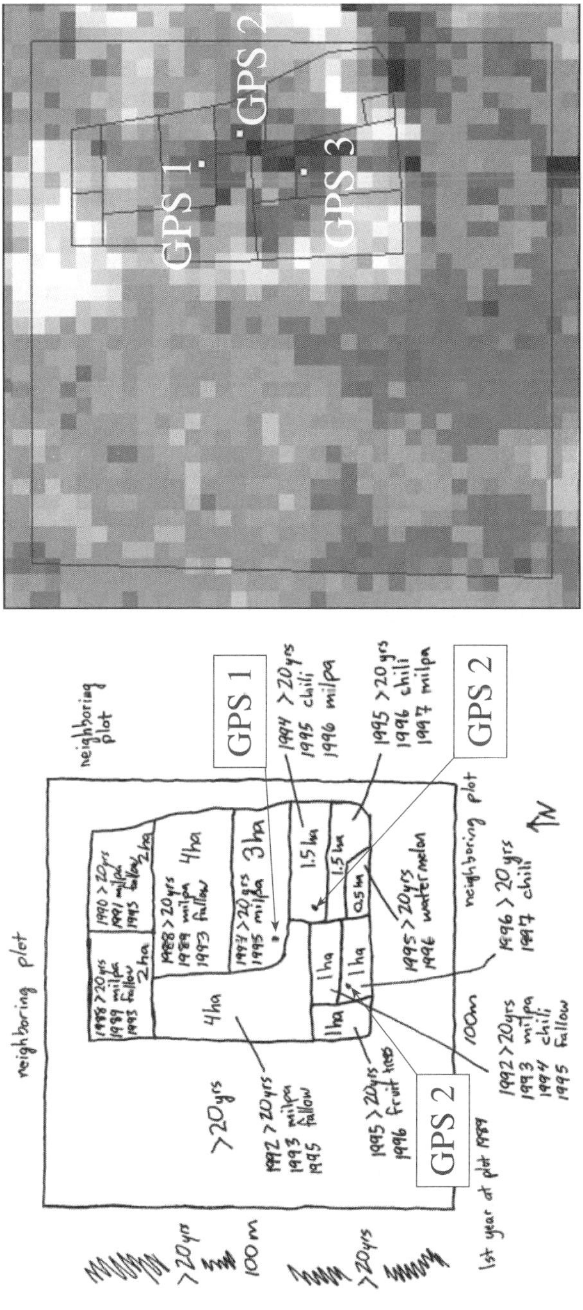


Land-use history (last 20 years) via sketch map

Parcels linked to imagery by GPS

**Fig. 9.6 a, b** Sketch maps of recent parcel use matched tc TM imagery.

year before to those with sufficiently old successional growth (*c.*25 years old) to mimic *mediano* in terms of species abundance.[7] These observations must be evaluated in terms of the new complexities in cultivation noted above and detailed for chili in the next chapter. For example, if chili is incorporated into the swidden cycle, then the amount of land cultivated increases and presumably the swidden cycle is altered slightly.

## 2. *Other Practices*

A number of cropping experiments are underway in the region, many sparked by NGOs and involving agroforestry and replanting hardwoods (Haggar 1999; Primack *et al.* 1998). Omitting these last types of projects, which seek to retain forest and rebuild the suite of species within them, various agencies have promoted and various individuals experimented with specialty crops on a small-scale basis. Throughout the region, small orchards of citrus fruits, primarily oranges, and bananas can be found, and in some locales, plots of tomatoes and watermelon. Promotion of some of this activity appears to be based on impressions or interpretations of past Maya and indigenous land uses, and have less grounding in demonstrated economic rewards. A few experiments with goat production exist, as do cattle projects, which continue to be supported by various state and federal programs, despite no proven successes as found in the Gulf Coast states of Mexico. All these activities pale in comparison to the amount of lands taken to swidden and chili. They bear watching in terms of their future impacts on land use and deforestation, however.

[7] Above we noted that maize yields, as reported in various surveys, had fallen from 1969 to present, yet the average amount of land taken to maize swidden (*milpa*) by households has remained more or less constant. Calculations made by 1997–8 SYPR-project survey indicate that deficit producer households constitute only 26% of the sample, making up that deficit by purchasing maize. We assume that these are either households that sell their labor to other farmers, as in the case of harvesting chili (Ch. 10), produce cash crops themselves (e.g. fruits, chili), or gain income off-the-farm. The survey revealed major distinctions in household material well-being, but no instances of loss of land, selling land (still illegal in most *ejidos*), or indicators that households were suffering from significant consumption shortfalls.

## 3. *The Evolving Swidden*

In some sense, the maize swidden or *milpa* of yesteryear remains very much alive in the southern Yucatán peninsular region. Slash-and-burn techniques are the dominant mode of cultivation, and all farmers grow maize on lands prepared this way. The average size of the consumption-oriented swidden has remained more or less constant during modern cultivation, despite the fact that some maize has been sold since at least the late 1960s. The significant shift underway in regard to cultivation is the incorporation of chili into the swidden cultivation cycle and the increasing movement towards intensive chili cultivation on 'mechanized' plots. The longer-term impacts of disking and chemical inputs as well as the expansion of the area dominated by invasive species (Ch. 6) bear watching.

The role of swidden cultivation in understanding deforestation and land change in the region is thus shaped by the conditions and forces that have begun to provide adjustments and alternatives to it. Over the near term, basic subsistence swidden will remain a principal component of understanding land dynamics. Incorporated with it, however, is chili cultivation and added to it is mechanized chili cultivation (Ch. 10). Taken as a package, these systems increase the total area cut from the forest and planted annually. Understanding how many and which households are engaged and are likely to engage in different packaging of these and other options enhances the potential to explain and project land dynamics in the region.

## REFERENCES

Acopa, D., and E. Boege (1998). 'The Maya Forest in Campeche, Mexico: Experiences in Forest Management at Calakmul', in R. B. Primack, D. Bray, H. A. Galletti, and I. Ponciano (eds.), *Timber, Tourists, and Temples: Conservation and Development in the Maya Forests of Belize, Guatemala, and Mexico*. Washington DC, Island Press, 81–97.

Bernstein, R. H., and R. W. Herdt (1977). 'Towards Understanding Milpa Agriculture: The Belize Case'. *Journal of Development Studies* 11: 373–92.

Chamberlain, R. S. (1948). *The Conquest and Colonization of Yucatan, 1517–1550*. Washington DC, Carnegie Institution of Washington Publication 582.

Cook, O. F. (1921). 'Milpa Agriculture: A Primitive Tropical System'. *Annual Report of the Smithsonian Institution, 1919*, 302–26.

Denevan, W. M. (1992). 'Stone vs Metal Axes: The Ambiguity of Shifting Cultivation in Prehistoric Amazonia'. *Journal of the Steward Anthropological Society* 20: 153–65.

Emerson, R. A. (1953). 'Preliminary Survey of the Milpa System of Maize Cultivation as Practiced by the Maya Indians of the Northern Part of the Yucatan Peninsula'. *Annals of the Missouri Botanical Gardens* 40: 51–62.

Emerson, R. A., and J. H. Kempton (1935). 'Agronomic Investigations in Yucatan'. *Yearbook of the Carnegie Institution of Washington Publications* 34: 138–42.

Ewel, J. J., C. Berish, B. Brown, N. Price, and R. Raich (1981). 'Slash and Burn Impacts on a Costa Rican Wet Forest Site'. *Ecology* 62: 816–29.

Garcia Espinoza, R., R. Quiroga Madrigal, and N. Granados Alvarez (1994). 'Agroecosystems for Sustained Corn Production in Hot Wet Regions of Mexico', in H. D. Thurston, M. Smith, G. Abawi, and S. Kearl (eds.), *Tapado—Slash/Mulch: How Farmers Use It and What Researchers Know About It*. Ithaca, NY, Cornell International Institute for Food, Agriculture, and Development, Cornell University, 61–74.

Gómez-Pompa, A. (1987). 'On Maya Silviculture'. *Mexican Studies/Estudios Mexicanos* 3: 1–17.

Haggar, J. (ed.) (1999). *Manual Agroforestal para la Península de Yucatán*. Chetumal, International Center for Resources, Agriculture, and Forestry (ICRAF)—Mexico.

Hernandez Xolocotzi, E., E. Bello Baltazar, S. Levy Tacher (compilers) (1994–5). *La Milpa en Yucatán: Un Sistema de Producción Agrícola Tradicional*. Vols. 1–2. Motecillo, Texcoco, Colegio Postgraduados.

Juo, A. S. R., and A. Manu (1996). 'Chemical Dynamics in Slash-and-Burn Agriculture'. *Agriculture, Ecosystem, and Environment* 58: 49–60.

Kauffman, J. B., R. L. Sanford, D. L. Cummings, I. H. Salcedo, and E. V. S. B. Sampaio (1993). 'Biomass and Nutrient Dynamics Associated with Slash Fires in Neotropical Dry Forests'. *Ecology* 74: 140–51.

Keys, E. (1999). 'Kalchikel Gardens: Women, Children, and Agricultural Education among the Maya of Highland Guatemala'. *CLAG Yearbook* 25: 89–100.

Killion, T. W. (ed.) (1992). *Gardens of Prehistory: The Archaeology of Settlement Agriculture in Greater Mesoamerica*. Tuscaloosa, University of Alabama Press.

Lundell, C. L. (1933). 'The Agriculture of the Maya'. *Southwest Review* 19: 65–77.

Nations, J. D., and R. B. Nigh (1980). 'The Evolutionary Potential of Lacandón Maya Sustained-Yield Tropical Agriculture'. *Journal of Anthropological Research* 36: 1–30.

Popenoe, H. (1957). 'The Influence of the Shifting Cultivation Cycle on Soil Properties in Central America', in *Proceedings of the Ninth Pacific Science Congress of the Pacific Science Association*. Bangkok, 72–7.

Primack, R. B., D. Bray, H. A. Galletti, and I. Ponciano (eds.) (1998). *Timber, Tourists, and Temples: Conservation and Development in the Maya Forests of*

*Belize, Guatemala, and Mexico*. Washington DC, Island Press.

Reina, R. (1967). 'Milpas and Milperos'. *American Anthropologist* 69: 1–20.

Reina, R. E., and R. M. Hill (1980). 'Lowland Maya Subsistence: Notes from Ethnohistory and Ethnography'. *American Antiquity* 45: 74.

Rojas Rabiela, T. (1985). 'La Technología Agrícola Mesoamericana en el Siglo XVI', in T. Rojas Rabiela and W. T. Sanders (eds.), *Historia de la Agricultural Epoca Prehispánica-Siglo XVI*. México DF, Instituto Nacional de Antroplogía e Historia, 129–232.

——(1991). 'La agricultura en la época prehispanica', in T. Rojas Rabiela (ed.), *La agricultura en tierras mexicanas desde sus orígenes hasta nuestros días*. México DF, Editorial Grijalbo S.A. de C.V., 15–138.

Roys, R. L. (1943). *The Indian Background of Colonial Yucatan*. Washington DC, Carnegie Institution of Washington.

Steggerda, M. (1941). *Maya Indians of Yucatan*. Washington DC, Carnegie Institution of Washington.

Turner, B. L., II (1974). 'Prehistoric Intensive Agriculture in the Mayan Lowlands'. *Science* 185: 118–24.

——(1983). *Once Beneath the Forest: Prehistoric Terracing in the Río Bec Region of the Maya Lowlands*. Boulder, Westview Press.

Turner, B. L., II, and S. B. Brush (eds.) (1987). *Comparative Farming Systems*. New York, The Guilford Press.

Turner, B. L., II, S. Cortina Villar, D. Foster, J. Geoghegan, E. Keys, P. Klepeis, D. Lawrence, P. Macario Mendoza, S. Manson, Y. Ogneva-Himmelberger, D. Perez-Salicrup, R. Roy Chowdhury, B. Savitsky, L. Schneider, B. Schmook, and C. Vance (2001). 'Deforestation and Agricultural Change in the Southern Yucatán Peninsular Region: Integrative Land Change for Global Change Studies'. *Forestry, Ecosystems, and Management* 54: 343–70.

Urrutia, V. M. (1967). 'Corn Production and Soil Fertility Changes under Shifting Cultivation in Uaxactun'. MA Thesis, University of Florida.

Vogeler, I. K. (1970). 'Frontier Settlements in South-Eastern Campeche: A Report for the 1970 National Geographic Society-Tulane University Archaeological Project at Becan, Campeche, Mexico'. Unpublished.

——(1974). 'The Cultural Ecological Setting of Southeastern Campeche', in R. E. W. Adams (compiler), *Preliminary Reports on Archaeological Investigations in the Rio Bec Area, Campeche, Mexico*. New Orleans, Middle American Research Institute, No. 31, Tulane University, 110–12.

——(1976). 'The Dependency Model Applied to a Mexican Tropical Frontier Region'. *Journal of Tropical Geography* 43: 63–8.

Wagner, H. O. (1968). 'Subsistence Potential and the Population Density of the Maya on the Yucatan Peninsula and Causes for the Decline in Population in the Fifteenth Century', in *Verhandlungen des XXXVIII International Amerikanisten Kongresses*. Munich, 179–96.

Watters, R. F. (1971). *Shifting Cultivation in Latin America.* Rome, Food and Agricultural Organization of the United Nations.

Wilhelmy, H. (1981). *Welt und Umwelt der Maya: Aufstieg und Untergang einer Hochkultur.* Munich, R. Piper.

Wilken, G. C. (1971). 'Food-Producing Systems Available to the Ancient Maya'. *American Antiquity* 36: 432–48.

Wilken, G. C. (1987). *Good Farmers: Traditional Agricultural Resource Management in Mexico and Central America.* Berkeley, University of California Press.

# 10

# Jalapeño Pepper Cultivation: Emergent Commercial Land Use

*Eric Keys*

## 1. *The Emergence of Chili Cultivation*

After maize, the commercial cultivation of jalapeño chili (*Capsicum annuum* L.; henceforth, jalapeño or chili) is the most important land use in the southern Yucatán peninsular region in terms of the numbers of farmers engaged in the practice and the total area devoted to it. Chili surpasses all other land uses in the way that it ties the region to the national economy and the way that it has altered household economies and land use. Open-backed semi-trailers and large pick-up trucks crowd the paved roads of the region between October and the end of January laden with large bags of jalapeños taken on the way to central gathering areas and, ultimately, the central wholesale market in Mexico City (Fig. 10.1). The flow of chili at this time is so large that even the casual observer would have little problem concurring with Uc Reyes's (1999: 4.24) claims that 'su cultivo [chile] constituye la más importante fuente de ingreso para los productores hortícolas de la entidad, además es generador de empleos para los trabajadores del campo.' (Its cultivation constitutes the most important source of income for horticultural producers, and generates jobs for field workers.)

The spread of the jalapeño describes how farmers adopt new crops, change their agriculture, and ultimately change their livelihood. It also teaches how land covers—forest, savanna, and farm field—can travel a new trajectory over a relatively short period of time, changing not only the way a region or landscape looks but the way it is viewed (Gudeman 1978; Watts and Goodman 1997).

**Fig. 10.1.** Coyote trucks loaded with jalapeños for transport to Mexico City.

These changes began in 1975 when colonizing farmers arrived from a traditional chili-growing zone—the Chiapas/Veracruz border—and introduced its commercial cultivation. After the first year they recruited national intermediaries or middlemen (in Spanish: *coyote*) to purchase and transport their product to the national market in Mexico City. Observing the success of the first jalapeño pioneers, other smallholders began to adopt the practice. Today, the south-central portion of the study region is called the *zonas chileras* (chile-growing zones); about 85 per cent of the smallholders there cultivate chili on as much as 7,500ha, usually on small plots, about 1.5ha on average (mean = 1.42 ha). Given the semi-subsistence or semi-commercial nature of the smallholder household (Ch. 11), however, chili cultivation is not a substitute for swidden and its maize. Chili cultivation involves the expansion of cultivated land as well as an intensification of cropping. Chili is monocropped, a practice that makes mechanized cultivation attractive and, coupled with the preference to cut older successional growth and mature forests for chili fields, signals potential stress on the region's forest lands and plans for forest conservation. Monocropping also fosters pests, especially in the tropics, and hence the purchase of pesticides, complete with apparent increased human health risks and unknown impacts on local ecosystems (Salabarrias A. 1999).

The overall impacts of chili cultivation of the southern Yucatán peninsular region are so large that this land use warrants special attention. It forms a new land-use strategy with implications to change the basic economic character of the agricultural frontier, especially within the overall plans to conserve forests as part of biospheres and archaeo-ecotourism programs (Chs. 1 and 3). Here the story of chili is examined as it takes place in the southern portion of the Municipio of Calakmul, the *zonas chileras* of the region. It is based largely on work undertaken in 1999–2000, involving lengthy open-ended interviews with farmers and middlemen in sixteen *ejidos* in the *zonas*, and a formal survey delivered to 160 households in five *ejidos*—Ricardo Payro Gene (aka Polo Norte), Manuel Castillo Brito, Cristobol Colón, Narciso Mendoza, and El Civalito.

## 2. *The Origins of Chile Cultivation in the Region*

Describing how chili spread throughout Calakmul and the region illustrates not only the local and extra-local dynamics that contribute to frontier agricultural change, but also the important role of smallholders as independent agents of change and how rapidly they can shift to new forms of production (Netting 1993). No commercial chili cultivation existed in the region in 1975, but by 1985 it was prevalent throughout. This rapid adoption reflects, in part, the paucity of viable commercial alternatives for land use, and the synchronicity of chili introduction with government and NGO programs pushing alternatives to subsistence cultivation: for example, cattle, agroforestry, and apiculture. Different migrants, of course, brought with them different experiences from their homelands. And, those from the Gulf Coast and southern states of Tabasco, Veracruz, and Chiapas brought chili cultivation (Long-Solis 1998).

According to farmer and *coyote* sources, the first market-oriented chili cultivation in the region began in 1975 in the *ejido* of Ricardo Payro Gene (Polo Norte), approximately 28km south of Xpujil. Local history claims that a man named Miguel Martinez initiated this activity in Polo Norte before moving on to other communities. Other sources, however, cite three farmers from the Juarez Municipio, Chiapas. Regardless of the number involved, all sources report that chili was initiated as an individual action involving a critical agent. In the early and mid-1970s no paved road between Ricardo Payro and Highway 186 (paved east–west road; Fig. 1.1) existed and most cultivators dried their chili and moved the produce to Xpujil on horseback. Dissatisfied with this transportation requirement, farmers from Ricardo Payro invited a buyer (*coyote*), known to them from their source communities, to establish business in the region. The buyer traveled to *ejidos* and bought dried chili for sale in the Mexico City market, *La Merced*, or in his home market of Puebla. He, in turn, began traveling to other parts of the region, informing smallholders that he would buy chili if the crop were available.

Chili seeds spread from community to community via different means. Some informants report that they attained seed by working for chili cultivators in Ricardo Payro, while others claim that early *coyotes* distributed seed with the understanding that future sales would be dedicated to them. Only one respondent out of 160 reported that he began farming because of the urging of a *coyote*. It is noteworthy that about 17 per cent of households interviewed claimed to have practiced chili cultivation in their former homelands, and thus potentially brought seed from source communities. Regardless, as farmers saw that those in Ricardo Payro and nearby *ejidos* were able to earn income from chili cultivation, they too began to adopt the crop (Table 10.1).

After cultivation began in Polo Norte, the next two communities to adopt chili were Cristobal Colón and Narciso Mendoza in 1978. Chile cultivation spread from *ejido* to *ejido* along the southern road, with generally a one-year lag between one community beginning and the next following. In 1997, the community furthest south on the road, Arroyo Negro

**Table 10.1.** Reasons given by farmers for engaging in commercial chili cultivation (n = 160)

| Reason for engaging in commercial chile cultivation | Number responding 'yes' | Per cent of sample responding 'yes' |
|---|---|---|
| Neighbor engaged in practice | 125 | 78.1 |
| Someone encouraged them | 67 | 41.9 |
| *Coyote* urged them or gave credit to them to engage | 1 | 0.6 |
| Government credit | 2 | 1.3 |
| Experiment | 77 | 48.1 |
| Household experience in this cultivation | 34 | 21.3 |
| Most of community engaged | 130 | 81.3 |
| Someone in family engaged | 108 | 67.5 |
| Other | 2 | 1.3 |

(abutting Petén, Guatemala) entered into chili cultivation. Figure 10.2 demonstrates that within 15 years of its introduction, over one-half of the households in the chili survey (southern Calakmul) had tried or were engaged in commercial chili cultivation. By 1998, 92 per cent of households in the chili survey had attempted cultivation at least once. Of these, 85 per cent reported that they continued to cultivate chili.

## 3. *The Practice and Rewards of Chili Cultivation*

Basically, chili production takes two forms—mechanized and non-mechanized (swidden)—although variations in both forms can be found in regard to capital inputs, hired labor, and basic attention given to the crop. Approximately 30 per cent of southern Calakmul chili farmers use mechanized land, while the remainder use swidden. Mechanized refers to the basic disking of the land in preparation of soils. Almost all subsequent practice through to the harvest is undertaken by hand. Mechanized farmers tend to cultivate more land in chili than non-mechanized farmers (1.85ha v. 1.17ha), and they report, on average, higher yields and higher net and gross profits, and undertake greater risks due to higher amounts of labor and capital invested in the cultivation process than swidden farmers.

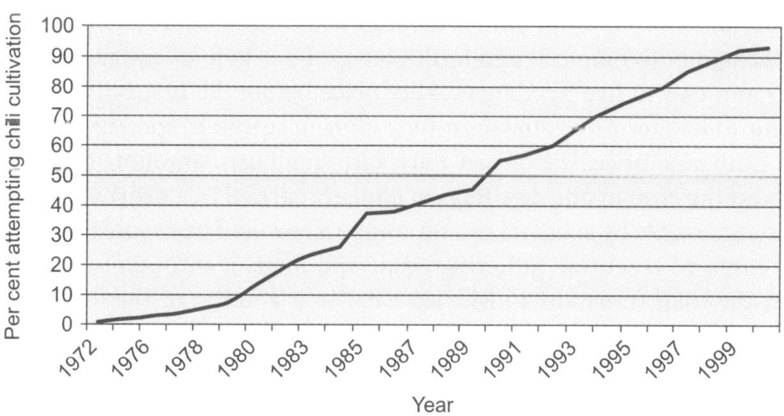


**Fig. 10.2.** Historical growth of chili cultivation, southern Calakmul, 1972–99. The figures are cumulative in that each year includes those that attempted chili cultivation previously.

Access to tractors and lands suitable for mechanized preparation seems to be related to the length of residency in the region, connectivity to government programs, and levels of wealth. Nearly all farmers interviewed expressed a desire to have their land mechanized. Government and non-governmental programs also encourage mechanization with the hopes that it will encourage 'sedentarization' and intensification of agriculture. It is not clear, however, that these processes are the end product for multiple reasons discussed below.

### 3.1. *Cropping Strategies*

Mechanized and swidden chili cultivation share similarities but also differ in some key ways. The description that follows indicates these distinctions, but combines the kinds of cultivation in their shared parts. Selection of chili plots and the cutting and burning of forest (if the plot has forest) is similar to those steps described for swidden (Ch. 9). The same land characteristics desired for maize production are sought for chili plots. If farmers have access to a tractor, the plot in question is sufficiently level, and the burn does not involve the first year of cultivation after the cut of older growth in successional or mature forest, the soil is turned with a disk. Following the first rains, the household or hired labor plants the field using a dibble or digging stick and last year's seed or seed purchased from farmers or other sources. Shortly thereafter, the farmer returns to fumigate the crop against weeds and insect pests (Fig. 10.3). The spraying of agricultural chemicals is universal among chili farmers in the region and represents a significant cost to them in both financial and health terms. Four to five fumigation sprayings and two to five weedings taking place before the first harvest, on or about 1 October. Given the sensitivity of the jalapeño to spoilage, affecting the farm gate price, the farmer must hire significant amounts of labor to harvest the crop in one day. Bags of chili are carried to a nearby collection point accessible by small truck and sold to a *coyote*. The *coyote* then moves the cargo to a central gathering point and loads it into semi-trailers for over-the-road transport to Mexico City. In a good year, the farmer may repeat fumigation and harvesting up to five times until the last cutting in late December or early January. When prices for green (fresh) chili are low, he may choose to smoke dry his last crop and sell it as *chipotle* (smoked and dried chili).

**Fig. 10.3.** Farmer spraying a chili field in southern Calakmul.

## 3.2. *Costs, Gains, and Favored Lands*

The costs of chili production in the region are high, but so too can be the rewards (Table 10.2). For swidden and mechanized chili, respectively, the costs are about N$5,000 and N$9,500, and the profit approaching N$5,000 and N$18,000. The discrepancies follow largely from the difference in yields, 12 to 15mt for mechanized chili and 6 to 10mt for swidden chili. It would appear that the disking of land and, perhaps, the higher levels of inputs and care given to a mechanized investment contribute to yield differences.

Contrary to findings from the general survey (Ch. 8), chili farmers in southern Calakmul claim to prefer to take chili plots from secondary growth forest more so than from old growth (49.3% v. 18.4%). And, as Table 10.3 shows, this preference is reflected in what lands are actually turned to chili; about 53 per cent of the chili plots in 1999 was taken from previously used lands and only 12 per cent from old growth. Farmers offer various reasons for choosing what parcels to plant in chili. Those claiming

**Table 10.2.** Inputs, outputs, and profit for 'average' year estimated by chili farmers

| Expense category | Mechanized farmer | Non-mechanized farmer |
|---|---|---|
| Labor expenses | N$4,453.00 (s.e. 682.07) | N$2,534.48 (s.e. 318.78) |
| Material costs | N$3,699.48 (s.e. 666.79) | N$1,885.01 (s.e. 218.80) |
| Transportation costs | N$1,316.26 (s.e. 230.75) | N$574.69 (s.e. 63.70) |
| Total costs | N$9,468.74 (s.e. 1089.55) | N$4,994.18 (s.e. 464.70) |
| Total income | N$26,706.89 (s.e. 5091.08) | N$9,352.19 (s.e. 894.12) |
| Price/kilogram | N$2.12 (s.e. 0.038) | N$1.94 (s.e. 0.065) |
| Production per farmer | 12.1 mt (s.e. 2.15) | 4.49 (s.e. 0.42) |
| After cost profit | N$17,956.41 (s.e. 4,621.95) | N$4,680.81 (s.e. 694.83) |

*Note*: N$ = new peso; there are about N$10 to $US1.

**Table 10.3.** Prior year land cover/land use of chili plots, 1999

| Land-use/land-cover | Swidden farmers (n = 84) | | Mechanized farmers (n = 49) | | All chili farmers (n = 133) | |
|---|---|---|---|---|---|---|
| | % | No. | % | No. | % | No. |
| Old growth | 12.6 | 11 | 10 | 5 | 11.8 | 16 |
| Secondary growth | 51.7 | 45 | 54 | 27 | 52.9 | 72 |
| Chili field | 21.8 | 19 | 14 | 7 | 19.1 | 26 |
| Maize field | 8.0 | 7 | 8 | 4 | 8.1 | 11 |
| Other agriculture | 2.3 | 2 | 12 | 6 | 5.9 | 8 |

*As noted in the text, virtually no old growth is taken directly into mechanized chili production. Farmers with mechanized chili, however, may also cultivate some chili by way of swidden absent disking.

to prefer older growth forest claim these parcels have less weed and pest problems. Those preferring successional growth, however, point out that these lands dry more quickly, and are less difficult to clear because of fewer roots and stumps. Nearly all farmers (85%), however, concur that mechanized land is optimal for cultivation, an opinion borne out by production figures. Many farmers express a desire to mechanize their plots through government programs or access to private tractors, and most (70%) felt that mechanization was best attained through political connections (*palancas*)—a proxy to social capital.

To mechanize older successional or forest growth, farmers generally clear a piece of land and cultivate it as a maize swidden. The following year, after the cut stumps have had time to soften and rot, a tractor owner is contracted to disk (turn) the soil (Fig. 10.4). Thus prepared, the land is planted in chili. While mechanization in the region is relatively recent—no farmer reported having mechanized land before 1993—observed fallow or reports

**Fig. 10.4.** Tractor turning up *sascab* (decayed limestone bedrock) soils underlying thin rendzina topsoils in Ricardo Payro Gene, Calakmul.

of fallow on these plots is nil. Only two farmers with mechanized parcels reported that they would let such land rest in 2000. Thus, it seems that mechanization leads to permanent or long-term parcel cropping where forest regrowth is unlikely. Thirty-one per cent of the chili farmers in southern Calakmul had mechanized land. Government aid programs, good prices for chili, and increasing numbers of tractors make it likely that mechanization will increase in the coming years. The land-cover effects of this change from swidden to more permanent cultivation are not clear.

## 4. *Instability in Production and Sales*

Although jalapeños are the dominant cash crop in the Municipio de Calakmul and the region in general, the production and sale of the peppers varies significantly from year to year. Linked to the national market through *coyotes*, chili farmers complain of inconsistent prices for their production. For example, the average price per kilo in 1998 was N$1.43, while in 1999 it was N$2.09. These price fluctuations are linked to the state of chili production elsewhere in Mexico. States in the north and Bajio region (colonial heartland) of Mexico have historically dominated chili production in the country and have long-standing ties to food-processing companies and intermediaries, and greater access to agricultural technologies, such as mechanization, refrigeration, and chemical applications, which tend to produce a superior jalapeño than that produced in the southern Yucatán peninsular region. All else equal, these areas constitute the primary and secondary suppliers; the Calakmul and larger region is best seen as a tertiary supply region whose crop increases in value when demand is not met by the other areas, due to bad weather (e.g. killing frost), pest outbreaks, and so forth. The production in the primary regions of the country was high in 1998, and Calakmul chile prices were low. In 1999, frosts and pest attacks in the north of Mexico lowered the total national supply, and the prices were relatively high in Calakmul.

Likely due to the insecurity of the price structure, chili farmers cultivate more land than other smallholders in the region, perhaps in an effort to expand their agricultural portfolio by increasing possible income from lower value but more securely priced crops (i.e. squash and beans). Figure 10.5 illustrates this strategy by way of a comparison of the total amount of land cultivated by non-chili, chili-swidden, and mechanized chili-swidden households. Note the increase in total area devoted to the range of crops moving from the non-chili to mechanized households. Chili farmers cultivate more land than non-chili farmers, and mechanized farmers more than non-mechanized farmers.

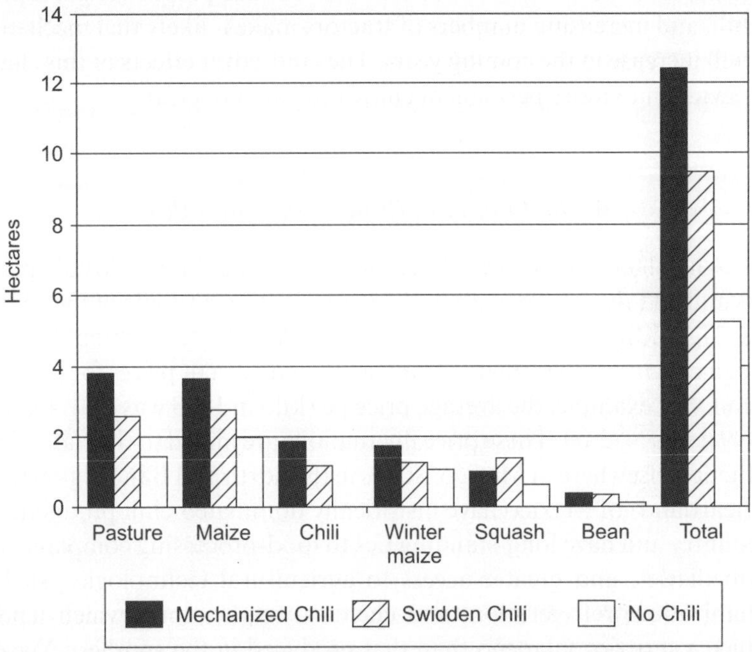


**Fig. 10.5.** Land devoted to different crops in southern Calakmul.

Because chili farmers cultivate more land, they might be expected to have less land in old growth forest. Figure 10.6 demonstrates the different land covers reported by farmers in the 1999 chili survey. Regrowth is any land that had been previously cultivated but is in some state of succession. Old growth is any land that the farmer believed never to have been cultivated. Long-term agriculture describes lands that are either mechanized or pasture, and are unlikely to revert to forest in the near future. And, short-term agriculture constitutes lands that are currently part of a swidden crop–fallow cycle and will be allowed to regrow in the near future. Mechanized chili farmers report holding the least amount of old growth land on their plots (Fig. 10.4) and have the most land in uses not likely to convert back to forest in the near future. They also reported the most amount of land in regrowth and short-term agriculture.

The reasons that chili farmers tend to use more land are varied but interrelated. Because they earn more money than most other farmers, they are more likely to hire labor from outside their household to engage in agricultural labor, such as weeding, fumigation, clearing, and harvesting. This increased wealth enables chili farmers to expand their agricultural portfolios. Expansion of the portfolios allows the farmer the opportunity to hedge against price fluctuations for their products. Increased wealth also affords them better connections to various government programs that

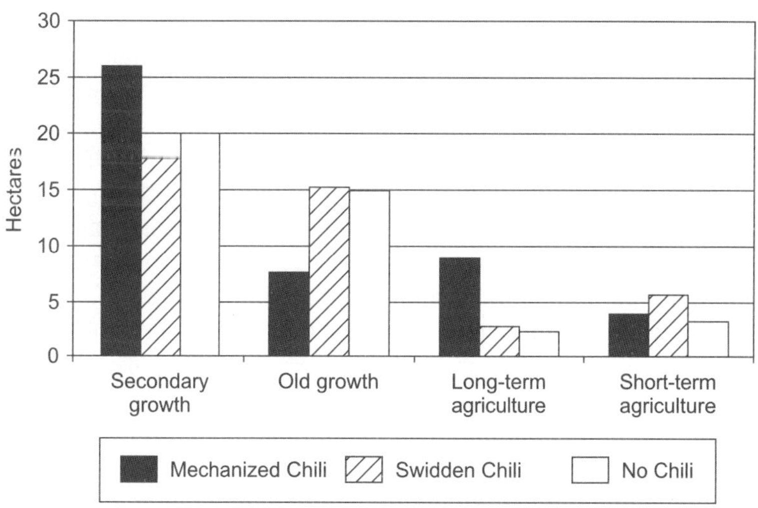


**Fig. 10.6.** Land covers reported by different types of farmers.

promote access to subsidies of various kinds and to the latest 'experiments' in cropping.

## 5. *Summary: Changing Economies and Ecologies*

The arrival of migrants from Tabasco, Veracruz, and Chiapas set in motion a chain of events that significantly shape the land-use and land-cover dynamics of the southern Yucatán peninsular region today. Some these migrants arrived in the region with experience in a special kind of land use not present before—commercial jalapeño production. Their arrival overlapped with the decline in large-scale projects, and the need of the *ejido* household to find alternative ways to improve material lifestyles beyond what could be gained from swidden and the sale of maize. While NGOs offered experimental alternatives, none were lucrative enough to engage so many households as was chili. Absent initial programmatic subsidies and support, a few smallholders took it upon themselves to use their homeland connections to *coyotes* to entice them to come to the *municipio*. Thus began the chili story in the region—the last part of a chain that supplies the national market with fresh and dried jalapeños. As the last link, it is most precarious economically and appears to give rise to year-to-year booms and busts. The fact that this production has swept through the region like wildfire, however, indicates its perceived comparative advantages over alternative land uses.

The future of chili in the region and its environmental implications, however, are not so obvious. The boom-bust nature of the farm gate prices suggest that a profitable but more stable substitute, should it be found, might readily shift commercial practices. Beyond this, the novelty of 'intensive' and 'capital' input cultivation of this kind opens many questions about its ecological and economic sustainability. Pests and soil impacts appear to be growing, requiring more investment in land inputs. And, despite intensification, the more commercial and capital the household becomes, the more of its lands are taken into cultivation and the less remain in older growth forest. Chili cultivation, therefore, has triggered a new set of economic and ecological relationships that must be taken into account in understanding and projecting deforestation and land change in the southern Yucatán peninsular region.

The emergence of commercial chili cultivation in this frontier region is critical for understanding current and future land-use and land-cover change, not only owing to its direct effects on the physical environment, but its indirect effects registered through household economic and lifestyle changes. To wit, within the last twenty-five years chili cultivation has:

- spread from one source community and a few hectares to every community and more than 50 per cent of the smallholders in the region; 85 per cent of farmers in the southeastern part of the Municipio de Calakmul contributed to the 7,500ha of chili cultivated in the *zonas chileras* in 2000;
- changed the SYPR from a region dominated by subsistence agriculture and the sale of surplus maize, to one involved with and affected by national produce markets, agrochemical industries, and market intermediaries;
- become increasingly intensive and non-swidden in terms of modern inputs and labor requirements and has increased the annual area cultivated among the more successful chili farmers;
- helped to foster insect and microbial pests, compromising human and environmental health of the region;
- created a growing class of short-distance, itinerant farm laborers; and
- led to significant increases in wealth and status among some smallholders.

## REFERENCES

Gudeman, S. (1978). *The Demise of a Rural Economy*. Boston, Routledge & Kegan Paul.

Long-Solis, J. (1998). *Capsicum y cultura: la historia del chilli*. Mexico City, Fondo de cultura económica.

Netting, R. M. (1993). *Smallholders, Householders: Farm Families and the Ecology of Intensive, Sustainable Development*. Stanford, Calif., Stanford University Press.

Salabarrias A., R. (1999). 'Retos y oportunidades del Sistema Agro-industrial del chile jalapeño en el estado de Quintana Roo'. MA Thesis, Universidad de Cholula.

Uc Reyes, C. (1999). 'Chile jalapeño', in J. Haggar (ed.), *Manual Agroforestal para la Península de Yucatán*. Chetumal, Quintana Roo, International Center for Research on Agroforestry, 4:24–27.

Watts, M. J., and D. Goodman (eds.) (1997). *Globalising Food: Agrarian Questions and Global Restructuring*. Boston, Routledge.

# 11

# The Semi-Market and Semi-Subsistence Household: The Evidence and Test of Smallholder Behavior

*Colin Vance*

Understanding household farming behavior among smallholders is an essential element of land-change studies inasmuch as a considerable portion of the world is dominated by land-users of this kind. Smallholders (peasants in some literature) are especially important within the tropical forests of Mexico, and the southern Yucatán peninsular region is no exception. This region, as elsewhere in the tropics, is characterized by underdeveloped markets and the consequent partial engagement of frontier farmers as market participants. Sparse exchange opportunities resulting from remoteness, low population density, and poorly developed infrastructure constrain these farmers to maintain a strong focus on consumption production, especially in terms of staple foods. Indeed, until the late 1960s, households in the region were totally subsistence-based and had virtually no experience with the agricultural market. Today, smallholder farmers retain consumption production, though a growing proportion also produce crops for sale. While this dual position in the market and in subsistence is an increasingly prevalent feature of smallholder farmers throughout the developing world, studies of deforestation commonly ascribe to them a wholly commercial orientation by employing profit-maximizing theoretical structures as a basis for econometrically modeling their land-use decisions (e.g. Chomitz and Gray 1996; Cropper, Griffiths, and Mani 1999; Cropper, Puri, and Griffiths

The views expressed in this chapter are those of the author and do not necessarily represent the official views of the US Environmental Protection Agency.

2001; Nelson, Harris, and Stone 2001; Nelson and Hellerstein 1997; Panayotou and Sungsuwan 1994; Pfaff 1999). In essence, the assertion of profit-maximization rests on the assumption that agents are fully engaged in markets, from which it follows that production, being strictly a function of farm technology and exogenously given input and output prices, is entirely independent of consumption and labor supply (Barnum and Squire 1979). This chapter explores the implications of relaxing the perfect-markets assumption for the modeling of semi-subsistence and commercial land-use decisions. By introducing variables measuring the consumption side of the colonist household, evidence is presented to suggest that, consistent with mixed or hybrid production themes (e.g. Singh, Squire, and Strauss 1986; Turner and Brush 1987), farmers operating in a context of thin product and/or labor markets do not exhibit behavior corresponding to that of a commercially oriented profit-maximizing farm.

The agricultural household model employed in economics provides the theoretical framework for identifying and interpreting this empirical finding. In recent years, this model has established itself as an important analytical tool for investigating the consequences of different agricultural policy measures in less-developed countries (Saha 1994), but there are few examples of its application to the question of land use. Given the model's ability to explain the sometimes sluggish—and even perverse—response of smallholders to changes in relative prices, it can provide useful insights into the implementation of appropriate price-based policy responses to deforestation.

## 1. *Smallholder Cultivation in the Region*

Semi-subsistent agricultural households simultaneously represent two distinct forms of organizing production: the profit-maximizing family farm and the subsistence-oriented household. This dichotomy implies that the household's allocation of resources may not be consistently in accord with the tenets of marginal value product equalization as measured by market prices; changes in relative prices do not solely determine the adjustment of inputs and outputs in production. In the face of few opportunities for market exchange, smallholder farmers[1] maintain a portfolio of commercial

[1] 'Smallholder' has taken on a meaning in the literature beyond the immediate indicator of 'small holdings'. It implies a peasant mode of agriculture in which cultivation is geared for both internal consumption by the producer and variously for the market. Capital inputs are usually low, and minimal risk is taken in regard to the consumption crop. 'Smallholder' is

and subsistence land uses that are selected to stabilize their fulfillment of basic food needs and cash income.

The dominance of *milpa* (basically for subsistence; see Ch. 9) and the increasing expansion of jalapeño chili cultivation (for market; see Ch. 10) reflect a combination of the household's capital endowment, its ability and willingness to sustain risk, and its reliance on markets. The jalapeño market is dominated by private traders with substantial market power in price setting. Nevertheless, few other market crops appear to be as profitable currently as is the jalapeño, recognizing the significant year-to-year variability in their price (Ch. 10). Farmers also occasionally sell surpluses of maize beyond that consumed by the family, but under very different market conditions than those that govern chili. The government sets a minimum guarantee price at which maize can be purchased or sold to CONASUPO (Compañía Nacional de Subsistencias Populares), a parastatal marketing agency. With only two CONASUPO outposts servicing the region, however, prohibitively high transport costs combined with a relatively low unit price for maize (1 peso/kilo) can easily defeat attempts to produce the crop commercially.[2] Opportunities to market maize locally are also few, particularly given highly correlated fluctuations in output across households within a locale and insufficient economic specialization to support a robust village market. Despite these marketing constraints, roughly 41 per cent of households reported to sell some portion of their maize output in 1997.

## 2. *Theoretical Antecedents*

Two broad classes of agricultural household models—separable and non-separable—have been developed to explain the differing production strategies of farm households depending on market opportunities (Singh, Squire, and Strauss 1986; Vance 2000). The separable model applies when all input, output, and insurance markets fully function. These conditions make it optimal for the utility-maximizing farmer to apply a production

employed owing to the negative connotations applied by some to the term 'peasant'. It is noteworthy that the 'smallholders' in question here may have usufruct rights to 80ha, although no more than 5–10ha are cultivated at any one time.

[2] During the year of the survey in 1997, the exchange rate was approximately N$10 pesos to the dollar ($US1).

logic that is independent of consumption considerations. The intuition behind this result rests on the assumption of market-purchased commodities in the household's utility function. As the acquisition of such commodities requires income, which itself is a function of exogenously given prices for output and labor, the household will attempt to maximize its net income just as would a profit-maximizing firm. That is, the household will supply its product to the market until the point at which the marginal costs of production equal the price it receives, and it will demand labor so as to equate the marginal revenue product to the wage. Whether the household is a net buyer or seller of the good or factor has no effect on this result, since, as a price taker, its valuation for both is determined exogenously. Models with this feature are sometimes referred to as *recursive* because there is a one-way effect of the optimizing production decision on consumption via the level of income obtained. Statistically, this implies that an econometric specification based on profit-maximization is sufficient to model the household's land allocation decisions.

The non-separable model relaxes the assumption of complete markets and thereby reconciles the theoretical assertion of profit-maximization with the sluggish or perverse price response that is often said to characterize smallholder farm systems (Medani 1975; Medellin, Apedaile, and Pachico 1994). It does so by explicitly incorporating various constraints on market participation that result in the violation of the separable property. Sources of such constraints include seasonal limitations on the amount of labor for hire, prohibitive costs of transporting farm output to the market, and excessive price variability coupled with aversion to risk. When the household chooses not to participate in the market, its production and consumption decisions cannot be analytically separated: production decisions will incorporate subsistence and/or leisure considerations (de Janvry, Fafchamps, and Sadoulet 1991). Statistically, this implies that a profit-maximizing empirical specification comprising only prices and agricultural technology will yield biased estimates. Correcting this bias necessitates incorporating variables that also measure the consumption side of household decision-making.

In what follows, a non-separable theoretical model is developed for analyzing the choice between competing land uses in a multi-crop economy characterized by the presence of underdeveloped markets. The analysis is motivated by a model developed in Ahn, Singh, and Squire (1981) to describe the land-use choices of a semi-commercial agricultural household producing cash crops and wage income. DeShazo and DeShazo (1995) apply a virtually identical framework to investigate the economics of deforestation in an agricultural frontier. While both these studies invoke the assumption of a complete set of markets for all the variables that the individual household can choose, the model presented here is structured to accommodate various constraints on market participation. It does so by explicitly distinguishing between the choice set of semi-subsistence and commercial land uses and by introducing marketing constraints on the output of the semi-subsistent crop. Since the solution of the model essentially reduces to the separable result derived by Ahn, Singh, and Squire given that the market environment constraints are not binding, it provides a more general representation of household choices with respect to land use.

The model is intended to describe the behavior of a family farm that employs land and labor to produce two goods: a cash crop, the entire output of which is sold on the market, and an agricultural staple, which the household consumes but which can also be bought or sold depending on whether the consumption target is met. In order to focus on the important features, various complexities are omitted from the framework. For example, the household is assumed to be a price taker on the labor market and can either hire labor or sell its own labor off the farm, thereby augmenting income from the sale of cash crops and surplus agricultural staples. Moreover, hired labor and family labor are viewed as perfect substitutes for performing on-farm tasks. Production, a twice-differentiable concave function of land and labor, is assumed to be riskless. The planning horizon of the model is one agricultural cycle over which the household maximizes a utility function:

$$\text{Max}U = U(M, X_a, X_l) \tag{1}$$

subject to

$$A_{ic} + A_{ia} \leq \hat{A}_i \quad \text{for } i = 1, \ldots k \tag{2}$$

$$X_l + F = T \tag{3}$$

$$pM + p_a X_a = p_c \sum_{i=1}^{k} Q_{ic}(A_{ic}, L_{ic}) + p_a \sum_{i=1}^{k} Q_{ia}(A_{ia}, L_{ia}) -$$

$$w\Big(\sum_{i=1}^{k} L_{ic} + \sum_{i=1}^{k} L_{ia} - F\Big) - \sum_{i=1}^{k} d_{ia} A_{ia} - \sum_{i=1}^{k} d_{ic} A_{ic} \tag{4}$$

$$\sum_{i=1}^{k} Q_{ia} + E_a \geq X_a \tag{5}$$

where

- $M$ is a market purchased manufactured good
- $X_a$ is an agricultural staple consumed by the household, but which also may be purchased or sold depending on consumption relative to production
- $X_l$ is family time allocated to leisure
- $A_{ic}$ is commercial land-area on land type *i*
- $A_{ia}$ is agricultural staple land-area on land type *i*
- $\hat{A}_i$ is the maximum available quantity of the $i^{th}$ type of land
- $F$ is family time allocated to labor (on- and off-farm)
- $T$ is the family's total time endowment
- $p$ is the price of the market purchased manufactured good
- $p_a$ is the price of the agricultural staple
- $p_c$ is the price of the commercial crop
- $Q_{ic}$ is the commercial crop output *c* on land type *i*
- $L_c$ is labor (family and hired) input for commercial crop *c* on land type *i*
- $Q_{ia}$ is the agricultural staple output *a* on land type *i*
- $L_a$ is labor (family and hired) input for staple *a* on land type *i*
- $w$ is the market wage rate
- $d_{ia}$ is the clearance cost for agricultural staple *a* and land type *i*
- $d_{ic}$ is the clearance cost for commercial crop *c* and land type *i*
- $E_a$ is the initial endowment of the agricultural staple

Thus, the family farm is assumed to maximize a utility function comprised of arguments for a market-purchased good (M), an agricultural staple ($X_a$), and leisure ($X_l$). The first constraint, given by equation (2), highlights the model's exclusion of a market for land, requiring the farm to allocate land of a specific type (or quality) across commercial ($A_{ic}$) and semi-subsistence ($A_{ia}$) uses without exceeding exogenously given endowments ($\hat{A}_i$). The family also faces a time constraint, given by equation (3), preventing it from allocating more time to leisure ($X_l$) plus on- and off-farm employment (F) than the total time available (T).

Equation (4) is a standard budget constraint equating expenditure on the sum of purchases of manufactured goods and agricultural staples with net revenue generated by the sale of the commercial crop, agricultural staple, and family labor both on and off the farm. The absence of a term capturing an explicit rental rate for land use in this expression reflects the lack of a property rights regime that would support a market for land, a characteristic common to frontier areas of developing countries.[3] The cost of land acquisition for use is instead captured by the term d, representing the annualized clearance costs incurred in preparing the land for cultivation (DeShazo and DeShazo 1995; Reis and Guzmán 1994).

Equation (5) incorporates the assumption of inefficient commercialization channels on trade in agricultural staples, requiring that the household's supply $(\sum_{i=1}^{k} Q_{ia})$ plus exogenously given initial endowments ($E_a$) must be greater than or equal to its demand ($X_a$). The constraint, written as such, essentially renders the staples non-tradable goods, so that $\sum_{i=1}^{k} Q_{ia} + E_a < X_a$ are perceived as irrelevant alternatives and therefore reside outside of the household's effective choice set.[4] Following Thornton and Eakin (1992), the constraint could be written in a more general form as $\sum_{i=1}^{k} Q_{ia} + E_a - X_a \geq G_a$, where $G_a$ is a market restriction parameter in order to accommodate situations where there may be some degree of market participation. In the context of the present model, the theoretical results obtained by introducing this complication remain the same.

After solving for F in equation (3) and substituting into equation (4), the Langrangian for this constrained optimization problem is expressed as:

$$\begin{aligned}
\text{Max}\,\zeta = {} & U(M, X_a, X_l) - l[pM + p_a X_a - p_c \sum_{i=1}^{k} Q_{ic}(A_{ic}, L_{ic}) - \\
& - p_a \sum_{i=1}^{k} Q_{ia}(A_{ia}, L_{ia}) + w(\sum_{i=1}^{k} L_{ic} + \sum_{i=1}^{k} L_{ia} - T + X_l) + \sum_{i=1}^{k} d_{ia} A_{ia} + \sum_{i=1}^{k} d_{ic} A_{ic}] + \\
& \delta[(\sum_{i=1}^{k} Q_{ia}(A_{ia}, L_a)) + E_a - X_a] + \sum_{i=1}^{k} \phi_i (\hat{A}_i - A_{ic} - A_{ia}) \qquad (6)
\end{aligned}$$

[3] Until changes to Article 27 of the Mexican Constitution in 1994, *ejidatarios* were prevented from selling or renting land. Today such transactions enjoy legal sanction, but there remains little evidence of a formal land market within the *ejido* sector of the southern Yucatán peninsular region (Ch. 8).

[4] An alternative approach to incorporating market imperfections into the model is by introducing a term for transaction costs, t, that is added and subtracted to $p_a$ on the left and right hand sides of Equation 4, respectively (e.g. Omamo 1998; de Janvry, Sadoulet, and Gordillo de Anda 1995). This approach gives rise to a price band in which ($p_a + t$) is the decision price faced by net buyers while ($p_a - t$) is the decision price faced by net sellers.

where $\lambda$, $\delta$, and $\phi_i$ are the Lagrange multipliers associated with the constraints on income, the agricultural staple, and land of type *i*, respectively.

The first-order Kuhn-Tucker conditions are:

$$\partial \mathcal{L}/\partial M = \partial U/\partial M - \lambda p = 0 \tag{7}$$

$$\partial \mathcal{L}/\partial X_a = \partial U/\partial X_a - \lambda p_a - \delta = 0 \tag{8}$$

$$\partial \mathcal{L}/\partial X_1 = \partial U/\partial X_1 - \lambda w = 0 \tag{9}$$

$$\partial \mathcal{L}/\partial \lambda = pM + p_a X_a - p_c \sum_{i=1}^{k} Q_{ic}(A_{ic}, L_{ic}) - p_a \sum_{i=1}^{k} Q_{ia}(A_{ia}, L_{ia})$$

$$+ w(\sum_{i=1}^{k} L_{ic} + \sum_{i=1}^{k} L_{ia} - T + X_1) + \sum_{i=1}^{k} d_{ia} A_{ia} + \sum_{i=1}^{k} d_{ic} A_{ic} = 0 \tag{10}$$

$$\partial \mathcal{L}/\partial \delta = \sum_{i=1}^{k} Q_{ia}(A_{ia}, L_a)) + E_a - X_a \geq 0 \text{ and} \tag{11a}$$

$$\delta[\sum_{i=1}^{k} Q_{ia}(A_{ia}, L_a)) + E_a - X_a] = 0 \tag{11b}$$

$$\partial \mathcal{L}/\partial L_{ic} = \lambda p_c(\partial Q_{ic}/\partial L_{ic}) - \lambda w = 0 \tag{12}$$

$$\partial \mathcal{L}/\partial A_{ic} = \lambda p_c(\partial Q_{ic}/\partial A_{ic}) - \lambda d_{ic} - \phi_i \leq 0 \text{ and} \tag{13a}$$

$$A_{ic}[\lambda p_c(\partial Q_{ic}/\partial A_{ic}) - \lambda d_{ic} - \phi_i] = 0 \tag{13b}$$

$$\partial \mathcal{L}/\partial L_{ia} = \lambda p_{ia}(\partial Q_{ia}/\partial L_{ia}) - \lambda w + \delta(\partial Q_{ia}/\partial L_{ia}) = 0 \tag{14}$$

$$\partial \mathcal{L}/\partial A_{ia} = \lambda p_{ia}(\partial Q_{ia}/\partial A_{ia}) - \lambda d_{ia} + \delta(\partial Q_{ia}/\partial A_{ia}) - \phi_i \leq 0 \text{ and} \tag{15a}$$

$$A_{ia}[\lambda p_a(\partial Q_{ia}/\partial A_{ia}) - \lambda d_{ia} + \delta(\partial Q_{ia}/\partial A_{ia}) - \phi_i] = 0 \tag{15b}$$

$$\partial \mathcal{L}/\partial \phi_i = \hat{A}_i - A_{ic} - A_{ia} \geq 0 \text{ and} \tag{16a}$$

$$\phi_i(\hat{A}_i - A_{ic} - A_{ia}) = 0 \tag{16b}$$

Equations (7), (9), and (10) represent standard first-order conditions of consumer behavior. The ratios of marginal utilities from the manufactured good and leisure to the shadow value of income are seen to equal the market given price and wage rate, respectively. These conditions arise from the assumption of existent markets for the two goods in question. A different condition is seen, however, to apply to the agricultural staple, for which there is a marketing constraint. Rearranging equation (8) as $\partial U/\partial X_a = \lambda(p_a + \delta/\lambda)$ suggests that the utility maximizing price for the staple diverges from the observable market price by an amount equal to $\delta/\lambda$. Thornton and Eakin (1992: 1020) formalize a corresponding finding with the construct of virtual prices, which they define as 'the set of prices that would induce the [owner-operated farm] to make the same choices in a perfect markets environment as it actually makes when there are one or more imperfect markets'. Viewed alternatively, the shadow price of the market constraint, $\delta$, can be regarded as the price that the household would be willing to pay to have the corresponding constraint relaxed by one unit. Moreover, because $\delta$ is determined by the solution to the constrained utility-maximization problem, the virtual price emerges as an endogenous variable. Thus, the household's perception of the scarcity of the good is not solely determined by the market-given price but also by a subjective valuation represented by $\delta$.

Moving to equations (12) through (16) on the production side of the model, it is first noted that if equation (16) is binding for the $i^{th}$ type of land, then $\phi_i \geq 0$ represents the shadow price for that type of land. Equation (13) indicates that if for commercial crop, $\lambda p_{ic}\partial Q_{ic}/\partial A_{ic} - \lambda d_{ic} < \phi_i$, then that crop will not be grown on land type *i*. If, however, $\lambda p_{ic}\partial Q_{ic}/\partial A_{ic} - \lambda d_{ic} = \phi_i$, then the $i^{th}$ type of land will be completely allocated to the commercial crop. The model thus produces the standard result of complete specialization by land type (Ahn, Singh, and Squire 1981), thereby accommodating cases in which farmers allocate land to a multitude of uses. Equation (15) is similar to Equation (13) except that here the comparison with $\phi_i$ involves an extra term, $[\delta\partial Q_{ia}/\partial A_{ia}]$, representing the shadow price of the constraint times the marginal product of type *i* land for the staple crop. This result underscores the violation of separability since it indicates that the production side of the model cannot be solved independently of the value of $\lambda$ (the marginal utility of income).

To further illustrate this point, consider the outcome of the model with and without the market constraint. Given the absence of binding market constraints, the determination of crop choice would involve a comparison of $\lambda p_{im}\partial Q_{im}/\partial A_{im} - \lambda d_{im}$ and $\phi_i$ for each crop *m* (be it a staple or commercial crop). In this case, the comparison is not affected by the value $\lambda$ since $\lambda$ is a scalar which can be canceled out. Finding the household's utility-maximizing choices would thus involve decomposing the first-order conditions into two sets for production and consumption and then solving recursively. Upon including the possibility of a binding market constraint on the staple crop, however, the additional term in the comparison,

[$\delta\partial Q_{ia}/\partial A_{ia}$], precludes cancellation of the term $\lambda$. The model may not, therefore, be treated as a recursive one, in which production decisions are independently determined by profit maximization with consumption subsequently determined given the level of profit achieved. Rather, these two choices are determined simultaneously. The empirical consequence of this feature of the model is that demographic variables, to the extent that they measure the consumption requirements of the household, will be seen to be significant in explaining the household's production behavior.

### 3. *The Applicability of the Non-Separable Model to the Deforestation Question*

Two related considerations are of primary importance in assessing the extent to which non-separable models can foster understanding of deforestation processes in frontier regions of the developing world. The first is the nature of political economic relations believed to characterize the region under study, and the second is the time frame of the analysis.

Regional studies of the deforestation issue commonly assume that the penetration of capitalist production is an inexorable dynamic that drives the frontier's expansion. This assumption serves as a justification for the application of the separable model to study deforestation processes, particularly over the long run. Angelsen (1995: 1716), for example, argues that a separable or profit-maximizing framework is more relevant for the study of long-term effects because 'traditional agrarian societies are increasingly being integrated in the larger market economy, and commercialization of village life creates increased cash income needs'. DeShazo and DeShazo (1995) and Ozório de Almeida (1992) concur with this view. While acknowledging an initial subsistence orientation of frontier colonists, they argue that the household will eventually shift its production focus toward maximizing profits pending sufficient familiarity with local ecological and economic conditions and the progression of 'capital penetration into the countryside' (Ozório de Almeida 1992: 138). Hyde, Amacher, and Magrath (1996: 223), addressing the policy implications that emerge from the assumption of existent markets, note that 'deforestation will induce price increases and investments in forestry well before deforestation attains its physical limits', thus justifying a focus on prices and responses to resource scarcity.

While the majority of frontiers have at least some sectors that are linked to the larger market economy, it is important to identify whether the continued encroachment of the market is, as often assumed, a unidirectional process that ultimately recruits all local economic systems into a capitalist ordering. Agricultural frontiers in the tropical world are often characterized by costly access, a dearth of personal and economic services, significant environmental constraints, and poorly defined property rights, all of which create adverse settling grounds for capitalist enterprises in the absence of sustained incentives from the government. Such regions commonly witness an initial infusion of capital followed by an abrupt exodus with the removal of government supports or the short-term liquidation of the natural resource endowment. Cleary (1993), for example, observes that since the late 1980s informal sector miners have displaced capitalist mining and ranching ventures in many parts of the Amazon, and attributes this to the inability of the government to sustain the flow of financial incentives and infrastructural development attracting investors. In the Philippines, Kummer and Turner (1994) note how forested areas initially exploited by logging companies are subsequently settled by subsistence farmers taking advantage of the access afforded by the logging roads. Capital penetration of the southern Yucatán peninsular region has also occurred along discontinuous trajectories (see Ch. 3). Most public and private development efforts have given rise to isolated pockets of boom and bust that have led the majority of the region's inhabitants to maintain an at least partial basis in subsistence production. Following over fifty years of sporadic government-backed development efforts, including wet rice projects, cattle ranching schemes, and massive logging concessions, what little industry exists in the southern Yucatán peninsular region today is far from sufficient to support a viable wage-earning class.

These examples call into question an appeal to the long run as a basis for imposing a profit-maximizing theoretical structure to understand the causes of deforestation. If anything, they indicate that subsistence production is, under some conditions, a more persistent form of organizing production than capitalist ventures. This is especially the case in fragile tropical ecosystems, since the resource base in these areas often cannot withstand the continuous application of capital intensive inputs required for commercially competitive production (Humphries 1993). Of course, examples exist of frontiers that have extensive and viable linkages to the larger economy, but these are regions that have successfully exploited local production possibilities where they have a comparative advantage. The southeastern portion of the Brazilian Amazon, initially devoted to the production of coffee for foreign markets, and the American Corn Belt are examples of export-propelled frontiers where capitalistic relations of production prevail (Katzman 1975; Ozório de Almeida 1992). Generally speaking, however, those regions that have not been incorporated into the international market economy on the basis of a comparative advantage in the production of some internationally traded product are not likely to be regions predominated by profit-maximizing agents, at least over the long term.

## 4. *An Empirical Application*

The theoretical model developed above indicates that the determinants of the farm household's land allocation decision depend on, among other factors, its relationship with the market. If a complete set of markets exists, the optimal land use portfolio is one that equalizes the marginal revenue product of the land across all uses. Since the determination of marginal revenue product is a function of exogenously given market parameters and is independent of household characteristics, measures of demographic composition should have no statistically significant effect in explaining land use for households that participate in all markets. This observation is tested here. Two econometric models are specified, the first for maize, a staple crop cultivated by the entire sample, but sold by slightly less than half; and the second for jalapeño, cultivated by approximately half the sample but strictly for the market (Ch. 10).

The determinants of maize cultivation are contingent on the decision of whether the household participates as a seller in the market. As the decision to sell maize is itself an endogenous choice, however, there may be unobservable variables that affect both the probability of being a seller as well as the area of maize planted. To account for the potential simultaneity bias arising from the existence of such variables, the empirical specification employs a two-stage, switching regression model with endogenous switching (Lee 1982; Maddala 1983). The first stage uses probit analysis on the entire sample of observations to identify the determinants of seller status and, based on the estimates from the model, to calculate the inverse Mill's ratio for each household. This ratio is derived from the density and distribution functions of the standard normal distribution and serves to capture that component of association with the seller and non-seller categories not accounted for by the included exogenous variables. The probit analysis employs all available observations; the dependent variable equals one if the household reported to sell maize and zero otherwise. The second stage estimates two separate regressions of area cultivated in maize for the sellers and non-sellers. These regressions include the inverse Mills ratio as an explanatory variable in order to control for potential biases arising from sample selectivity. Because the residuals of the second-stage regressions are heteroskedastic, they are estimated by weighted least squares using the Huber/White estimates of variance.

As in the case of maize, estimating the area cultivated in jalapeños requires distinguishing between the determinants of market participation and those of area cultivated. The decision to produce jalapeño simultaneously represents a decision to participate in the market as the entire output is sold, but because 48 per cent of the sample produced no jalapeño whatsoever, a modified version of the switching regression, the Heckman sample selection model is used (Heckman 1974). Like the switching regression, the model comprises an equation determining sample selection and a regression model. These two components are estimated jointly using the maximum likelihood technique, with estimates of the inverse Mills ratio used as starting values in the iteration process.

To identify both empirical models requires the selection of variables that uniquely determine the discrete decision (i.e. whether to participate in the market) but not the continuous decision (i.e. area cultivated). In the present example, this selection can be informed by the work of Goetz (1992) and Key, Sadoulet, and de Janvry (2000), where these authors demonstrate that the market participation decision depends on both fixed and proportional transaction costs of participating in the market while the extent of participation depends on proportional transaction costs only. Identification of the model can therefore be achieved by including in the selector equation variables that proxy for fixed transaction costs. Five such variables are included: a dummy variable for vehicle ownership, the on-road distance from the *ejido* to the nearest market and its square, a dummy variable indicating one of the sampled *ejidos* that is situated on the main highway and serves as a depot for local produce, and a dummy variable indicating whether the head of the household is a native Spanish speaker.[5] Following the work of Goetz (1992), these variables are intended to control for the fixed costs of information gathering and of accessing the market. It is expected that the dummies for vehicle ownership, the favorably situated *ejido*, and first language in Spanish positively affect the probability of

[5] Of household heads, 31% spoke an indigenous language, usually of Maya origin, as their native language.

market participation as they reduce information and access costs. Conversely, distance is predicted to have a negative effect, with the squared term included to allow for potential non-linearities.

Among the other exogenous regressors included in the model are variables theoretically expected to affect the continuous land-use decision: demographic indices; travel costs to the plot, which serves as a proxy for the farm-gate price; time allocated to off-farm labor; available liquidity; farm capital (human and physical); and ecological factors. Expected effects with regard to most of these variables are intuitive and have been examined extensively elsewhere in the literature. The present discussion focuses on the demographic variables, as it is respect to these indices that differential effects are expected according to seller status as is predicted by the non-separable model. Specifically, it is hypothesized that the sellers of maize and jalapeño, having made the decision to participate in the market, will reach their production decision for each of these two crops with reference to the associated farm-gate prices. Consequently, there should be no statistically significant relationship between the area of maize or jalapeño planted and the consumption requirement of the household. Such a relationship, however, is expected among households producing maize solely for subsistence. As their valuation of that crop is determined endogenously as a function of demographic composition, a positive effect of the household consumption requirement on the area of maize cultivated is expected.[6] In addition, both models test if a labor market constraint leads to non-separability by including a measure of the labor endowment of the household.

To distinguish between household supply of labor and consumption demand effects, demographic structure is decomposed into two mutually exclusive variables that were constructed on the basis of educational attainment and age. The first variable measures the number of household members greater than or equal to 12 years of age who have less than nine years of schooling, and is intended to measure the potential pool of domestic

[6] Strictly speaking, the separable model predicts the insignificance of demographic variables on maize cultivation for both the seller and buyer categories of households. While 26% of the households in the sample reported themselves to have occasionally purchased maize for the survey year, the analysis merges these households with the self-sufficient producers under the assumption that the two categories apply an identical operating logic to the productive resources available. Maize cultivation is the principal occupation of most household heads, particularly among the buyers. As a result, it is assumed that any purchases of maize were more likely to be the result of unexpected shortages relative to domestic consumption rather than of planned production targets.

on-farm labor (Sadoulet, de Janvry, and Benjamin 1998).[7] The second variable, comprising children under 12 and members with nine or more years of schooling, measures the remainder of the household. Since young children and highly educated members are assumed not to contribute to field labor, this variable is taken to be a consumption measure that is purged of an on-farm supply-side effect.[8] No a priori prediction is made for the coefficient on the labor force measure other than to exclude a negative sign. Given the existence of shallow labor markets, the measure would be expected to have a positive coefficient, reflecting non-separability in production and consumption as a consequence of the household's endogenous valuation of its time endowment.

## 5. *Econometric Results*

### 5.1. *Stage I: The Determinants of Seller Status for Maize*

Results from the first stage probit analysis used to identify the determinants of seller status are presented in Table 11.1.[9] Among the identifying variables, the signs on most of the coefficient estimates are consistent with intuition. The dummy variables indicating the favorably situated *ejido* and vehicle ownership are both positive and statistically significant in increasing the probability of being a seller, likely reflecting the lower costs of information gathering and increased accessibility. The other measure of information gathering costs, the Spanish language dummy, unexpectedly has a negative and statistically significant coefficient. The project has no immediate interpretation of this result and continues to explore possible answers. Distance to market has a non-linear effect on the probability of seller status, first increasing and then decreasing the probability of being a seller.

The signs on most of the remaining variables are also consistent with intuition. Access to credit and a larger land endowment positively affect the

[7] In a study of labor markets in Mexico's *ejidatario* sector, these authors present persuasive theoretical and empirical evidence to support the proposition that skilled labor (defined by nine or more years of schooling) only works in off-farm employment.

[8] Following Pichón (1997), children were weighted by one-third to approximate their consumption requirement relative to adults.

[9] The data used in this model and those that follow have been weighted to reflect the survey design. See Vance (2000) for details.

**Table 11.1.** Probit and weighted least squares models of seller status and hectares planted in maize

| Variable | Probit of seller status | WLS of hectares planted in maize | |
|---|---|---|---|
| | | Sellers | Non-sellers |
| Consumption index | −0.030 | −0.443 | 1.134 |
| | (−0.357) | (−0.732) | (3.447)* |
| Labor index | −0.041 | 0.257 | 0.384 |
| | (−1.822) | (2.086)* | (2.914)* |
| Age of head | −0.002 | 0.055 | 0.004 |
| | (−0.256) | (2.347)* | (0.157) |
| Education of head | −0.466 | 0.168 | −0.051 |
| | (−1.144) | (1.137) | (−0.542) |
| Distance to plot | 0.024 | −0.101 | −0.078 |
| | (1.678) | (−1.729) | (−4.044)* |
| Area of plot | 0.004 | 0.002 | 0.011 |
| | (3.109)* | (0.229) | (1.650) |
| Elevation | 0.002 | −0.007 | −0.002 |
| | (1.831) | (−1.208) | (−0.550) |
| Good soil | −0.163 | 1.938 | 0.777 |
| | (−0.480) | (2.336)* | (1.978)* |
| Access to chainsaw | −0.030 | 0.107 | −0.527 |
| | (−0.171) | (0.088) | (−0.665) |
| Number of cattle | −0.015 | −0.026 | 0.057 |
| | (−1.902)* | (−0.987) | (0.860) |
| Months of off-farm labor | 0.005 | 0.180 | −0.067 |
| | (0.753) | (1.909)* | (−0.865) |
| Government credit | 0.061 | 0.530 | 0.332 |
| | (2.642)* | (2.510)* | (2.857)* |
| First language Spanish | −0.548 | | |
| | (−2.134)* | | |
| Depot *ejido* | 1.212 | | |
| | (1.985)* | | |
| Distance to nearest market | 0.074 | | |
| | (2.524)* | | |
| Distance squared | −0.001 | | |
| | (−2.045)* | | |
| Vehicle ownership | 0.504 | | |
| | (2.812)* | | |
| Selectivity term (sell = 1) | | 0.447 | |
| | | (0.273) | |
| Selectivity term (sell = 0) | | | −2.274 |
| | | | (−2.612)* |
| Constant | −2.210 | −0.352 | 1.394 |
| | (−2.565)* | (0.19) | (0.92) |
| R-square | | 0.45 | 0.37 |
| Observations | 187 | 78 | 109 |

*Note*: *denotes significance at the 10% level or below. T-statistics are in parentheses.

probability of seller status, with the latter result likely due to the fact that with more land, farmers are able to extend the fallow period, thereby increasing the land's productivity for future use in the cycle. Cattle ownership decreases the probability of seller status, perhaps a consequence of pasture competing for lands otherwise in the fallow cycle or because investment in cattle indicates an economic status in which the low rewards for selling maize are not considered. Finally, the labor endowment and the distance from the household to the plot are statistically insignificant.

## 5.2. *Stage II: The Determinants of Area Cultivated in Maize*

With the inverse Mills ratio included as a regressor to control for selectivity bias, the next two models in Table 11.1 test hypotheses concerning the determinants of maize cultivation for the sellers and non-sellers of that crop, respectively. Sample selectivity appears to be an issue only in the non-seller regression, with the negative coefficient on the inverse Mills ratio likely reflecting the influence of unobservable variables that lead a non-seller to plant less maize than a household selected at random (Goetz 1992). The most noteworthy distinction revealed by the model is the asymmetric effect of the consumption index. This variable is significant in the non-seller regression, while it is insignificant among the sellers, thereby lending support to the hypothesis that the latter reach their production decision independent of consumption considerations.[10]

Other than the consumption index, however, there are few variables that suggest a distinctly different behavior between the two regimes. The dummy variable indicating favorable soil and the credit index are positive and significant in both models. Likewise, the labor index is also seen to be positive and significant in the two models. Although in this case there were

[10] One caveat regarding this finding is that no distinction was made between the *ex ante* decision of how much maize to plant and the *ex post* decision of how much to sell. The latter decision is partly determined by the quantity of realized output, an amount the farmer cannot be sure of at the time of planting. Hence, the observance of market participation *ex post* is not a sufficient condition for the specification of a non-recursive (i.e. profit-maximizing) model, since ex ante the farmer may still be reaching his decision on the basis of the consumption requirements of the household. A high level of risk aversion could exacerbate this problem by leading households to plant an area far larger than that required to feed the family in an average year, thereby increasing the likelihood that they have a surplus for sale. One empirical implication of this feature is that the seller category may consist partly of households that are reaching their decisions with reference to consumption considerations, which would increase the chance of falsely rejecting separability for this group.

no a priori predictions, this finding likely reflects the household's endogenous valuation of its own labor due to some combination of few opportunities for off-farm employment, low availability of labor for hire, and potentially higher monitoring costs associated with hired labor.[11] Differential effects are seen with respect to the variables measuring the total family months allocated to non-farm labor in the previous year, the age of the household head, and the distance separating the household from its agricultural plot. The former two are positive and statistically significant for the sellers only while the latter is negative for both groups but significant for the non-sellers only.

### 5.3. *Jalapeño Chili*

The final model investigates the determinants of commercial jalapeño chili production with an application of Heckman's sample selection model. To allow for qualitative comparisons, the specification includes the same variables in the selector and regression equations that were used in the model for maize. The results presented in Table 11.2 suggest that many of the factors determining both the participation and production decisions for jalapeño are different from those of maize. With regard to participation indicators, the most curious result is the significance of the consumption index as a determinant of jalapeño cultivation. This finding may relate to the fact that the harvest for jalapeño occurs over a highly concentrated time span during which those households marginally engaged (small hectarages of jalapeño) must harness their entire stock of labor.[12] In this regard it is noted that a joint chi square test of the significance of the consumption and labor index (not presented) rejects the null that the coefficients are simultaneously zero. Included among the other statistically significant determinants of the market participation decision for jalapeño are the age and education of the household head, months allocated to off-farm labor, and the elevation of the plot, none of which were statistically significant determinants of the market participation decision for maize. The remaining two statistically significant variables, cattle ownership and credit, have negative and positive signs, respectively, as for the case of maize.

[11] Such market imperfections are readily incorporated into the theoretical model by introducing an additional constraint on hired labor.

[12] In contrast, households strongly committed to chili production and possessing large hectarages invariably hire labor to pick the peppers (Ch. 10).

**Table 11.2.** Heckman model of producer status and hectares planted in jalapeño chili

| Variable | Selector equation | Regression equation |
|---|---|---|
| Consumption index | 0.277 | −0.310 |
| | (1.669)* | (−1.095) |
| Labor index | 0.075 | 0.092 |
| | (1.146) | (1.827)* |
| Age of head | −0.030 | 0.008 |
| | (−2.522)* | (0.629) |
| Education of head | −0.147 | 0.090 |
| | (−2.361)* | (1.006) |
| Distance to plot | 0.029 | 0.009 |
| | (1.225) | (0.409) |
| Area of plot | −0.001 | 0.003 |
| | (−0.332) | (1.179) |
| Elevation | −0.006 | 0.001 |
| | (−2.260)* | (0.577) |
| Good soil | 0.171 | 0.048 |
| | (0.484) | (0.175) |
| Access to chainsaw | 0.092 | 0.547 |
| | (0.324) | (1.868)* |
| Number of cattle | −0.019 | −0.029 |
| | (−1.781)* | (−1.305) |
| Months of off-farm labor | −0.056 | 0.030 |
| | (−2.785)* | (0.822) |
| Government credit | 0.105 | 0.122 |
| | (2.351)* | (1.592) |
| First language Spanish | 0.184 | |
| | (0.681) | |
| Depot *ejido* | −0.371 | |
| | (−0.484) | |
| Distance to nearest market | −0.015 | |
| | (−0.501) | |
| Distance squared | 0.000 | |
| | (0.205) | |
| Vehicle ownership | 0.612 | |
| | (1.351) | |
| Constant | 2.455 | −0.264 |
| | (2.022)* | (−0.327) |
| Selectivity term | | −0.548 |
| | | (−1.620)* |
| Log likelihood | | −30248 |
| Wald chi-square | | 35.76* |
| Observations | | 187 |

*Note*: *denotes significance at the 10% level or below. Z-statistics are in parentheses.

It is notable that distance to market and its square are included among the statistically insignificant coefficient estimates. This may result from differences in the marketing networks of maize and jalapeño. Whereas maize is generally transported by the farmers themselves, jalapeño is transported by marketing intermediaries or *coyotes* who typically purchase the harvest at or near the site of the farmer's field. While none of the identifying variables are individually significant, the overall appropriateness of the selector equation specification is supported by a significant estimate for the sample selection parameter, or Mills ratio.

Only two variables in the regression are statistically significant in explaining the area planted in jalapeño: the labor index and ownership of a chainsaw, the former of which was also significant for the case of maize. Consistent with the expectation that the household's valuation of jalapeño is determined exogenously by the market price, the consumption index has no statistically significant effect in determining the area cultivated. That the chainsaw dummy is positive and statistically significant, contrasting with the case of maize, apparently supports the much larger annual hectarage cleared by those engaged in jalapeños over those who are not (Ch. 10).

## 6. *Conclusions*

Does the basic structure of the agricultural household model hold for the *ejidatarios* of the southern Yucatán peninsular region, or must it be modified according to the mixed or hybrid household model? By including a potentially binding constraint on market participation for the subsistence crop, maize, the Kuhn-Tucker conditions of the model revealed that the production of this crop is analytically non-separable. Specifically, the planting decision is not made strictly with reference to exogenously given market parameters and technology, as under the profit-maximization paradigm. Rather, this decision also incorporates the consumption requirements of the household.

The empirical results highlight the analytical advantages of distinguishing the determinants of land-use decisions according to the household's relationship with the market. By controlling for the endogeneity of the market participation decision and subsequently estimating separate regressions by seller status, the switching regression model revealed a number of differences in land-use determinants that otherwise would not have been detected. The most general finding is that demographic variables matter for explaining how these frontier cultivators reach both semi-subsistence and commercial land-use decisions when other theoretically supported controls are present. The statistical significance of the consumption index for maize likely follows from market imperfections, while this index for jalapeño was insignificant. In both the models of maize and jalapeño, the household's labor endowment was positive and significant, likely reflecting the existence of thin markets for labor.

These results indicate that purely market-based approaches to estimate deforestation and agricultural change in the region will miss the mark, as long as the current conditions (local political economy) hold. They also suggest that policy instruments relying on market-based approaches may not have the intended effect on peasant land-use choices if these policies are not complemented by those that also reduce the transaction costs of market participation that apparently constrains them to reach their production decisions with reference to household-specific supply and demand conditions.

## REFERENCES

Ahn, C. Y., I. Singh, and L. Squire (1981). 'A Model of an Agricultural Household in a Multi-Crop Economy: The Case of Korea'. *Review of Economics and Statistics* 63: 520–5.

Angelsen, A. (1995). 'Shifting Cultivation and Deforestation: A Study from Indonesia'. *World Development* 23: 1713–29.

Barnum, H., and L. Squire (1979). *A Model of an Agricultural Household.* Washington DC, World Bank.

Chomitz, K. M., and D. A. Gray (1996). 'Roads, Land Use, and Deforestation: A Spatial Model Applied to Belize'. *World Bank Economic Review* 10: 487–512.

Cleary, D. (1993). 'After the Frontier: Problems with Political Economy in the Modern Brazilian Amazon'. *Journal of Latin American Studies* 25: 331–49.

Cropper, M., C. Griffiths, and M. Mani (1999). 'Roads, Population Pressures, and Deforestation in Thailand, 1976–1989'. *Land Economics* 75: 58–73.

Cropper, M., J. Puri, and C. Griffiths (2001). 'Predicting the Location of Deforestation'. *Land Economics* 77: 172–86.

de Janvry, A., M. Fafchamps, and E. Sadoulet (1991). 'Peasant Household Behavior with Missing Markets: Some Paradoxes Explained'. *The Economic Journal* 101: 1400–17.

de Janvry, A., E. Sadoulet, and G. Gordillo de Anda (1995). 'NAFTA and Mexico's Maize Producers'. *World Development* 23: 1349–62.

DeShazo, R. P., and J. R. DeShazo (1995). 'An Economic Model of Small Holder Deforestation: A Consideration of the Shadow Value of Land on the Frontier', in O. Sandbukt (ed.), *Management of Tropical Forests: Towards an Integrated Perspective*. Oslo, Center for Development and the Environment, 153–66.

Goetz, S. J. (1992). 'A Selectivity Model of Household Food Marketing Behavior in Sub-Saharan Africa'. *American Journal of Agricultural Economics* 74: 444–52.

Heckman, J. J. (1974). 'Shadow Prices, Market Wages, and Labor Supply'. *Econometrica* 42: 679–94.

Humphries, S. (1993). 'The Intensification of Traditional Agriculture among Yucatec Maya Farmers: Facing up to the Dilemma of Livelihood Sustainability'. *Human Ecology* 21: 87–102.

Hyde, W. F., G. S. Amacher, and W. Magrath (1996). 'Deforestation and Forest Land Use: Theory, Evidence, and Policy Implications'. *The World Bank Research Observer* 11: 223–48.

Katzman, M. (1975). 'The Brazilian Frontier in Comparative Perspective'. *Comparative Studies in Society and History* 17: 266–285.

Key, N., E. Sadoulet, and A. de Janvry (2000). 'Transaction Costs and Agricultural Household Response'. *American Journal of Agricultural Economics* 82: 245–59.

Kummer, D., and B. L. Turner, II (1994). 'The Human Causes of Deforestation in Southeast Asia'. *BioScience* 44: 323–8.

Lee, L. F. (1982). 'Some Approaches to the Correction of Selectivity Bias'. *Review of Economic Studies* 49: 355–72.

Maddala, G. S. (1983). *Limited-Dependent and Qualitative Variables in Econometrics*. Cambridge, Cambridge University Press.

Medani, A. I. (1975). 'Elasticity of the Marketable Surplus of a Subsistence Crop at Various Stages of Development'. *Economic Development and Cultural Change* 23: 424.

Medellin, M. A., L. P. Apedaile, and D. Pachico (1994). 'Commercialization and Price Response of a Bean-Growing Farming System in Colombia'. *Economic Development and Cultural Change*: 795–816.

Nelson, G., and D. Hellerstein (1997). 'Do Roads Cause Deforestation? Using Satellite Images in Econometric Analysis of Land Use'. *American Journal of Agricultural Economics* 79: 80–8.

Nelson, G., V. Harris, and S. Stone (2001). 'Deforestation, Land Use, and Property Rights'. *Land Economics* 77: 187–205.

Omamo, S. W. (1998). 'Transport Costs and Smallholder Cropping Choices: An Application to Siaya District, Kenya'. *American Journal of Agricultural Economics* 80: 116–23.

Ozório de Almeida, A. L. (1992). *The Colonization of the Amazon*. Austin, University of Texas Press.

Panayotou, T., and S. Sungsuwan (1994). 'An Econometric Study of the Causes of Tropical Deforestation: The Case of Northeast Thailand', in K. Brown and D. Pierce (eds.), *The Causes of Tropical Deforestation: The Economic and Statistical Analysis of Factors Giving Rise to the Loss of Tropical Forests.* London: UCL Press, 192–210.

Pfaff, A. S. P. (1999). 'What Drives Deforestation in the Brazilian Amazon? Evidence from Satellite and Socioeconomic Data'. *Journal of Environmental Economics and Management* 37: 26–43.

Pichón, F. J. (1997). 'Colonist Land—Allocation Decisions, Land Use, and Deforestation in the Ecuadorian Amazon Frontier'. *Economic Development and Cultural Change* 44: 127–64.

Reis, E., and R. Guzman (1994). 'An Econometric Model of Amazon Deforestation', in K. Brown and D. Pierce (eds.), *The Causes of Tropical Deforestation: The Economic and Statistical Analysis of Factors Giving Rise to the Loss of Tropical Forests.* London: UCL Press, 172–91.

Sadoulet, E., A. de Janvry, and C. Benjamin (1998). 'Household Behavior with Imperfect Labor Markets'. *Industrial Relations* 37: 85–108.

Saha, A. (1994). 'A Two-Season Agricultural Household Model of Output and Price Uncertainty'. *Journal of Development Economics* 45: 245–69.

Singh, I., L. Squire, and J. Strauss (1986). 'The Basic Model: Theory, Empirical Results, and Policy Conclusions', in I. Singh, L. Squire, and J. Strauss (eds.), *Agricultural Household Models: Extensions, Applications, and Policy.* Baltimore, The Johns Hopkins University Press, 17–47.

Thornton, J., and B. K. Eakin (1992). 'Virtual Prices and a General Theory of the Owner Operated Firm'. *Southern Economic Journal* 58: 1015–29.

Turner, B. L., II, and S. Brush (eds.) (1987). *Comparative Farming Systems.* New York, Guilford Press.

Vance, C. (2000). 'Land Use Change, Deforestation, and Peasant Farm Systems: A Case Study of Mexico's Southern Yucatan Peninsular Region'. Ph.D. Dissertation, Clark University.

# PART IV

## *Spatial Modeling of Land Change: Empirical Approaches in Data-Sparse Environments*

The environmental impacts of the spatial dimensions of land-use and land-cover change, such as ecosystem fragmentation, have been addressed in the ecological sciences for some time, including the development of various modeling techniques applicable to the problem. *Spatial modeling* of land-use change within the domain of the social sciences is less well developed, at least in regard to fine-tuned spatio-temporal specificity. Both modeling domains require *explanatory* models that can address *where*, *when*, and *why* land changes occur. Explaining human-induced land change beyond correlations to specific activities or proximate causes (e.g. road building), however, requires that human behavior be addressed (e.g. farming decision-making) and, ideally, that this behavior be linked to distal and mediating factors impinging on that behavior (e.g. national policy and local tenure). Various behavioral theories of land-use change exist that differentially consider distal and mediating factors. Unfortunately, serious data limitations inhibit tests of them in situations akin to the southern Yucatán peninsular region. Remotely sensed data greatly facilitate the observation of *where* and *when* land-change takes place, but observations from space do not provide much insight into *why* land-use change occurs, save in those few cases where established and distinctive pattern-process relationships exist. It is in the need to understand the *why* that the SYPR project, as do similar studies attempting to explain land change, links spatially explicit social data with biophysical and remotely sensed data.

Given the paucity of data in frontiers, the SYPR project has collected and created a myriad of spatially explicit data sets. Joining these data raises a number of theoretical, empirical, and methodological issues concerning how best to use them to produce robust land-use and land-cover change models. The project takes two separate, but complementary approaches in developing spatially explicit land-change models for the region as well as *ejidos* and households. One approach is based on statistical techniques, where theoretically informed economic models are tested using econometric approaches. The second approach is based on simulations procedures (or what is commonly labeled 'integrated assessment models' in the literature), which allow for a potentially broader array of social behaviors and spatial interactions, but relies heavily on the statistical models or rules of thumb (which the SYPR project seeks to avoid) for parameter estimates. Both approaches yield powerful insights into the land dynamics at play, although each is significantly restricted by data limitations. Such results point to the potential for improved explanation and projections of spatially explicit land change as the spatial specificity and the range of data improve. With such improvements, the project should be able to determine the level of understanding gained, for example concerning changes in carbon stocks or biodiversity, by the added level of model specificity.

*The Editors*

# 12

# Spatially Explicit, Statistical Land-Change Models in Data-Sparse Conditions

*Jacqueline Geoghegan, Laura Schneider, and Colin Vance*

## 1. *Frontiers of Land Modeling*

A range of research interests beyond global environmental change science increasingly calls for advances in land-change models and, specifically, models that have fine-grained locational outputs. The rationale for such modeling about land change has been articulated elsewhere in this book (Ch. 1; Introduction to Part IV) and need not be reiterated here. It is important to note, however, that advances in question are assisted by the advances in the analytical sophistication of geographical information systems, hardware (GPS) that permits geographical coordinates to be established easily in the field, and for land-change studies, increasing temporal and spatial resolution of satellite imagery. Much of the first phase of land-change models that incorporate these systems and data has been empirical-based, time series assessments, such as Markov-chain models (e.g. Turner 1988), that let the record of land change determine future projections, or the spatial level of assessment has been large-grain (e.g. counties, states, regions). The SYPR project seeks a different approach demonstrated here: to test theories of land change in regard to their ability to explain fine-grained land change in the region at different spatial scales of assessment.

Two complementary econometric modeling approaches are used here to investigate the factors that affect deforestation at the regional and household scales of analysis. Both approaches use the individual satellite pixels as the data on land-use change, from the classification of TM imagery described in Ch. 6. A regional model spans the entire study area of agricultural *ejidos*, and links the satellite imagery with publicly available geophysical data and socio-demographic government census data. The second model focuses exclusively on the parcels associated with the household survey data collected specifically for this project, discussed in Part III, especially Ch. 8. This latter approach uses the same geophysical data of the aggregate approach, but uses the much richer socio-demographic data derived from the linkage of individual farm plots and the satellite imagery via the sketch mapping exercise described in Chs. 8 and 9.

While both models take a theoretical approach of individual maximization, they differ in a number of ways, the most important of which is the role of time in the decision-making process. This is not to suggest that *individual* behavior is necessarily different according to the spatial extent or temporal scale of the analysis, but rather that the types of theoretically derived, testable hypotheses and the empirical methodologies used to test them have to be supportable by the data that have been created and collected through this project. Indeed, in both models, the choice of explanatory variables and derivation of testable hypotheses concerning land-use change are based on a broad array of frameworks from social science theory. Reviews of these can be found elsewhere (Irwin and Geoghegan 2001).

The regional model estimated in this chapter focuses exclusively on the location of land-use change, using as the dependent variable the results from a single GIS change analysis created from two different time periods of TM images, while the household model investigates both the location and timing of land-use change by tracking the land use associated with each particular pixel through multiple TM images over time. This dual approach is taken because the existing, albeit small, spatially explicit economic literature on land-use change has bifurcated into these two separate modeling approaches. The current literature that has focused exclusively on the *location* of land-use change has analyzed deforestation in tropical countries: Belize (Chomitz and Gray 1996); northern Mexico (Nelson and Hellerstein 1997); Brazil (Pfaff 1999); Thailand (Cropper, Griffiths, and Mani 1999; Cropper, Puri, and Griffiths 2001); and Panama (Nelson, Harris, and Stone 2001). Models that include the temporal dimension as well as location of land-use change have focused on urban fringe development in the United States (Geoghegan and Bockstael 2000; Irwin and Bockstael 2001), but this latter approach can be modified to analyze tropical deforestation as well. Therefore, one of the goals of this chapter is to compare the results of these two approaches.

## 2. *Conceptual Framework of Land-Change in the Region*

A rich literature exists on theory relevant to the land changes underway in the southern Yucatán peninsular region, especially deforestation and agriculture. In economics, this theory has been built around principles of a maximizing land manager, either of profit maximization if the land manager is engaged in the market, or utility maximization in the case of thin markets and subsistence cultivation as discussed in Ch. 11. An alternative literature exists beyond economics that attempts to incorporate the variability controlled for in the economic models (e.g. Turner and Ali 1996) but these tend to be analytically weak. The modeling effort here draws not only on the various principles that have been advanced elsewhere, but also on the lessons learned in the SYPR project about the various factors affecting land-use decisions and the conditions in which those decisions are made—that is, the integrated land-change science advanced by the project.

The SYPR project is guided by the principle that land managers (in this case, primarily *ejidatarios*) are the basic agents of change in land use, whose actions are strongly influenced by the institutions and policies in which they operate. As demonstrated in Ch. 11, these agents in the region operate within a mixed decision domain that results in actions that are neither fully consistent with market nor subsistence behavior. The controlling usufruct tenure system, thin markets, and the relative marginality of landscape for capital-intensive cultivation have heretofore inhibited market cultivation and helped to sustain subsistence production. Assuming a profit-maximixing behavior with full market integration is not sufficient for modeling, as the various chapters in the volume inform us, owing to a series of other factors and institutions, besides markets that strongly influence land manager decision-making. Among these are, to name a few: various subsidies (Chs. 7, 8) that not only promote maize production beyond subsistence needs but pasture creation absent livestock to graze; non-market controlled land access; significant variation in access to middlemen (markets) owing to variation in roads and the paucity of them; and ethnicity and education that affect the social capital of the land manager (e.g. ability to access subsidies, gain information, entice middlemen to pick up their commercial crops). In addition, individual land managers have different quality lands to cultivate.

Chapters 3 and 7 demonstrate that the political economic conditions in which land managers make decisions are not only complex but have changed significantly in the region's contemporary development. These findings do not diminish the applicability of land-change concepts that focus on the land manager—who, after all, makes the decisions about land—but inform us of those factors that should be accounted for in the overall assessment. Unfortunately, in data-poor conditions, incorporating all the known factors that should be considered is highly restricted and, as in all modeling approaches, some simplifications must be made and potential proxies used for variables in hypothesis testing.

## 3. *The Models: Theoretical and Empirical Specifications*

As almost all forms of land-use change reflect a calculation of comparative costs and benefits, the theoretical models developed in this chapter are founded on the theory of economic choice: that individuals choose the best option available to them, given their constraints. These constraints can consist of such things as income and time available to a household, as well as the institutional framework within which a household operates. For example, it is important to recognize that the land allocation mechanism in Mexico's *ejido* sector is not market-based, but rather politically and institutionally based (Ch. 7). Consequently, the classic spatial von Thünen model of land use, (e.g. Chomitz and Gray 1996; Nelson and Hellerstein 1997), in which there is a market equilibrium resulting from a trade-off of farm-gate prices (agricultural returns minus transportation costs) and agricultural land prices, is mostly irrelevant as a theoretical modeling structure as there is no land market in the *ejido* context (i.e. land tenure reforms in the Mexican Constitution have had little, if any, impact yet in the region). Therefore, this theoretical framework cannot be relied upon for understanding the location of the agricultural activity of an individual *ejidatario*. Other hypotheses, similar to von Thünen, can be tested, however (Chisholm 1962). Among these: did older *ejidos* obtain lands with better access to markets; were original members of an *ejido* allocated the 'best' lands first; does access and transportation costs affect land-use decisions, given the location of each *ejidatario*'s parcel? While we currently do not have data to test the first two hypotheses concerning how a particular *ejidatario* gained access to a particular parcel of land, we are able to test different specifications of the third hypothesis in this chapter.

The simplest characterization of the land-use decision for an individual land manager is based on profit maximization, applicable when the farmer is engaged in product and factor markets. A more general characterization, that of utility maximization, is required to model the behavior of a subsistence-based farmer (Ch. 11). The general case of utility maximization is developed here. The first theoretical model considered focuses exclusively on the *location* of land-use change. The land manager will choose land use in period $T$ that maximizes expected net utility. Let $A(i,T)$ represent the present value of the future stream of utility to pixel $i$ that is cleared in time period $T$. Let $F(i,T)$ represent the value of the future stream of utility for leaving pixel $i$ in forestry use in each time period, and let $C(i,T)$ be the one-time clearing costs associated with clearing the pixel in time period $T$. Then $V(i,a,T)$ represents the *net* benefits to pixel $i$ of deforestation at time $T$ and the static conversion rule for the individual land manager is to clear pixel $i$ when the net future stream of benefits to agricultural use are greater than the stream of benefits to forest use:

$$V(i,a,T) > V(i,f,T) \tag{1}$$

Given that all factors that affect $V$ are not observable, this statement can be rewritten in terms of the *probability* of parcel $i$ being cleared, with an error term $e$, representing the unobserved random component to net returns. This reformulates the decision rule into clear parcel $i$ if:

$$\text{Probability}\,(V(i,a,T) + e(i,f,T)) > \text{Probability}\,(V(i,f,T) + e(i,f,T)) \tag{2}$$

Let the characteristics of pixel $i$ be $X(i)$, including the characteristics of the land manager of that pixel, and assume that the net returns are a linear function of these characteristics, the decision rule for clearing is now:

$$\text{Probability}\,(X(i,a,T)\beta_i + e(i,f,T)) > \text{Probability}\,(X(i,f,T)\beta_i + e(i,a,T)) \tag{3}$$

By assuming an extreme value distribution for the error terms, the parameters $\beta$ of this model can be estimated using a logit specification (for these econometric models, see Madalla 1983). Specifically, the regional model estimates the probability that a pixel that begins the time period of observation in forest use will be converted for agricultural use or remains in forest use. A preliminary application of a logit model estimation for a portion of the study region can be found in Geoghegan and colleagues (2001).

The second theoretical model is slightly more sophisticated than that given above as it focuses on both the *location* and *timing* of land-use change and is based on Irwin and Bockstael (2001). The empirical specification of this model is used to estimate the factors affecting deforestation at the individual plot level, using the household survey data for information on socio-economic factors. In this model, let $A(i, t)$ be the net benefits to agricultural use for *each time period* after pixel $i$ is cleared in time period $T$. Let $F(i, t)$ be the net benefits to the farmer for leaving pixel $i$ in forestry use in *each time period*, and let $C(i,T)$ be the one-time clearing costs associated with clearing the pixel in time period $T$ and let $\delta$ be the discount rate. Then the net benefits to the farmer of clearing pixel $i$ in period $T$ are:

$$\sum_{t=0}^{\infty} A(i,T+t)\delta^{T+t} - \sum_{t=0}^{\infty} F(i,T+t)\delta^{T+t} - C(i,T) \quad (4)$$

For $T$ to be the optimal time period for clearing, the following two conditions must hold:

$$\sum_{t=0}^{\infty} A(i,T+t)\delta^{T+t} - \sum_{t=0}^{\infty} F(i,T+t)\delta^{T+t} - C(i,T) > 0 \quad (5a)$$

$$\begin{aligned}&\sum_{t=0}^{\infty} A(i,T+t)\delta^{T+t} - \sum_{t=0}^{\infty} F(i,T+t)\delta^{T+t} - C(i,T) > \\ &\sum_{t=1}^{\infty} A(i,T+1)\delta^{T+t} - \sum_{t=1}^{\infty} \mathrm{F}(\mathrm{i},\mathrm{T}+1)\delta^{\mathrm{T+t}} - \delta\mathrm{C}(\mathrm{i},\mathrm{T}+1)\end{aligned} \quad (5b)$$

The first condition, equation (5a) requires that net benefits to clearing are positive, and is intuitively the same as the optimal condition in the first model found in equation (1). The second condition considers that although clearing may yield net positive benefits at time $T$, there may still be benefits to waiting because of the potential for even higher benefits at some future date. Such a circumstance could arise, for example, in anticipation of improved technologies that reduce clearing costs.

Let the characteristics of pixel $i$ be $X(i)$. The optimal time for clearing this pixel then is the first time period in which the following holds:

$$A(X(i),T) - F(X(i),T) - \delta C(X(i),T+1) \geq 0 \quad (6)$$

Add an error term to equation 6 to account for unobservable characteristics:

$$A(X(i),T) - F(X(i),T) - \delta C(X(i),T+1) - \varepsilon(i) \geq 0 \quad (7)$$

The probability that pixel $i$ will be deforested in period $T$ is the hazard rate for period $T$:

$$h(i,T) = \frac{G[W(i,T+1)] - G[W(i,T)]}{1 - G[W(i,T)]} \quad (8)$$

where $G$ is the cumulative distribution function for the error term, and

$$W(i,T) = A(X(i),T) - F(X(i),T) - \delta C(X(i),T+1) \quad (8a)$$

$$W(i,T+1) = A(X(i),T+1) - F(X(i),T+1) - \delta C(X(i),T+2) \quad (8b)$$

As $h(I,T)$ is a hazard rate, a hazard model is used to estimate the parameters in $X(i)$. Hazard models typically estimate the conditional probability of exiting a state given that the state has been occupied for some length $t$. The dependent variable, the duration, is the length of time that elapses from the beginning of the state until its end or until measurement is taken and therefore truncates the observation. For this chapter, the duration of interest is the length of time that an individual pixel remains in forest before being converted to cropland or pasture. Let $P_{it}$ be the probability that deforestation occurs to pixel $i$ in interval $t$ given that the pixel was not deforested in any earlier periods and let $t_i$ be the time period in which pixel $i$ is deforested. Then, if it is assumed that the error terms follow an extreme value distribution, as in the logit model above, the empirical specification can be derived as a complementary log-log model, where the $\beta_i$ terms are parameters to be estimated:

$$\log[-\log(1-P_{it})] = X(i,T)\beta_i \quad (9)$$

This specification is well suited for these data. The complementary log-log model accommodates cases in which the objective is to model continuous time processes using discrete time data. Deforestation occurs continuously through time across the landscape. The event of forest clearance itself, however, is only observed at discrete times, as the temporal resolution of the TM data is rather coarse, due to the difficulty of obtaining cloud-free data for the tropics. Therefore, the exact timing of the deforestation event is not known, only that it occurred during some relatively long time period (see also Allison 2000).

## 4. *Data Sources*

Before moving to the complete empirical specifications of the models developed above and the specific testable hypotheses, several issues related to data must be explicated. The data for the dependent variable, land-use change, are derived from the satellite imagery. As was described in Ch. 6, there are seven land-cover classes for the region: wetland forest, upland forest, seasonal inundated savannas, 4–15-year secondary growth, agriculture (including pasture and 1–3-year secondary growth) and the invasive *Pteridium aquilinum* (bracken fern). For the purposes of modeling deforestation, specifically the change in forest cover into agricultural use, these land-use classes were aggregated into a forest category consisting of wetland forest, upland forest, inundated savannas, 4–15-year secondary growth, and an agricultural category consisting of pasture, cropland, and bracken fern. Therefore the change category of 'deforestation' consisted of a transition from the general forest category into the general agriculture category. However, in order to test hypotheses concerning the potential differential effect of initial land-cover on deforestation, as may emerge for example from the trade-offs between higher soil quality associated with older vegetation and the lower clearing costs of younger vegetation, the initial forest-cover type of primary forest, either wetland or upland, of each pixel is included as an explanatory variable in both econometric models.

The data for the independent variables come from a number of sources. Both models use: elevation and slope from a digital elevation model; soil types digitized from a1:250,000 Mexican government (INEGI) map; digitized road network from INEGI 1:50,000 topographic maps; rainfall data (interpolated to cover the region) from the Mexican government (SARH). The soil maps for the region are complex in their distribution, offering many geographical 'clusters' of soils based on taxonomic order and type, in addition to other characteristics. For modeling purposes, the various types and subtypes have been aggregated into two simple categories that reflect the basic conditions facing farmers: rendzina and other mollisols ('upland soils', preferred by farmers); litosols and vertisols (all other soils). Only four precipitation-temperature stations exist in the region. To these, sixteen others lying just outside the region were added to create twenty stations or precipitation points and used to create an interpolated map (krigging analysis) of average regional annual rainfall (averaged over thirty years). The road layers were used to calculate distances from each individual pixel to roads and markets.

For the regional model, which spans the entire spatial extent of the agricultural *ejidos*, Mexican government census data are used for the explanatory socio-demographic variables, with the unit of observation being the *ejido*. Using the map of *ejido* boundaries (e.g. Fig. 7.1), the census data are linked with the pixels associated with each of the *ejidos* via a uniform distribution, so that each pixel in the *ejido* was assigned the value of the variable from the census data for that *ejido*. The data include information on number of households, language and literacy, and structural characteristics of houses, such as electricity and water service. While the level of detail is not ideal, the data do span the entire spatial extent of our study region, and constitute the only current source of demographic data for *ejidos* not covered in our survey.

For the household model, the survey data, as described extensively in Ch. 8, are used for the explanatory socio-demographic variables, together with the biophysical and distance variables used in the regional model described above.

## 5. *Variables Used in the Empirical Models and Associated Hypotheses*

As suggested in the theoretical models presented above, the land-clearance decision will be based on a comparison of the net benefits from forest and non-forest land uses. There are several testable socio-economic and environmental factors, denoted by the $X(i)$ variables above, that could influence this comparison. For both the regional logit model and the household complementary log-log model, the variables in the $X(i)$ vector are qualitatively similar, consisting of socio-economic, biophysical, and spatial variables. Also, in both models, the unit of observation is the TM pixel, but for the regional model all that is known about the management unit of each individual pixel is that it is in a particular *ejido*, while for the household model, the land manager associated with that pixel as well as the total amount of land available is known. To some extent, this important difference leads to different specifications for the socio-economic and spatial variables in the two models. For example, the total area of the *ejido* is known (Fig. 7.1) for the regional model, but it is not known how much of that land is actually available for agricultural use (e.g. some lands are for settlement, others for community forest reserves). Therefore, it is assumed that all pixels have a potential for agricultural use. However, some of the contiguity of agricultural land uses is controlled for by included variables on the total number of agricultural pixels in a $5 \times 5$ pixel window around each pixel as well as a measure of the distance from each pixel to the nearest agricultural pixel. The hypothesis is that the greater and closer the number of agricultural pixels, the greater the likelihood for a pixel to be cleared for agricultural use.

Summary statistics for the variables used in the models can be found in Table 12.1 for the regional model and Table 12.2 for the household model.

### 5.1. *Demographic Information*

For the household model, the number of males older than 11, number of females older than 11 and total number of children under 12 are included as exogenous variables, this information having been collected during the field survey. These indices are measured as the average number of members in each age/sex category over the corresponding time interval of the imagery and accordingly vary across households and through time. For the regional model, by contrast, only total number of males and females for the *ejido* in 1990 can be included given that the census data contains information on total population size only by gender, not by age. The expectation in both cases is that population pressure will increase the probability of deforestation, and as demonstrated in Ch. 11, the labor–consumer ratio influences the amount of land cultivated, consistent with peasant economic behavior.

### 5.2. *Education and Ethnicity*

For the household model, other socio-economic variables include education of household head as well as the number of household members with

**Table 12.1.** Summary statistics for regional aggregate census logit model

| Variable | Definition | Mean | Std. Dev. | Min. | Max. |
|---|---|---|---|---|---|
| exten | Forest Extension (dummy) | 0.20 | 0.40 | 0 | 1.00 |
| male | Male population (per *ejido*) | 316.49 | 368.91 | 9 | 1529.00 |
| female | Female population (per *ejido*) | 209.13 | 192.26 | 11 | 725.00 |
| span | Spanish speaking (per *ejido*) | 97.91 | 157.94 | 0 | 636.00 |
| m_prim15 | Higher education (per *ejido*) | 24.73 | 31.47 | 0 | 109.00 |
| num_pixl | *Ejido* size (no. of pixels) | 284549.60 | 252115.8 | 8063 | 791775.00 |
| primary | Primary forest (dummy) | 0.89 | 0.32 | 0 | 1.00 |
| elevation | Elevation (m amsl) | 170.63 | 65.87 | 0 | 350.00 |
| rain | Rainfall (mm) | 1123.27 | 86.44 | 0 | 1315.00 |
| slope | Slope (degrees) | 1.11 | 2.25 | 0 | 67.00 |
| freq_ag | Number ag pixels (5 × 5) | 0.43 | 1.80 | 0 | 24.00 |
| d_roads | Distance to roads (m) | 5198 | 3991.32 | 0 | 25750.00 |
| agdist | Distance to nearest ag land (m) | 860.99 | 966.73 | 30 | 7252.00 |
| d_market | Distance to market (m) | 30458.98 | 15751.22 | 30 | 77420.00 |
| pwatr | Percent with water (hh/*ejido*) | 31.72 | 39.32 | 0 | 100.00 |
| pelec | Percent with electricity (hh/*ejido*) | 47.37 | 37.63 | 0 | 95.65 |
| p_credit | Percent receiving credit (hh/*ejido*) | 20.03 | 16.01 | 0 | 76.79 |
| goodsoil | Good soil (dummy) | 0.68 | 0.47 | 0 | 1.00 |

**Table 12.2.** Summary statistics from household survey

| Variable | Definition | Mean | Std. Dev. | Min. | Max. |
|---|---|---|---|---|---|
| male | Average no. of males > 11 over interval | 1.61 | 1.03 | 0 | 7.00 |
| female | Average no. of females > 11 over interval | 1.44 | 1.04 | 0 | 7.00 |
| kids | Average no. of children < 12 over interval | 1.60 | 1.79 | 0 | 6.75 |
| primary | Primary forest | 0.79 | 0.41 | 0 | 1.00 |
| soil | Upland soil | 0.76 | 0.43 | 0 | 1.00 |
| dem | Elevation | 168.04 | 71.09 | 0 | 310.00 |
| slope | Slope | 1.23 | 2.78 | 0 | 58.67 |
| dry | Precipitation | 56.10 | 4.29 | 49 | 72.00 |
| totpixl | Plot size (no. of pixels) | 1321.92 | 1051.02 | 6 | 4819.00 |
| saw | Percent of interval owning chainsaw | 0.25 | 0.37 | 0 | 1.00 |
| vehic | Percent of interval owning vehicle | 0.13 | 0.30 | 0 | 1.00 |
| edhead | Education of household head | 3.31 | 3.87 | 0 | 14.00 |
| ed9 | No. of household members w/> 8 years education | 1.05 | 1.57 | 0 | 7.00 |
| span | Native Spanish speaker | 0.83 | 0.38 | 0 | 1.00 |
| credit | Percent of interval receiving government credit | 0.34 | 0.36 | 0 | 1.00 |
| dis | Distance household to plot | 9.07 | 7.62 | 0 | 28.60 |
| nearmkt | Distance *ejido* to nearest market | 21.23 | 19.22 | 0 | 82.63 |
| yr | Duration of occupancy | 20.23 | 12.82 | 0 | 58.00 |
| yrsq | Duration of occupancy squared | 573.54 | 674.30 | 0 | 3364.00 |

more than eight years of education, while for the regional model, education is measured by the population older than 15 who has attended post-elementary school. Education can have two conflicting effects on land-use decisions. It can increase off-farm employment opportunities, so thereby leading to a decrease in deforestation. However, higher education can also be hypothesized to increase management skills and access to capital, thereby increasing the probability of market engagement, for example, in chili production, and leading to an increase in deforestation. In both models, native speakers of Spanish are controlled for either by a dummy variable if the household head is a native speaker, as in the household model, or by a measure of the total number of native speakers of Spanish in the *ejido*, as in the regional model. It is hypothesized that non-Spanish-speaking households, typically those of less assimilated ethnic groups, have less social capital (e.g. links to subsidies, information, and non-agricultural labor opportunities) than Spanish speakers. The non-Spanish speakers are much more likely to pursue a more subsistence-oriented production behavior.

### 5.3. *Subsidies/Credit*

A variable measuring credit received is included in both models but with different specifications. In the household model, this variable is measured by the percentage of the time interval over which the household received government agricultural credit, while in the regional model it is measured by the percentage of the *ejido* population that received government agricultural credit. The expectation in both models is that a greater level of agricultural credit will lead to expansion of agricultural output, thereby increasing deforestation, with some qualitative evidence that the amount of pasture created in response to specific government programs is subsidizing this activity (see Ch. 7).

### 5.4. *Wealth and Infrastructure*

Because of data limitations, other potential wealth effects and physical capital are included in different ways. In the household model, the percentage of each observed time interval during which the household owned a chainsaw and a vehicle are included, both of which are expected to increase deforestation. For the regional model, the only vaguely similar variables available were the percentage of the households in the *ejido* that had either electricity or running water in their houses. Wealth, as with the education variables, can have two conflicting effects. On the one hand, wealthier communities likely have more off-farm opportunities, decreasing land pressure, while on the other hand, they have more social capital, leading to access to subsidies and capital inputs for chili production. So, it is not clear which effect will dominate in the model; it is an empirical question.

### 5.5. *Tenure*

Land tenure issues are also controlled for in different ways, as different issues arrive with the two data sets. In the household survey data, one of the *ejidos* has a communal system of land tenure as opposed to recognized plot boundaries (usufruct) within the *ejido*. This *ejido* is indicated by a tenure dummy that other *ejidos* are compared against in the household model. The assumption here is that differential access to land affects the amount and location of land cultivated. This type of information is not available at the regional level. At this level, however, another tenure issue unfolds—the distinction between agricultural *ejidos* and forest extension *ejidos*. In most of the last type, strong regulations prohibit cultivation or severely limit it, as described in Chs. 3 and 7. Therefore, a dummy is included to indicate if a pixel is within a forest extension *ejido*.

### 5.6. *Biophysical Information*

In both models, the biophysical variables are the same: dummy variables indicating whether the pixel began the time period of observation in primary forest and whether the pixel registered upland soils. Also included are the slope and elevation of the pixel, and the average annual rainfall associated with the pixel. The potential effect of primary forest is unclear. Primary forests are associated with higher fertility soils, but have much higher clearing costs than secondary forests. Farmers prefer upland soils, so it is expected that this variable will have a positive effect on the probability of deforestation. It is hypothesized that steeper slope and higher elevation decrease the probability of deforestation, because the former characteristic increases labor and soil loss and the latter, largely crowns of hills, have thinner soils. An increase in average rainfall is expected to increase the probability of clearing for agricultural land use because this average tends to correlate with less variability in the timing and amount of precipitation.

### 5.7. *Amount and Location of Land*

Land availability is controlled for by the land area associated with each household in the survey in the household model and by the total land area of the *ejido* in the regional model. The expectation in both cases is that with a larger land endowment, the probability of deforestation for an individual pixel will decrease. The spatial variables associated with each model are slightly different. For the household model, the distance measures include distance from the household to the plot and distance from the *ejido* that the household is in to the nearest market. For the regional model, these measures are the distance from each pixel to the nearest road as well as distance to the nearest market. In each case, it is expected that an increase in accessibility will lead to an increase in deforestation.

### 5.8. *Temporal Issues*

To capture the effect of unobserved, inter-temporal factors affecting land-use choices, the household model also includes a variable that measures the duration of the household's occupancy as of the end of the time interval. The square of this variable is also included to allow for non-linearities. A similar variable does not exist from the census data for the regional model. Finally, both models include dummy variables for each TM scene used to create the dependent variable in each model to control for the fact that the TM scenes are of differing lengths (Allison 2000).

## 6. *Estimation Results*

The results from the regional logit model can be found in Table 12.3 and the results from the household hazard model can be found in Table 12.4. The size of the estimated coefficients for the two models cannot be easily tested to determine if they are statistically significantly different from each other, as they are different statistical models using different datasets. It is possible to compare the sign of the estimated coefficients of similar variables between the two models, however. Specifically, it is of interest to identify how each similar explanatory variable increases or decreases the probability or risk of deforestation in both of the model specifications.

With respect to the demographic variables in the hazard model, two of the three indices, males over 11 and children under 12, are seen to be positive and statistically significant determinants of the hazard of deforestation. The coefficient on females is not statistically significant. As in the hazard model, the logit model indicates that male population has a positive and statistically significant effect on increasing the probability of deforestation, while female population is negative and statistically significant. While the estimated coefficients on the female population variable differ between the models, the overall results are similar, an unsurprising result given that the majority of households in the region are semi-subsistent producers, for which family members simultaneously represent a source of labor as well as demand for outputs from the agricultural plots.

All the biophysical variables are statistically significant and all but one has the same sign across the two models. The primary forest variable is negative in both models, suggesting that farmers prefer to clear secondary forest, all else being equal, perhaps because clearing costs are much lower,

**Table 12.3.** Econometric results from regional logit model

| Dependent variable: deforestation or not | Estimated coefficient | t-statistic |
|---|---|---|
| primary | −0.8773 | −150.74 |
| elevation | −0.0019 | −91.57 |
| rain | 0.0012 | 22.24 |
| slope | 0.0145 | 13.15 |
| goodsoil | 0.3281 | 53.37 |
| freq_ag | 0.0592 | 67.11 |
| agdist | −0.0021 | −201.97 |
| d_roads | −0.0001 | −109.03 |
| d_market | −4.45e-06 | −12.72 |
| female | −0.0004 | −2.22 |
| male | 0.0035 | 21.73 |
| num_pixl | −2.01e-06 | −88.26 |
| m_prim15 | −0.0098 | −39.26 |
| span | −0.0041 | −69.04 |
| pwatr | −0.0048 | −33.82 |
| pelec | −0.0006 | −4.49 |
| exten | −0.0795 | −5.96 |
| p_credit | 0.0038 | 14.60 |
| constant | −2.099 | −30.35 |
| Pseudo $R^2$ | 0.2264 | |
| Number of observations | 3944875 | |

even though the weeding costs associated with agricultural use on secondary growth is higher (Dvorak 1992).[1] Superior upland soils as well as greater average rainfall increase the hazard or probability of deforestation in both models, as expected. Higher elevation is estimated to lower the likelihood of deforestation in both models, as higher elevated lands tend to be related to more rugged terrain, shallow soils and caprock, and less soil moisture, all factors inhibiting yields. The only biophysical variable that differs between the two models is the slope of the pixel. The logit model yields the counterintuitive result that increased slope increases the probability of deforestation. Thus the hazard model might yield the more consistent result that at the small scale of an individual parcel, slope decreases the probability of deforestation, but at the regional level, the more important factors of soil type and forest type better capture the clearing decision.

[1] This finding warrants further investigation for this reason. Many *ejidatarios* have surplus land, and yet it is not clear that they rotate their plots across all of it. We are not yet certain if the focus on certain parcels—those already cleared and/or in fallow—reflects perceptions about soils, the presence of *bajo* or other non-preferred conditions to cultivate, the distance to forested lands, or rules about forest reserves with *ejidos*.

**Table 12.4.** Econometric results from individual household hazard model

| Dependent variable: deforestation or not | Estimated coefficient | t-statistic |
|---|---|---|
| male | 0.0224 | 2.558 |
| female | 0.0105 | 1.141 |
| kids | 0.0152 | 3.026 |
| primary | −0.6696 | −38.924 |
| soil | 0.2598 | 11.911 |
| dem | −0.0104 | −54.039 |
| slope | −0.0165 | −4.277 |
| dry | 0.0369 | 7.938 |
| totpixl | −0.0002 | −22.513 |
| saw | 0.2101 | 6.994 |
| vehic | 0.4566 | 13.250 |
| edhead | 0.0213 | 8.268 |
| ed9 | 0.0379 | 6.489 |
| span | −0.1022 | −4.401 |
| credit | −0.2056 | −7.311 |
| dis | −0.0455 | −29.551 |
| nearmkt | −0.0038 | −4.105 |
| yr | −0.0397 | −13.099 |
| yrsq | 0.0005 | 8.933 |
| constant | −0.0628 | −0.215 |
| Likelihood Ratio $Chi^2(25)$ | 17725.93 | |
| Number of observations | 115017 | |

With respect to the variables measuring the land endowment, each model yields negative and statistically significant coefficients, so that all else being equal, the larger the land endowment, the less the pressure on forested lands for agricultural use. In the household hazard model, the time-varying parameters of ownership of a chainsaw and vehicle both increase the hazard of deforestation, which is consistent with the idea of lower labor costs in forest clearance and in access to the plot. For the regional logit model, the variables that attempt to measure a level of *ejido* wealth or connectivity with political forces, the percent of houses with running water and electricity, both decrease the probability of deforestation. These results may reflect a correlation with unobserved variables such as access to more off-farm opportunities, as would characterize the *ejidos* closer to the market centers where electricity and running water are more common.

The education variables are all statistically significant in each model, but with different signs: in the hazard model education has a positive effect on deforestation, while in the logit model it has a negative effect. To the extent that higher education implies a higher opportunity cost of on-farm labor due to increased wage-earning potential, the expectation is for these variables to carry negative coefficients. However, higher education could also imply better agricultural management skills, leading to an increase in potential returns to agricultural use. Perhaps it is this difference in the interpretation of the education variables that leads to the difference in the empirical results from the two scales of the model. Or these results may also reflect the different spatial scales of analysis involved. For example, as the education variable for the logit model is the average education level for the entire *ejido,* the variation in this variable could also be a proxy for other variables not included in the logit model, such as duration of occupancy or age of *ejido*, which is controlled for in the hazard model, leading to the change in sign in the estimated coefficient. Being a Spanish speaker decreases the hazard and probability of deforestation in both models, which may result from a greater reliance of indigenous farmers on the resource base as opposed to off-farm wage-earning opportunities.

The credit variables are statistically significant in both models, but have differing signs. In the household model the estimated coefficient is negative, while in the regional model, it is positive. This outcome could be the result of the different way that the variable is defined in each of the models. However, if it can be argued that the two variables are the same basic measure, then perhaps the results show that at the household scale, the estimated negative coefficient on credit could be capturing the increased ability to purchase land-saving inputs, such as chemical fertilizers, thereby intensifying land use and decreasing the pressure on forests, while at the *ejido* scale, the greater the amount of government credit, the greater the ability to, for example, hire labor and clear more forest land for agricultural use. The estimated positive coefficient of the tenure dummy lends support to the hypothesis that insecure property rights increase incentives to clear forest as a means of establishing access in the household model, while in the regional model the negative coefficient on the forest extension dummy variable indicates that all else being equal, land is less likely to be cleared there for agricultural use.

For the location and access variables in both models, the estimated coefficients are negative and statistically significant. This result is consistent with the hypothesis that higher travel costs decrease the returns from agricultural land use, through a reduced farm-gate price of output for a market-oriented farmer, or the increased travel distance decreases the time available for agricultural activity for the subsistence farmer, especially moving the harvest to the homestead.

In the hazard model, different specifications of the variable measuring duration of occupancy were explored by means of a nested likelihood ratio test, where it was determined that the quadratic functional form was optimal in terms of fit and parsimony. The estimates indicate that the conditional probability of forest conversion decreases with the passage of time at a decreasing rate, with some evidence of a reversal after thirty-eight years. This result may reflect a confluence of factors, including fallow cycle strategies and adaptation to local market opportunities and ecological constraints, as well as household life-cycle issues. A similar variable is not available at the regional level, although future research will attempt to include a variable on the date of *ejido* formation to test similar hypotheses concerning the role of duration of occupancy on land-use change.

## 7. *Predictions*

Each of the estimated models was used to predict deforestation with its associated dataset. Prediction in discrete choice models, as in the two models estimated in this chapter, is different than in a regression model, in that the prediction for any observation is a *probability* of an event or choice occurring. In the models here, the probability or hazard of a particular pixel becoming deforested is calculated using the estimated model. These probabilities can range from zero to one and are calculated as follows:

$$\widehat{P}_i = \frac{\exp(\widehat{\beta}_i X_i)}{1+\exp(\widehat{\beta}_i X_i)} \tag{10}$$

In order to produce a measure of how well the estimated models of deforestation fit the actual deforestation that did or did not occur, some way of comparing the predicted probabilities of deforestation for each pixel to the actual deforestation that did or did not occur for each pixel is needed. This, in turn, requires some way of choosing a critical value of the predicted probability of deforestation to 'count' as deforested. Often researchers take 50 per cent as the critical threshold, that is, if the probability of a change occurring is greater than 50 per cent, then that observation is calculated as a change, in order to compare to the actual changes. A different approach is taken here, however, that has been developed in previous work of the project (Geoghegan *et al.* 2001). This approach can conceptually be divided into two steps for each model. Instead of imposing an ad-hoc percentage to consider a pixel a 'deforested' pixel, first the total number of actual pixels, *n*, that were deforested was identified. Secondly, the spatial distribution of highest probability pixels was sequentially identified until *n* had been selected. That is, each model does not attempt to predict the actual amount of deforestation, but takes that amount as given, and predicts the spatial distribution of that deforestation over the landscape. For the regional model, this critical threshold for the logit model is 21 per cent, so that any pixel with a predicted probability of deforestation of 21 per cent or higher was considered deforested for comparison with actual deforestation.

The maps of the results of this comparison of predicted deforestation to actual deforestation for the regional model are found in color at Pl. 3 for the entire region and Pl. 4 for a windowed area of the regional model: Nicolas Bravo for the eastern edge of the region on Highway 186.[2] The overall pattern of predictions in these figures shows the large number of pixels that were correctly predicted to remain in forest throughout the region. The correctly predicted deforested pixels are mainly found along the road network and contiguous to previous agricultural land use, showing the influence of these important variables on the predictions. The incorrectly predicted deforested pixels (i.e. the pixels that were predicted to be deforested, but actually remained in forest), however, also show a similar spatial pattern, suggesting that these two variables, distance to roads and distance to nearest agricultural land use, is over-predicting in the model. The incorrect forest predictions (i.e. those pixels that were predicted to remain in forest use, but actually were deforested) are much more randomly distributed throughout the landscape, suggesting that there are further spatial processes occurring in these areas that are not currently explained by the model. The most important information that is currently lacking at the regional level is further information on the location of alternative land uses to cultivation and pasture within an *ejido*, such as the communal forest lands, as mentioned previously, or individual agricultural plot boundaries, as is available for the survey data.

[2] The map at Pl. 3 also includes the borders of each different TM scene that was used in the analysis and their associated dates. The white areas on the maps (Pls. 3, 4) are pixels that were not included in the model because they were not part of an agricultural *ejido*, or they were originally not in forest or agricultural use, or their land use was unknown from the satellite data, due to clouds or other problems.

A similar procedure was completed for the household hazard model. The critical threshold for the hazard model is 29 per cent, so that all pixels with a greater than 29 per cent probability of deforestation are considered deforested for the predictions. This difference in probabilities between the regional model and the household model could be considered a measure of the improved fit of the household model, given the higher cut-off probability. Another way to evaluate the relative fits of the models is to compare the total number of correct and incorrect predictions. The regional model correctly predicts 33 per cent of the deforested pixels, while the household model correctly predicts 46 per cent of the deforested pixels. For the pixels that remain forested, the regional model correctly predicts 96 per cent, while the household model correctly predicts 89 per cent.

The improvement of the household model for predicting deforestation reflects the value of adding the richer household-level survey data that better captures the individual causes of deforestation than is possible at the aggregate level with the census data in the logit model. One possible reason for the better predictions on remaining in forest for the logit model over the hazard model is that for the latter, the agricultural plot boundaries for the individual household are known from the sketch-map work associated with the household survey. For the regional model, however, only the boundaries of each *ejido* are known, not the boundaries of the potential agricultural land within each *ejido*. Many of the *ejidos* contain forested land that is not available for agricultural use (e.g. communal forest reserves; Ch. 8) and the geographical boundaries of these reserves are not currently known. Therefore, included in the logit model are forested pixels that are assumed to be available for agricultural use, but in actuality are not available. It is highly likely that these forested pixels are correlated with values of variables that would predict a low probability of deforestation from the estimated model; for example, the forested pixel is far from the road. However, it might be that land far from the road that was specifically chosen as the forest reserve, therefore the model would predict the pixel remaining in forest for the wrong underlying cause.[3]

In order to compare the prediction results from the regional logit model with the prediction results from the household hazard model, we focus only on the observations that appear in both data sets. This totals 68,049 pixels,

[3] The individual parcels from the household survey are small and scattered over the region, rendering their presentation difficult. To display the results from even a few *ejidos* such that the individual parcels could be seen would require multiple pages. For this reason, maps at this level are not shown.

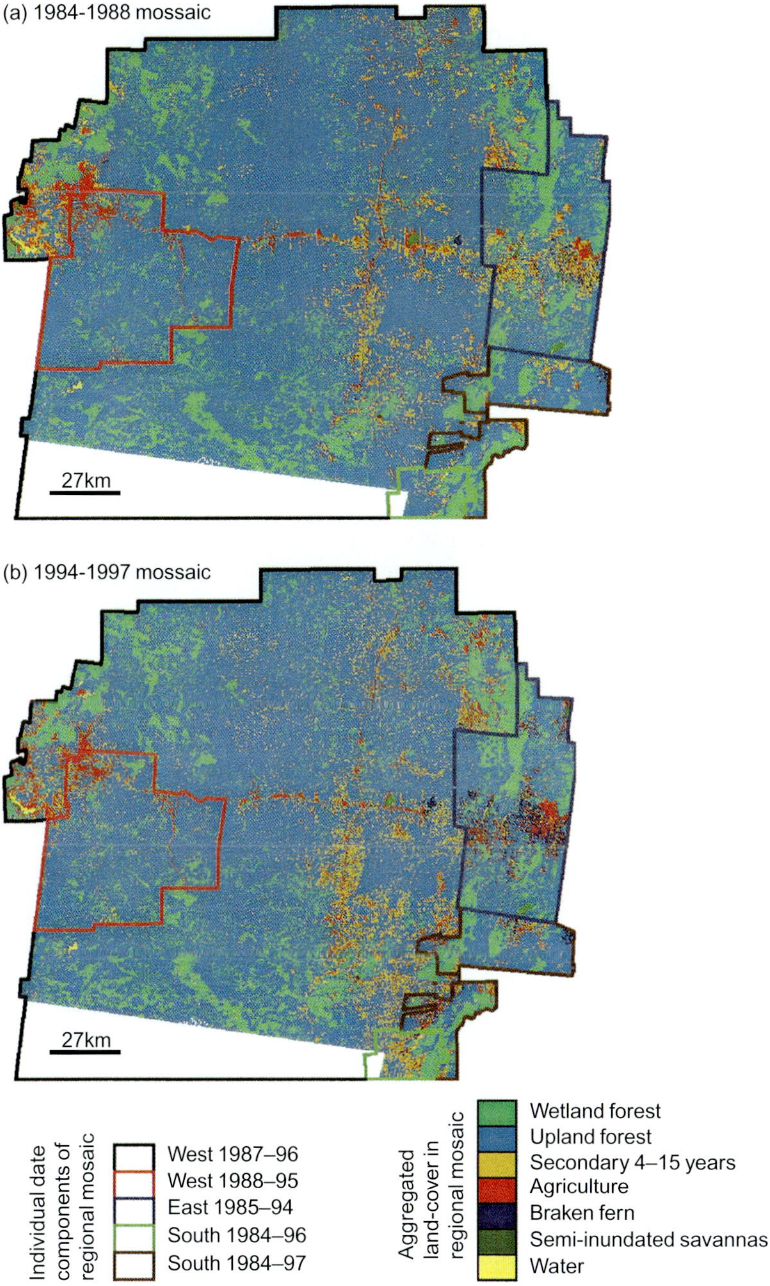


**Plate 1.** Regional land cover using Landsat TM classification.

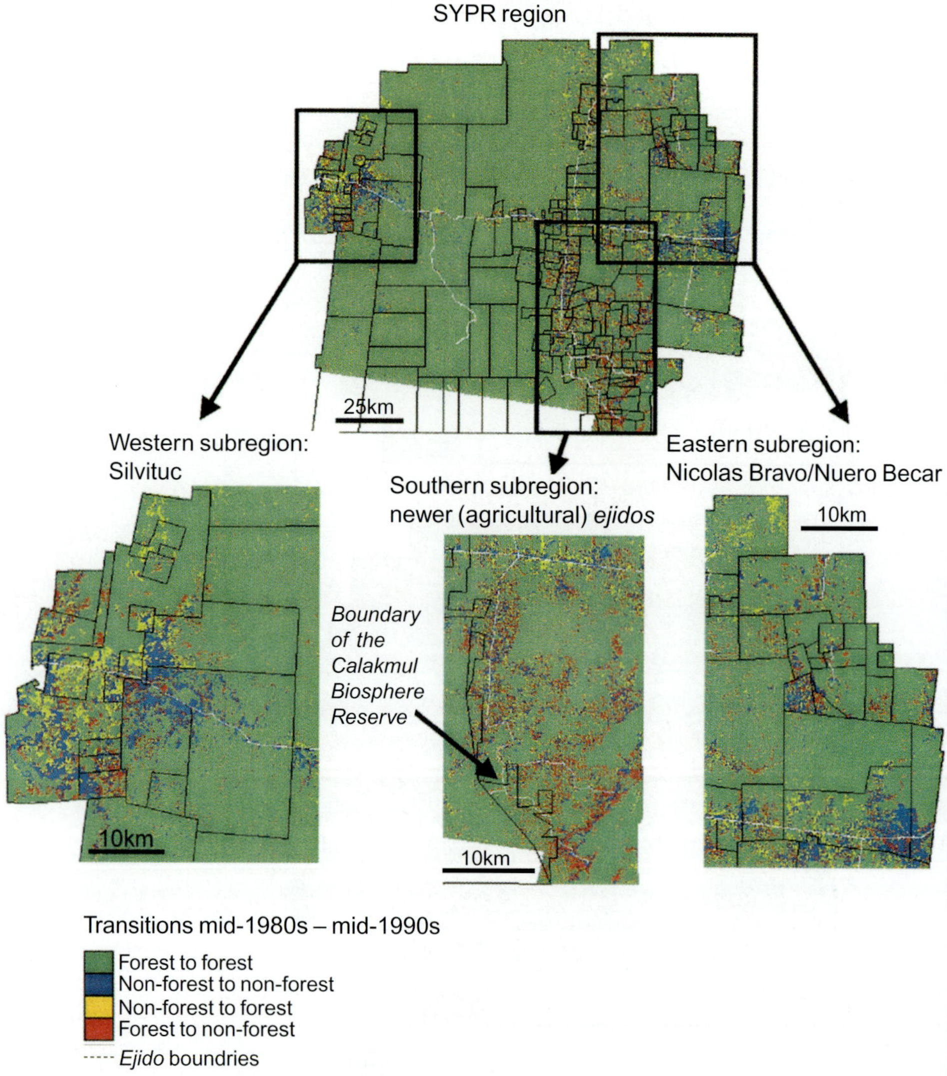


**Plate 2.** Land-use change and deforestation hotspots, 1987–97.

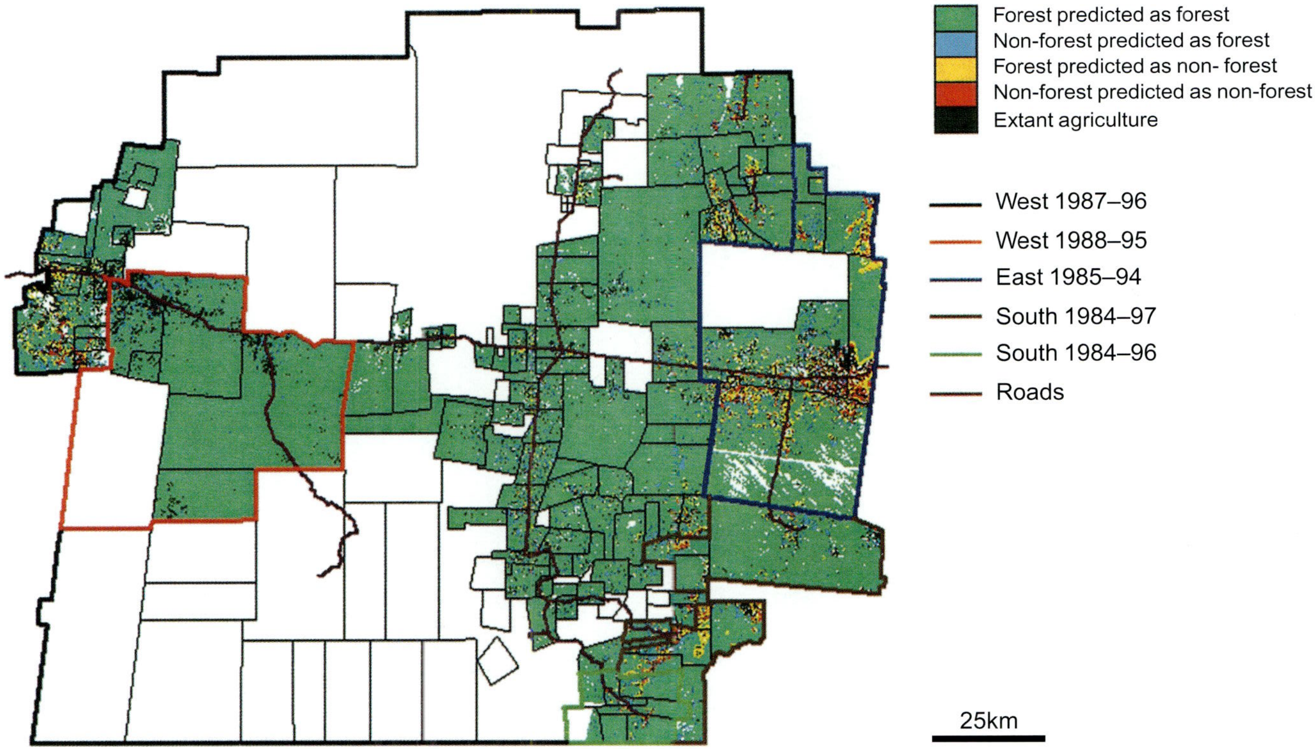


**Plate 3.** Regional model: Correct and incorrect predictions.

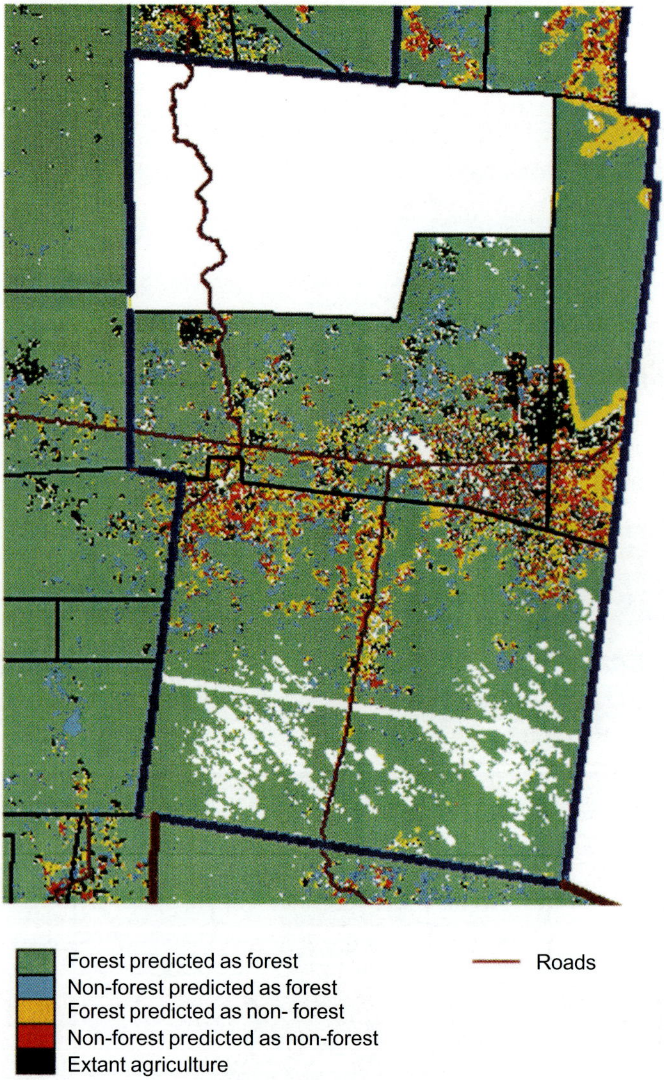


**Plate 4.** Regional model detail.

which is less than 2 per cent of the observations used in the regional logit model. For this comparison data set, 75 per cent of the total pixels were correctly predicted by both models; 7 per cent were correctly predicted by the logit model, but incorrectly predicted by the hazard model; 12 per cent were incorrectly predicted by the logit model, but correctly predicted by the hazard model; and the remaining 6 per cent were incorrectly predicted by both models. Both models incorrectly predict behavior, and future research will focus on further investigation of what modeling features drive these results.

## 8. *Conclusions*

This chapter has demonstrated the methodologies that have been and continue to be developed in this research project for combining different types of spatial data for modeling land-use change. Elsewhere, we have referred to this kind of effort as 'socializing the pixel' and 'pixelizing the social' (Geoghegan *et al.* 1998), a centerpiece of proposed work by the larger land-use/cover-change community of researchers. The aim of this body of work is to improve understanding and explanation of the variation in the regional dynamics of land change, including improvements in spatially explicit models that move beyond the magnitude of changes within a region to the locations of that change.

There remains much data and modeling work to be done to achieve these aims. As previously discussed, further information on the location of within-*ejido* settlement locations and *ejido* forest reserves, and variations in *ejido* usufruct rules would likely greatly improve the accuracy of the regional model. Further modeling work includes adding more temporal observations to the regional model, so that a hazard model approach could be used for those data and a more direct comparison between the regional model and the household model would be possible. At present, it is likely that at least some of the differences between the two models estimated in this chapter are artefacts of the different empirical specifications used and not necessarily a feature of the spatial or temporal scales of the data or the models.

Additional modeling considerations include modeling other land changes, such as forest regrowth, to attempt better to capture the dynamic of crop–fallow cycles. This effort will require the use of biophysical feedbacks related to the ecology work described in Chs. 4–5 on soil-vegetation nutrient cycling under different land uses to investigate how individual land-use histories affect land-use trajectories. Specifically, future work will investigate the feedbacks between land-use history and invasive species, such as the bracken fern that occurs throughout the region (Chs. 4 and 6). Other research will focus on the relationship between the spatial patterns of land-use change, such as fragmentation, and how that affects the biotic diversity of the landscape.

Another important future linkage between the econometric modeling work and the ecology research concerns carbon cycle issues, focusing specifically on the relationship between land use and carbon sequestration and how this relates to global warming policies: for example, using the econometric models to determine changes in the regional carbon stock and primary productivity under different sets of assumptions. Based in the recent history of land change that has occurred in the region, trajectories of land-use change will be developed from which the net impacts on future carbon stocks can be calculated. Moreover, in response to issues relating to the use of forests to sequester carbon (Watson *et al.* 2000), an analysis of the potential costs of carbon sequestration for the region can be performed. Recent work in Chiapas, Mexico (De Jong, Tipper, and Montoya-Gomez 2000) concludes that because agriculture in that region is only marginally profitable, the costs of carbon sequestration will be low, as it will take small payments to farmers to remove land from agricultural use. This Chiapan study, however, does not take into account the semi-subsistence nature of *ejido* agriculture, as detailed in Ch. 11, and it likely understates the 'value' of the land to the *ejidatario* and the costs of retiring land from agricultural use absent alternative employment opportunities. The true cost of carbon sequestration is likely to be much higher than estimated.

## REFERENCES

Allison, P. D. (2000). *Survival Analysis Using the SAS System: A Practical Guide*. Carey, NC, SAS Institute Inc.

Chisholm, M. (1962). *Rural Settlement and Land Use: An Essay on Location*. London, Hutchinson.

Chomitz, K. M., and D. A. Gray (1996). 'Roads, Land Markets and Deforestation: A Spatial Model of Land Use in Belize'. *World Bank Economic Review* 10: 487–512.

Cropper, M., C. Griffiths, and M. Mani (1999). 'Roads, Population Pressures, and Deforestation in Thailand'. *Land Economics* 75: 58–73.

Cropper, M., J. Puri, and C. Griffiths (2001). 'Predicting the Location of Deforestation'. *Land Economics* 77: 172–86.

De Jong, B. H. J., R. Tipper, and G. Montoya-Gomez (2000). 'An Economic Analysis of the Potential for Carbon Sequestration by Forests: Evidence from Southern Mexico'. *Ecological Economics* 33: 313–27.

Dvorak, K. (1992). 'Resource Management by West African Farmers and the Economics of Shifting Cultivation'. *American Journal of Agricultural Economics* 74: 809–15.

Geoghegan, J., and N. E. Bockstael (2000). 'Smart Growth and the Supply of Sprawl', Paper presented at the Association of Environmental and Resource Economists. *Workshop on the Effectiveness of Resource and Environmental Regulation*. La Jolla, Calif. Unpublished.

Geoghegan, J., L. Pritchard, Jr., Y. Ogneva-Himmelberger, R. Roy Chowdhury, S. Sanderson, and B. L. Turner II (1998). ' "Socializing the Pixel" and "Pixelizing the Social" in Land-Use and Land-Cover Change', in D. Liverman, E. Moran, R. Rindfuss, and P. Stern (eds.), *People in Pixels: Linking Remote Sensing and Social Science*. Washington DC, National Academy of Sciences Press, 51–69.

Geoghegan, J., S. Cortina Villar, P. Klepeis, P. Macario Mendoza, Y. Ogneva-Himmelberger, R. Roy Chowdhury, B. L. Turner II, and C. Vance (2001). 'Modeling Tropical Deforestation in the Southern Yucatan Peninsular Region: Comparing Survey and Satellite Data'. *Agriculture, Ecosystems, and Environment* 85: 25–46.

Irwin, E. G., and N. E. Bockstael (2001). 'Interacting Agents, Spatial Externalities, and the Endogenous Evolution of Residential Land Use Pattern'. *Journal of Economic Geography* 2: 31–54.

Irwin, E. G., and J. Geoghegan (2001). 'Theory, Data, Methods: Developing Spatially-Explicit Economic Models of Land Use Change'. *Agriculture, Ecosystems, and Environment* 84: 7–24.

Maddala, G. S. (1983). *Limited Dependent and Qualitative Variables in Econometrics*. Cambridge, Cambridge University Press.

Nelson, G. A., and D. Hellerstein (1997). 'Do Roads Cause Deforestation? Using Satellite Images in Econometric Analysis of Land Use'. *American Journal of Agricultural Economics* 79: 80–8.

Nelson, G., V. Harris, and S. Stone (2001). 'Deforestation, Land Use and Property Rights'. *Land Economics* 77: 187–205.

Pfaff, A. (1999). 'What Drives Deforestation in the Brazilian Amazon? Evidence from Satellite and Socioeconomic Data'. *Journal of Environmental Economics and Management* 37: 26–43.

Turner, B. L., II, and A. M. S. Ali (1996). 'Induced Intensification: Agricultural Change in Bangladesh with Implications for Malthus and Boserup', in *Proceedings, National Academy of Sciences*, 14984–91.

Turner, M. G. (1988). 'A Spatial Simulation Model of Land-Use Changes in a Piedmont County in Georgia'. *Applied Mathematics and Computation* 27: 39–51.

Watson, R. T., I. R. Nobel, B. Berlin, N. H. Ravindranath, D. J. Verardo, and D. J. Dokken (eds.) (2000). *Land Use, Land Use Change, and Forestry. A Special Report of the IPCC.* Cambridge, Cambridge University Press.

# 13

# The SYPR Integrative Assessment Model: Complexity in Development

*Steven Manson*

## 1. *Integrated Assessment*

Be it global environmental change or environment and development, land-use and land-cover change is central to the dynamics and consequences in question in the southern Yucatán peninsular region. Designing policies to address these impacts is hampered by the difficulty of projecting land use and land cover, not only because the dynamics are complex but also because consequences are strongly place-based. This chapter describes an *integrated assessment* modeling framework that builds on the research detailed in earlier chapters in order to project land-use and land-cover change in a geographically explicit way.

Integrated assessment is a term that describes holistic treatments of complex problems to assess both science and policy endeavors in global environmental change (Rotmans and Dowlatabadi 1998). The most common form of integrated assessment is computer modeling that combines socio-economic and biogeophysical factors to predict global climate. Advanced in part by the successes of these global-scale models, integrated assessment has expanded to structure knowledge and set research priorities for a large range of coupled human–environment problems. Increasing recognition is given to the need for integrated assessment models to address regional-scale problems that are masked by global-scale assessments (Walker 1994). Such models must address two issues to project successfully land-use and land-cover change at the regional scale. First, change occurs incrementally in spatially distinct patterns that have different implications for global change (Lambin 1994). Second, a model must account for the complexity of, and relationships among, socio-economic and environmental factors (B. L. Turner *et al.* 1995). The SYPR integrated assessment model, therefore, has a fine temporal and spatial grain and it places land-use and land-cover change at the intersection of land-manager decision-making, the environment, and socio-economic institutions. What follows is a description of an ongoing integrated assessment modeling endeavor of the SYPR project (henceforth, SYPR IA model).

## 2. *Conceptual Framework: Actors, Institutions, and Environment*

The depth and breadth of the SYPR project poses a challenge to the integrated assessment modeling effort since some unifying framework must reconcile a broad array of issues, theories, and data. The global change research community offers a general conception of how environmental change results from infrastructure development, population pressure, market opportunities, resource institutions, and environmental or resource policies (Stern, Young, and Drukman 1992). When applied to land-use and land-cover change, these causes and effects are apportioned among the conceptual foci of social systems, ecological systems, and land managers (B. L. Turner *et al.* 1995). The SYPR IA model recasts these foci as a three-component *actor-institution-environment* conceptual framework. The first part focuses on the decision-making of households and other actors that directly contribute to change in land use and land cover. The second component concerns socio-economic institutions that affect actor decision-making. Both actors and institutions interact with the third component, the biophysical environment.

Before examining the conceptual framework in detail, it is useful to understand its context. Use of separate, yet linked, actor and institutional components reconciles views of decision-making that range from narrowly defined individual rational choice to perspectives that focus on social and cultural structures. The actor component has roots in rational choice theory. Although criticized for its shortcomings, this framework has accommodated criticism and remains an analytically defensible technique when used in a manner that recognizes its limitations (Myers and Papageorgiou 1991). An extension to rational choice is bounded rationality, which relaxes assumptions regarding the generation of choice alternatives, such as perfect access to information, and attempts to account for factors such as learning over time (Simon 1997; Tversky and Kahneman 1990). Bounded rationality can allow for external influences in a manner consistent with social and cultural perspectives on decision-making (Jager *et al.* 1997), embodied here in the institutional component of the conceptual framework.

A variety of theoretical frameworks speak to actor decision-making in a land-change context (see also Agarwal *et al.* 2000; Kaimowitz and Angelsen 1998; Lambin 1994). The actor component of the conceptual framework concerns agrarian decision-making within the context of bounded rationality and resources available to actors. This view is based on agrarian and development literature that posits household behavior in terms of rational choice theory, where activity ranges from subsistence to market production (Singh, Squire, and Strauss 1986). An actor chooses production activities that maximize utility; under bounded rationality the actor satisfices utility. The utility afforded to the actor by production activities is subject to external factors and actor attributes. The latter can be treated as a *resource profile* that includes measures of access to information, ability to learn, and personal characteristics (Sen 1981).

Actor decision-making is affected by socio-economic institutions (Ch. 7). These are conceived as 'shared concepts used by humans in repetitive situations organized by rules, norms, and strategies' (Ostrom 1999: 37). Rules are shared prescriptions that are predictably enforced, while norms are those that are generally enforced through costs and inducements. Strategies are the regularized plans that individuals make given their understanding of rules, norms, and the likely behavior of others. Institutions are characterized by their members, geographical extent, degree of formalization, regulatory mechanisms, and administrative structure (Young 1994). Institutions are important to the SYPR IA model primarily because they influence actor decision-making, and they are a useful proxy for the external large-scale political economy.

Completion of the conceptual framework hinges on how the environment relates to institutions and actors. Actor production strategies in forests center on agriculture, timber, and non-timber forest products (Galletti 1998). These strategies are subject to biophysical configurations of precipitation, soil characteristics, secondary growth, pest infestations, and disturbances such as fire or hurricanes. Complex relationships, laden with feedback, result when these biophysical forces are in turn affected by production strategies that lead to changes, such as cover conversion, soil erosion, or nutrient depletion (Uhl 1987; Chs. 4 and 5).

The actor-institution-environment framework offers a general view on human–environment situations given its derivation from global environmental change research. Its key focus is how actor decision-making affects the environment and how it is influenced by the environment and by institutions. The framework is sufficiently flexible, however, to allow for the fact that actors are active participants in institutions. Similarly, institutions and the environment can be seen as affecting each other through the intermediation of agent decision-making. Finally, the conceptual framework complements work presented in previous chapters, particularly those concerning models of decision-making (Chs. 10 and 11). The dynamic linkage between actors and institutions is grounded in the rational choice framework that underlies SYPR project's economic research.

## 3. *Simulating Land-Use and Land-Cover Change*

While there are a number of spatiotemporally explicit modeling techniques, no single approach can implement the actor-institution-environment framework in order to create the SYPR IA model. It has been necessary, therefore, to create a unified software system that can combine a variety of approaches. While this simulation system has primarily been used to create the SYPR IA model, it can be treated as an entity unto itself.

The simulation system relies on scripts or text files written in a simulation language. Each script is a blueprint for building a simulation model from basic building blocks, such as control structures, variables, functions, and software objects. The simulation language provides three groups of software functions. The first is specialized simulation routines that allow the user to automate tasks and incorporate the functionality of other software applications. The second set of functions forms an *agent-based model*, a system of software objects that semi-autonomously manipulate information and perform actions. The third set of functions comprises a *cellular automata*, a two-dimensional grid where cell values change over time according to rules based on the value of adjacent cells or external inputs. The following sections examine these specialized functions within the context of the SYPR IA modeling effort.

Just as an assortment of tools and materials can be used to construct various buildings, the simulation system is a general toolbox that can be used to construct an array of models. When constructing a building, a general vision of the structure is embodied in a blueprint that in turn guides the application of tools and materials. The conceptual framework and integrated assessment principles provide the general intent and form of the SYPR IA model. A first draft blueprint is provided by the technique of *dynamic spatial simulation* as it has been applied to modeling deforestation (e.g. Gilruth, Marsh, and Itami 1995; Southworth, Dale, and O'Neill 1991). This method portrays the landscape as a two-dimensional grid where rules based on factors such as agricultural suitability determine how cells change. In an exhaustive review of deforestation models, this method is considered the 'most advanced modeling approach for a complex, dynamic and spatial problem such as tropical deforestation' (Lambin 1994: 92).

This basic architecture has been expanded by using cellular automata functions to model the environmental component of the conceptual framework and agent-based modeling is used for the actor and conceptual components. It is important to emphasize the difference between the uses to which the agent and cellular models are applied. Cellular automata models are used to represent actor decision-making when the actors are distributed in a manner consistent with the assumptions underlying cellular automata, such as tesselated actors reliant on local interactions (Benenson 1999). In cases where these assumptions do not hold, there is much to be gained by agent-based approaches instead of cellular automata (Torrens and O'Sullivan 2001).

Before continuing, it is important to review some terminology employed in modeling endeavors of this kind. Key is the distinction among the terms 'agent', 'actor', and 'institution'. The latter two refer to components of the conceptual framework in that 'actors' are land managers and 'institutions' are socio-economic and political frameworks, while an 'agent' is a software object used to represent individual actors and institutions. The correct term for an agent representing an actor or institution, respectively, is 'actor-agent' or 'institution-agent', but when examining obviously computational or methodological details of a model, the -agent suffix is assumed. Also important is the use to which the terms 'model' and 'simulation' are applied. The model is a framework of concepts and data that is embodied in simulation language scripts to become a 'simulation model'. In this sense, it is most appropriate to refer to the 'SYPR integrated assessment simulation model' but the 'SYPR IA model' suffices given the specificity of its application.

## 4. *SYPR IA Model Overview*

When set in motion, the SYPR IA model creates an artificial world, a simulated southern Yucatán peninsular region as it was in 1970. Time then starts in this simulated world, measured as a series of simulated years over forty 'model years' (1970 to 2010). Within each year, four general processes occur in the simulated study region. First, parameters exogenous to the simulated world are introduced or changed. Second, institutions change actor resource profiles or decision variables. Third, the environment, expressed primarily as variables related to land cover, evolves according to ecological rules and the impact of previous actor decision-making. Finally, each actor in the simulated southern Yucatán peninsular region makes land-use decisions that take into account its resource profile, exogenous parameters, institutional imperatives and environmental factors. The results of these decisions are registered in the actor's resource profile and as an environmental impact when appropriate. Upon completion, the simulation results are validated against observed data, as described below. The next six subsections explore the simulation procedure in detail.

Before continuing, however, a quick overview of data is in order. Much of the modern history of land-use and land-cover change in the region is accessible through government documentation, scientific research, the local populace, remotely sensed data, and the SYPR project. As described throughout this volume, the SYPR project provides spatial data and an actor survey of households (Chs. 8 and 10). Taken across eleven *ejidos*, the survey details an extensive set of socio-economic characteristics that bear on decision-making, including household statistics, production activities, labor factors, institutional affiliations, and spatiotemporally explicit land-use histories. In addition to these land-use histories, the project's spatial data includes satellite imagery, land-use and land-cover maps, aerial photography, and geographical information system layers for a host of socio-economic and biogeophysical phenomena.

### 4.1. *User Model Inputs*

Within each iteration, the model first initializes or updates information that feeds into later simulation stages. These user-specified simulation variables are stored in database tables and spatial layers where it may be accessed by model components. The bulk of user variables concern production activities, such as the rate at which land may be cleared or the effort required to maintain land under cultivation. This information is derived from research literature and project personnel. Scripts calculate production characteristics every year as a function of household characteristics. The rate at which land can be cleared, for instance, depends on labor and technology, such as number of family members or possession of a chainsaw. Miscellaneous data, such as the amount of food necessary to support an individual, affects agent decision-making. These are determined through consultation with SYPR project members, government publications, and research literature. The extent and depth of variables is typically dependent on the degree to which these data are necessary to other parts of the model.

### 4.2. *Institutions*

Critical institutions in the region include state-owned forests, forest concessions, the biosphere reserve, government subsidy programs, non-governmental organizations, and *ejidos* (Alcorn and Toledo 1998; Chs. 3, 7, and 10). The model focuses on large-scale agricultural institutions: land tenure, governmental subsidies, and economic markets. Land tenure includes *ejidos*, the biosphere reserve, state-owned forests, and timber concessions. Government payments through the PROCAMPO program offer direct payments to farmers (as described in Ch. 7).

The SYPR IA model uses agent-based model functions to represent both institutions and actors. This method is increasingly used to combine empirical and theoretical knowledge of decision-making. Agents are software programs that represent adaptive autonomous entities that extract information from their environment and apply it to functions such as perception, planning, and learning (Conte, Hegselmann, and Terna 1997). All three institutions have spatial characteristics that can be expressed as geographic information system layers that institution-agents combine with internal rules to determine how and when to influence actor-agent decision-making. The land-tenure agent guides actor decision-making by limiting land use according to location and actor characteristics. Subsidy institutions make available funds to targeted actors by changing their resource profiles. The economic market agent creates a layer for distance to market and makes it available to all agents and sets prices according to user-specified variables.

All three institutions act in a deterministic manner and, as such, are less interesting than actors as examples of how an agent-based model can be used. Given the flexibility of the modeling system, however, later iterations of the SYPR IA model can link to econometric research or to larger-scale economic models in a manner similar to that employed by other regional scale models (e.g. Riebsame *et al.* 1994; Yates and Strzepek 1998). It is also possible to develop more intentional and dynamic institution-agents that would have fuller participation of actors (e.g. Seror 1994).

All three institutions in the SYPR IA model (land tenure, subsidies, and the market) are important to land use in the southern Yucatán peninsular region. Land tenure is essential to land use and much of this resides in *ejidos*. Plot-level data, as gathered in the 1997 and 1998 survey (Ch. 8), are valuable given its spatial specificity, but acquiring it for the entire study region would prove resource intensive. Market and subsidy information is critical to harness the explanatory power offered by actor decision-making models based on economic theory. Finally, Ch. 7 describes how institutions not included in the present model can affect land use, and incorporating them is a long-term goal of the SYPR IA modeling effort.

### 4.3. *Environment*

The environmental component of the conceptual framework is based on cellular automata, which have a long history of being applied to physical and biological processes (Smith and Bull 1997). The raw material of cellular automata is a discrete lattice of uniformly shaped cells. This lattice is functionally identical to the geographic information system raster layers in which the SYPR project stores spatial data, since these are two-dimensional grids of cells. The SYPR IA model therefore uses these layers as a base for the cellular automata, defined as $[S, N, T]$ where $S$ is a set of finite cell states, $N$ is the set of states of neighboring cells, and $T$ is a set of transition rules that pair input-state tuples to an output state for the cell.

The prime example of states are land-use and land-cover classes, where states are mutually exclusive categories, such as those created by the image-processing efforts of the project (Ch. 6). Examining a forest/non-forest case, cells exist either in a forest state ($s_1$) or non-forest state ($s_2$), $S = \{s_1, s_2\}$. The neighborhood can be any configuration; for this example, the neighborhood is the cell itself and its four non-diagonal neighbors. At regular intervals, each cell in the layer is tested to see what state it will become in the next time-step. This future state is determined by $T$, where the number of possible rules that enumerate every permutation of $S$ and $N$ is $S^N$. In the case of the example here, there are two states (forest/non-forest) and a cell's five neighbors (four other cells and itself); there are $2^5$ different possible rules.

In theory, *T* is a fully enumerated set, but the simulation language allows users to specify rules that are more general. For example, one can specify that 'if the number of neighbors in a forested state is greater than three then change the state of the cell to forest'. By using compound conditional statements, complex ecological transition rules within a land-cover layer are created. Cellular automata can also refer to cells in other layers in order to incorporate other biogeophysical factors into state transition rules. This mechanism is codified as generalized cellular automata, which allows transition rules that do not reference adjacent neighborhood cells (Takeyama and Couclelis 1997). These other layers are treated as cellular automata where there are no transitions taking place (i.e. $T = \{\varnothing\}$) and the definition of states, *S*, is relaxed so that cells can take continuously varying values. These states do not need to be discrete since no state transformations take place. Other layers can then refer to these values in their transition rules. The ability to reference other layers is how cells affected by actor-agents are recognized as such and are accommodated by the environment. The simulation language also allows transition rules probabilistically to assess neighborhood states in order to make cell transitions more stochastic.

Despite the flexibility offered by the simulation system, only a few of the range of ecological relationships that exist in the region (Chs. 2, 4, and 5), are included in this version of the SYPR IA model. Current rule sets focus on modeling secondary succession, pest invasions, and soil fertility degradation. Future directions are suggested by work on rangeland dynamics (Li and Reynolds 1997), species composition (Silvertown *et al.* 1992), forest succession (Hogeweg 1988), and a host of biological models (Ermentrout and Edelstein-Keshet 1993; Gronewold and Sonnenschein 1998). Generalized cellular automata show potential for modeling large-scale phenomena such as fires or hurricanes as provided by project ecological research.

These potential extensions beg the question: how complex should the model environment be? Having an essentially unchanging environment is in keeping with deforestation models in general, is simple to implement, and may be adequate for short-term estimates of deforestation. A static environment is inadequate, however, if ecological processes such as succession are incorporated into observed actor decision-making; the extent to which this is true is still unclear. Moreover, ecological processes at the species composition scale are difficult to specify given species–area relationships, recruitment limitations, and succession dynamics.

## 4.4. *Actors*

Key actors in the region are land managers, ranchers, and smallholders whose production activities include swidden cultivation, agroforestry, modest logging, and market cultivation. A submodel probabilistically inserts actor-agents into the simulated study site at one of several hundred settlement locations according to exogenously given population densities. Actors are created in a given location, corresponding to a spatial layer cell, and make annual production decisions in the area around their respective locations. Actor-agents exist in a landscape represented by a collection of spatial layers contained either in the environment component or separately within the model. Actor-agents 'know' the surrounding landscape in so far as the agent software processes have access to layers representing spatial attributes such as land-cover type or distance to roads.

An actor makes decisions about production activities in order to maximize utility. In the simplest case, the actor is limited to a single strategy, such as *milpa* (swidden agriculture), and the actor's utility is tied to the amount and location of land under this form of cultivation. In a computational setting, this choice is played out on the landscape by the actor choosing which cells in the land-use layer will host production, a multicriteria evaluation problem (Eastman *et al.* 1995). An actor-agent determines the suitability, $S$, of each grid cell $c$, for a given production activity $l$, $S_{cl}$:

$$S_{cl} = \sum_{i=1}^{m} w_{il} v_{il} * \prod_{j=1}^{n} b_{jl} \qquad (1)$$

as a function of a set of continuously varying spatial factors $V = \{v_l, \ldots, v_m\}$, a set of factor weights $W = \{w_l, \ldots, w_m\}$, and a set of Boolean constraints $B = \{b_l, \ldots, b_n\}$. Factors $V$ are spatial layers of landscape attributes such as elevation. The importance for each use $l$ of any given factor $v_{il}$ is considered as a weight, $w_{il}$, which is determined by each agent, usually as a function of household resource profiles, in a manner considered below. Constraints $B$ are layers in which cells have the value of either one or zero. When calculating $S_{cl}$ for a given cell, if the corresponding cell in any Boolean layer is zero then $S_{cl}$ equals zero, regardless of the values of the other constraints and factors. A land-tenure institution-agent, for instance, can force an actor-agent effectively to ignore a whole swath of cells by marking their corresponding cells in a Boolean layer with the value of zero. Another use of Booleans is to limit the cells in which an agent can be interested in order to reflect some realistic situation, such as an agent not considering far-off cells, or in order to limit the actor's knowledge in keeping with the precepts of bounded rationality.

After determining $S_{cl}$ for all cells $C$, an actor $a$ selects a subset $C^a$ to maximize the sum of cell suitabilities for the production strategy. Selecting these cells is an intensive integer programming problem. Heuristic solutions either rank cells by suitability and select a given number of cells for the agent or lie in compactness and contiguity algorithms. Finally, when multiple production strategies are possible, multicriteria evaluation is used to pick cells for competing objectives and then these cells are allocated according to a multi-objective land allocation process (Eastman *et al.* 1995).

Multicriteria evaluation and multiobjective land allocation are heuristic processes that approximate the utility offered to the actor by production activities. Ideally an actor considers all permutations of cells in light of all pertinent decision-making variables in order to maximize utility or engage in some satisficing behavior. The present model, however, employs two computational short cuts. First, the number of cells selected for $C^a$ is a stochastic function of demographic characteristics, agent production characteristics, or area under cultivation by households per the actor survey. Second, area of land subject to different forms of production is either chosen directly by the user in order to force the actor to consider a fixed set of alternatives or via an actor-agent decision rule that chooses a mix of productive activities based on their past utilities (after Arthur 1993). Better estimation of the amount and distribution of land dedicated to production in later versions of the SYPR IA model should be possible through tighter integration with the econometric modeling and understanding of social processes explored by other portions of this volume. A heuristic that better reconciles actor utility with measures of suitability is also feasible given the similarity of this problem to linear programming approaches to multiobjective land allocation (Diamond and Wright 1988).

The SYPR IA model focuses on several methods of determining suitability ($S_{cl}$) for actors. The first is a rule-based or heuristic process. For example, an agent can determine suitability as a probabilistic function of road proximity and then use a plot for three years before returning it to fallow. This rule is similar to those used by other dynamic spatial simulations (e.g. Gilruth, Marsh,and Itami 1995; Southworth, Dale, and O'Neill 1991). Heuristics are applicable to situations where data on few landscape attributes are available. When remotely sensed imagery is the only source of information, for instance, then heuristics based on cells classified as roads can work well. At the same time, heuristics are dependent on the spatial extent of their underlying data and areas where roads missing due to misclassification or cloud cover will not be modeled well.

Second, an agent can approximate $S_{cl}$, the best example being estimated parameter models programmed into agents. Actors calculate cell suitability in this manner with the simulation system's ability to process equations that combine both spatial and aspatial data to produce agent-specific suitability rankings. Examples of estimated parameter models are offered in Ch. 12, by other research informed by economics (Chomitz and Gray 1996; Kaimowitz and Angelsen 1998), or by efforts based on other inductive alternative-oriented methods (Ludeke, Maggio, and Reid 1990; Mertens and Lambin 2000). The key value of this approach is the flexibility it offers to modeling actor decision-making and the opportunity to draw on theoretical frameworks tied to empirical data. A wide variety of estimated models may be accommodated methodologically and the conceptual framework can host a variety of theoretical backdrops.

The third means of determining suitability is with genetic programs. These are software programs that evolve in a Darwinian fashion according to their fitness in meeting measurable goals (Banzhaf *et al.* 1998; Koza 1992). Genetic programs are also bounded rationality representations of decision-making (Dawid 1999) and are therefore in keeping with the conceptual framework. Genetic programs evolve over simulation time to match surveyed actor land-use histories to spatial data and an array of decision variables from the actor survey. Alternatively, they can be used to parameterize empirically an equation linking a dependent land-use variable against independent variables in a manner functionally similar to that used to develop estimated parameter models noted above.

Genetic programs also offer a means of exploring agent communication because individual programs can be shared among agents. Similarly, genetic program evolution approximates learning over time (Beckenbach 1999). Genetic programming is a parallel means of searching complex solution spaces that is resistant to the pitfalls of local maxima and minima, discontinuities, and high dimensionality. This power comes at the cost of difficult implementation, interpretation, and the emerging nature of research linking social science perspectives on decision-making to genetic programming representations (Chattoe 1998).

The SYPR IA model thus far has focused largely on smallholders. Introduction of other kinds of actors offers greater variety in patterning and verisimilitude. Use of varied agents may also be useful for modeling communication among actors at the cost of over-specification and higher data needs. None the less, the actions of ranchers or the role of *coyotes* in encouraging chile cultivation, for instance, would be interesting model extensions (Ch. 10).

### 4.5. *Results and Validation*

The SYPR IA model runs as a series of Monte Carlo simulations whereby the model executes many runs, each one creating a simulated environment for each model year. For any given year, a set of simulated land-cover layers may be superimposed to create a single probabilistic image. Stochasticity enters the simulation through random queuing and placement of agents, heterogeneous actors, varying agent processes, and cellular automata transition rules. Snapshots of ecological and human-induced land-use/cover changes are validated against observed data where the latter exist. They are used to understand the ramifications of land-use/cover change.

The simulation model creates a variety of outcomes that vary according to model configuration. Figure 13.1 illustrates the difference in spatial outcomes between three simulations of land-use/cover change in a small part of the study area (Manson 2000). Each simulation has an identical initial configuration save for different models of actor decision-making. The environment is limited to cellular automata rules that mimic secondary succession in the absence of actor influence. Institutions, except for land-tenure, are ignored except as noted below.

The key difference between the three models lies in the agents' derivation of $S_{cl}$ in terms of equation 1, and the following values are used in this calculation:

1. Land use $l$, where actors are interested in a generic land use termed 'deforestation' that approximates deforestation for the purpose of cultivation.
2. Factors $V$ include (a) distance to roads; (b) distance to forest; and (c) initial land use (three forest types, agriculture, infrastructure).
3. Constraints $B$ denoting prohibited areas, which include (a) land-tenure; and (b) current infrastructure, such as roads and current settlement. Cells in each constraint layer are initialized to either zero or one, where zero denotes areas that are off limits to agent intervention.
4. Factor weights $W$ are determined by each agent according to one of three methods.

In order to determine factor weights, the first set of actors uses a heuristic to probabilistically choose cells as (1) a function of distance to roads, where weights are asymmetrical sigmoidal probabilistic function that has its inflection point at 50m and becomes asymptotic at 600m; (2) distance to forest edge, where weights probabilistically decline linearly to zero at 500m; and (3) time in current land use, where agents randomly leave cultivated land after one to three years and have a random chance of using recent secondary regrowth of 4–7 years in age. The second means of determining $W$ is a logit regression model applied to the three spatial data factors $V$ and accounting for $B$. This model is methodologically and conceptually similar to, but simpler than, the first decision-making model presented by Ch. 10 since it has fewer number of factors and it is run for separate, random subsets of agents within the modeled area. The area-wide 'cut-off threshold' is replaced by the number of cells required by each agent to enumerate cells in $C^a$ as specified in section 4.4 above. The third decision-making model is similar to the second in that it uses the same threshold for $C^a$, factors $V$, and constraints $B$. It differs in that it uses genetic programs that evolve over time. Each agent is stochastically imbued with three to twenty programs that represent different strategies to assess suitability. Each genetic program has a tree structure of varying size and shape composed of variables that correspond to factors $V$ and coefficients that correspond to factor weights $W$. Genetic program trees are reinterpreted by the agents as an equation that each agent solves to create a suitability surface.

While there are obvious visual differences between the results illustrated by Fig. 13.1, the simulation system offers several means of determining the difference in accuracy between probabilistic projected land-use/cover change and that observed. The first is Kappa Index of Agreement (KIA), which calculates how closely two categorical maps compare in terms of location while accounting for chance in creating seemingly correct outcomes (Pontius 2000). The second and third tests are fractal dimension (a measure of patch complexity) and contagion (the extent to which land uses are clumped) (O'Neill *et al.* 1988; M. G. Turner, Costanza, and Sklar 1989). Both of these provide a sense of how well the simulated surfaces reproduce landscape pattern. The fourth test is a multi-resolution goodness of fit metric (Costanza 1989). This compares the relative number of correctly classified cells within a moving window at multiple resolutions in order to balance the roles that location and quantity play as a function of scale. Finally, a Monte Carlo uncertainty analysis measures how confident

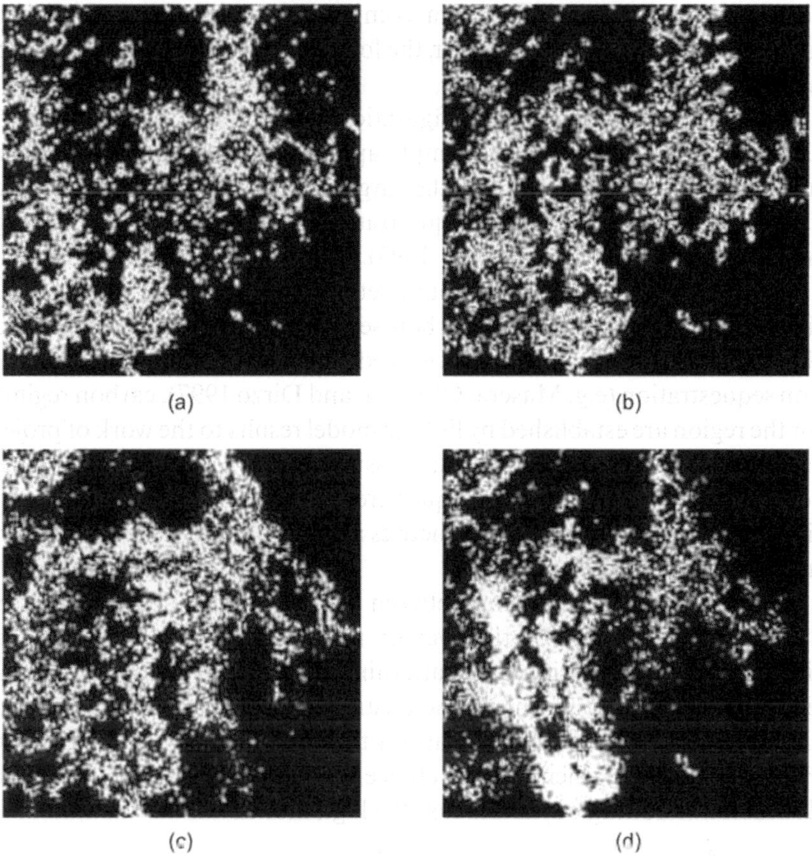


**Fig. 13.1.** Actual change (a) and change projected by agents using (b) heuristics; (c) logit model; and (d) genetic programming.

one can be about the predicted land-use/cover change transition of any given location (Ogneva-Himmelberger 1998).

When a single validation test is applied to different model configurations, some are by definition more correct than others. Over multiple tests, however, a model configuration that scores well on one test can score worse on another, such as when one configuration projects quantity correctly while another better approximates the spatial patterning of observed land use. With the KIA, for instance, the genetic program model has the highest correspondence to observed data compared to the other two models. Under the contagion test, however, the logit model scores best, followed by heuristics and genetic programs.

Differences among model configuration over validation metrics can be an advantage. Consider two example applications that link the regional integrated assessment effort to the larger global environmental change agenda. Carbon release and sequestration are implicit in land-change transitions over time (Houghton 1993). Each cell in land-use and land-cover layers experiences one of a finite set of state transitions that has a distinct effect on the amount of carbon sequestered at that location. Since these cell transitions can be characterized in terms of their net effect on carbon sequestration (e.g. Masera, Ordonez, and Dirzo 1997), carbon regimes for the region are established by linking model results to the work of project researchers and ecological literature. Assuming that a relationship between cell transitions and quantity of sequestered carbon exists, model configurations that score well on quantity metrics are appropriate, genetic programs in the case above.

Conversely, the relationship between habitat fragmentation, expressed as cell-transitions, and biotic diversity may be more reliant on the spatiotemporal configuration and patterning of change. Although the relationship between ecological characteristics and landscape measures is not simple, if it is possible to link pattern to biotic diversity (after Li 2000; Plotkin *et al.* 2000), then the best choice is a model scoring well on fractal dimension or contagion, in this case the logit model.

### 4.6. *Summary*

The SYPR IA model extracts key land-use/cover change interactions in the southern Yucatán peninsular region. Institutions interact with actors by changing resource profiles and other spatial and aspatial variables impinging on actor decision-making. The land-tenure institution, for instance, manipulates land-tenure grids that are referred to by agent processes that represent actor decision-making. Actors make production decisions in order to maximize utility in order to change their resource profiles, such as to grow food for consumption. These actions also change the state of cells in spatial layers that represents land use. As cellular automata transition rules rely on these cells to represent ecological phenomena, the modeled environment is affected by actors. The reverse also holds, however, as land used by an actor is subject to environmental events such as secondary succession that require continued actor intercession to counteract. As the simulation is iterative, there is constant interplay between actors and institutions, the effects of actor decision-making on the environment, and the effects of environmental transitions on actor decision-making.

## 5. *Regional integrated assessment*

The SYPR IA model addresses the distinct spatiotemporal patterns of land-use and land-cover change and the complexity of, and relationships among, socio-economic and environmental factors. By facilitating the management of information and relationships, the SYPR IA model improves understanding of the causes and mechanisms governing land-use/cover change. Being an integrated assessment model, it allows examination of the impacts of policy alternatives that cannot be easily explored in reality.

As illustrated throughout the chapter, there remains much to do in terms of knowledge integration and model refinement. Better linkage of ecological research to simulation outcomes remains an ongoing challenge, especially in regard to linking model results to larger issues, such as carbon sequestration and biodiversity. It is also necessary to encapsulate better more nuanced explorations of actors and institutions offered in other chapters. Agent definitions must be expanded to include a broader range of institutions, particularly those acting over less than the full spatial scale of the southern Yucatán peninsular region. A larger array of models can be introduced into the decision-making processes of agents representing actors, particularly in terms of stronger links to economics and econometrics. As the SYPR project moves into its second phase, the SYPR modeling efforts will improve and expand, seeking to offer a complex, integrated assessment model of land change.

## REFERENCES

Agarwal, C., G. L. Green, J. M. Grove, T. Evans, and C. Schweik (2000). 'A Review and Assessment of Land-Use Change Models: Dynamics of Space, Time, and Human Choice', in *Fourth International Conference on Integrating GIS and Environmental Modeling (GIS/EM4)* [CD-ROM]. Banff, Canada.

Alcorn, J. B., and V. M. Toledo (1998). 'Resilient Resource Management in Mexico's Forest Ecosystems: The Contribution of Property Rights', in F. Berkes and C. Folke (eds.), *Linking Social and Ecological Systems: Management Practices and Social Mechanisms.* Cambridge, Cambridge University Press, 216–49.

Arthur, W. B. (1993). 'On Designing Economic Agents that Behave like Human Agents'. *Journal of Evolutionary Economics* 3: 1–22.

Banzhaf, W., P. Nordin, R. E. Keller, and F. D. Francone (1998). *Genetic Programming: An Introduction.* San Francisco, Morgan Kauffman.

Beckenbach, F. (1999). 'Learning by Genetic Algorithms in Economics', in T. Brenner (ed.), *Computational Techniques for Modelling Learning in Economics*. Boston, Kluwer Academic Publishers, 73–100.

Benenson, I. (1999). 'Modelling Population Dynamics in the City: From a Regional to a Multi-agent Approach'. *Discrete Dynamics in Nature and Society* 3: 149–70.

Chattoe, E. (1998). 'Just How (Un)realistic are Evolutionary Algorithms as Representations of Social Processes?' *Journal of Artificial Societies and Social Simulation* 1 <www.soc.surrey.ac.uk/JASSS/1/3/2.html> accessed 1 January 1999.

Chomitz, K. M., and D. A. Gray (1996). 'Roads, Land Use, and Deforestation: A Spatial Model Applied to Belize'. *The World Bank Economic Review* 10: 487–512.

Conte, R., R. Hegselmann, and P. Terna (eds.) (1997). *Simulating Social Phenomena.* Berlin, Springer-Verlag.

Costanza, R. (1989). 'Model Goodness of Fit: A Multiple Resolution Procedure'. *Ecological Modelling* 47: 199–215.

Dawid, H. (1999). *Adaptive Learning by Genetic Algorithms: Analytical Results and Applications to Economic Models.* Berlin, Springer.

Diamond, J. T., and J. R. Wright (1988). 'Design of an Integrated Spatial Information System for Multi-Objective Land Use Planning'. *Environment and Planning B* 15: 205–14.

Eastman, J. R., W. Jin, P. A. K. Kyem, and J. Toledano (1995). 'Raster Procedures for Multi-Criteria/Multi-Objective Decisions'. *Photogrammetric Engineering and Remote Sensing* 61: 539–47.

Ermentrout, G. B., and L. Edelstein-Keshet (1993). 'Cellular Automata Approaches to Biological Modeling'. *Journal of Theoretical Biology* 160: 97–113.

Galletti, H. A. (1998). 'The Maya Forest of Quintana Roo: Thirteen Years of Conservation and Community Development', in R. D. Primack, D. B. Bray, H. A. Galletti, and I. Ponciano (eds.), *Timber, Tourists, and Temples: Conservation and Development in the Maya Forest of Belize, Guatemala, and Mexico*. Washington DC, Island Press, 33–46.

Gilruth, P. T., S. E. Marsh, and R. Itami (1995). 'A Dynamic Spatial Model of Shifting Cultivation in the Highlands of Guinea, West Africa'. *Ecological Modelling* 79: 179–97.

Gronewold, A., and M. Sonnenschein (1998). 'Event-Based Modelling of Ecological Systems with Asynchronous Cellular Automata'. *Ecological Modelling* 108: 37–52.

Hogeweg, P. (1988). 'Cellular Automata as a Paradigm for Ecological Modelling'. *Applied Mathematics and Computation* 27: 81–100.

Houghton, R. A. (1993). 'The Role of the World's Forest in Global Warming', in K. Ramakrshna and G. M. Woodwell (eds.), *World Forests for the Future: Their Use and Conservation*. New Haven, Yale University Press, 21–58.

Jager, W., M. B. A. van Asselt, J. Rotmans, C. A. J. Vlek, and P. Costerman Boodt (1997). *Consumer Behavior: A Modelling Perspective in the Context of Integrated Assessment of Global Change.* Amsterdam, Rijksinstituut voor Volksgezondheid en Milieu (Dutch National Institute of Public Health and the Environment.

Kaimowitz, D., and A. Angelsen (1998). *Economic Models of Tropical Deforestation: A Review.* Jakarta, Centre for International Forestry Research.

Koza, J. (1992). *Genetic Programming: On the Programming of Computers by Means of Natural Selection.* Cambridge, Mass., MIT Press.

Lambin, E. F. (1994). *Modelling Deforestation Processes: A Review.* Luxemburg, European Commission.

Li, B.-L. (2000). 'Fractal Geometry Applications in Description and Analysis of Patch Patterns and Patch Dynamics'. *Ecological Modelling* 132: 33–50.

Li, H., and J. F. Reynolds (1997). 'Modeling Effects of Spatial Pattern, Drought, and Grazing on Rates of Rangeland Degradation: A Combined Markov and Cellular Automaton Approach', in D. A. Quattrochi and M. F. Goodchild (eds.), *Scale in Remote Sensing and GIS*. New York, Lewis, 211–30.

Ludeke, A. K., R. C. Maggio, and L. M. Reid (1990). 'An Analysis of Anthropogenic Deforestation Using Logistic Regression and GIS'. *Journal of Environmental Management* 31: 247–59.

Manson, S. M. (2000). 'Agent-Based Dynamic Spatial Simulation of Land-Use/Cover Change in the Yucatán Peninsula, Mexico', in *Fourth International Conference on Integrating GIS and Environmental Modeling (GIS/EM4)* [CD-ROM]. Banff, Canada <http://www.colorado.edu/research/cites/banff/pubpapers/81/> Last accessed: January 1, 2002.

Masera, O. R., M. J. Ordonez, and R. Dirzo (1997). 'Carbon Emission from Mexican Forest: Current Situation and Long Term Scenarios'. *Climatic Change* 35: 265–95.

Mertens, B., and E. F. Lambin (2000). 'Land-Cover-Change Trajectories in Southern Cameroon'. *Annals of the Association of American Geographers* 90: 467–94.

Myers, G. M., and Y. Y. Papageorgiou (1991). '*Homo Economicus* in Perspective'. *The Canadian Geographer* 35: 380–99.

Ogneva-Himmelberger, Y. A. (1998). 'Exploring Empirical Diagnostic Modeling of Land-Use/Land-Cover: An Example from Southern Yucatán Peninsular Region'. Ph.D. dissertation, Worcester, MA, Clark University.

O'Neill, R. V., J. R. Krummel, R. H. Gardner, G. Sugihara, B. Jackson, D. L. DeAngelis, B. T. Milne, M. G. Turner, B. Zygmunt, S. W. Christensen, V. H. Dale, and R. L. Graham (1988). 'Indices of Landscape Pattern'. *Landscape Ecology* 1: 153–62.

Ostrom, E. (1999). 'Institutiomnal Rational Choice: An Assessment of the Institutional Analysis and Development Framework', in P. A. Sabatier (ed.), *Theories of the Policy Process*. Boulder, Westview Press, 35–71.

Plotkin, J. B., M. D. Potts, D. W. Yu, S. Bunyavejchewin, R. Condit, R. Foster, S. Hubbell, J. LaFrankie, N. Manokaran, L. H. Seng, R. Sukumar, M. A. Nowak, and P. S. Ashton (2000). 'Predicting Biodiversity in Tropical Forests'. *Proceedings of the National Academy of Science* 97: 10850–4.

Pontius, R. G. J. (2000). 'Quantification Error Versus Location Error in Comparison of Categorical Maps'. *Photogrammetric Engineering and Remote Sensing* 66: 1011–16.

Riebsame, W. E., W. J. Parton, K. A. Galvin, I. C. Burke, L. Bobren, R. Young, and E. Knop (1994). 'Integrated Modeling of Land-Use and Cover Change'. *BioScience* 44: 350–6.

Rotmans, J., and H. Dowlatabadi (1998). 'Integrated Assessment Modeling', in S. Rayner and E. L. Malone (eds.), *Human Choice and Climate Change*. Seattle, Battelle Press, 291–377.

Sen, A. (1981). *Poverty and Famine: An Essay on Entitlement and Deprivation.* New York, Oxford University Press.

Seror, A. C. (1994). 'Simulation of Complex Organizational Processes: A Review of Methods and their Epistemological Foundations', in N. Gilbert and J. Doran (eds.), *Simulating Societies: The Computer Simulation of Social Phenomena.* London, UCL Press, 18–40.

Silvertown, J., S. Holtier, J. Johnson, and P. Dale (1992). 'Cellular Automaton Models of Interspecific Competition for Space: The Effect of Pattern on Process'. *Journal of Ecology* 80: 527–34.

Simon, H. A. (1997). 'Behavioral Economics and Bounded Rationality', in H. A. Simon (ed.), *Models of Bounded Rationality*. Cambridge, Mass., MIT Press, 267–433.

Singh, I., L. Squire, and J. Strauss (eds.) (1986). *Agricultural Household Models: Extensions, Applications, and Policy*. Baltimore, The Johns Hopkins University Press.

Smith, G. C., and D. S. Bull (1997). 'Spatial and Temporal Ordering of Events in Discrete Cellular Automata: An Overview'. *Ecological Modelling* 96: 305–24.

Southworth, F., V. H. Dale, and R. V. O'Neill (1991). 'Contrasting Patterns of Land Use in Rondonia, Brazil: Simulating the Effects on Carbon Release'. *International Social Science Journal* 130: 681–798.

Stern, P. C., O. R. Young, and D. Drukman (eds.) (1992). *Global Environmental Change: Understanding the Human Dimensions.* Washington DC, National Academy Press.

Takeyama, M., and H. Couclelis (1997). 'Map Dynamics: Integrating Cellular Automata and GIS through Geo-Algebra'. *International Journal of Geographical Information Science* 11: 73–91.

Torrens, P. M., and D. O'Sullivan (2001). 'Cellular Automata and Urban Simulation: Where Do We Go from Here?'. *Environment and Planning B* 28: 163–8.

Turner, B. L., II, S. Skole, G. Sanderson, L. Fischer, L. Fresco, and R. Leemans (1995). *Land-Use and Land-Cover Change: Science/Research Plan.* Stockholm and Geneva, IGBP/HDP.

Turner, M. G., R. Costanza, and F. Sklar (1989). 'Methods to Evaluate the Performance of Spatial Simulation Models'. *Ecological Modelling* 48: 1–18.

Tversky, A., and D. Kahneman (1990). 'Rational Choice and the Framing of Decisions', in K. S. Cook and M. Levi (eds.), *The Limits of Rationality*. Chicago, University of Chicago Press, 60–89.

Uhl, C. (1987). 'Factors Controlling Succession Following Slash-and-Burn Agriculture in Amazonia'. *Journal of Ecology* 75: 377–407.

Walker, B. H. (1994). 'Landscape to Regional-Scale Responses of Terrestrial Ecosystems to Global Change'. *Ambio* 23: 67–73.

Yates, D. N., and K. M. Strzepek (1998). 'An Assessment of Integrated Climate Change Impacts on the Agricultural Economy of Egypt'. *Climate Change* 38: 261–87.

Young, O. (1994). *International Governances: Protecting the Environment in a Stateless Society.* New York, Cornell University Press.

# Retrospective: The Three Frontiers Revisited

Three frontiers serve as the conceptual and real-world basis of the SYPR project. To recapitulate, these frontiers are (1) the return to synthesis science, (2) the emergence of 'integrated land-change science', including the need for spatially explicit land-change models, and (3) the occupation and deforestation of the southern Yucatán peninsular region, the heretofore largest tract of seasonal tropical forest remaining in Middle America. The SYPR project was initially conceived as a synthesis approach to understanding the land-change dynamics underway in the region, sufficiently expansive but detailed in its research to lead to a new generation of spatially fine-tuned land-change models. While modest, mid-course adjustments in the project are underway, the project believes that its successes are substantial to date, pointing to the utility of the pursuit of the conceptual frontiers as a means of addressing the changes in process in this real-world frontier, or any other region of tropical deforestation.

In retrospect, the vast tracts of seasonal tropical forest and the pervasive *ejido* belie the complexity of land-change dynamics underway in the region. Land cover proved to be rather uniform in terms of species present across the region, but distinguishing between the two main forest types proved difficult where the upland–*bajo* boundary was not sharply delimited by elevational differences, and between older successional growth and *mediano* or mature upland forest. These difficulties were evident in forest transect work and in classification through TM Landsat imagery, the latter requiring more training sites aimed at this issue. The sharp north–south precipitation gradient across the region parallels gradients in nutrient dynamics and, presumably, variance in biophysical feedbacks on land uses. Invasive species, especially bracken fern, appear to be increasing in parts of the region, but their pattern of invasion and persistence is not well understood. And, despite local claims about sustainable crop–fallow cycles, nutrient evidence hints that successive cycling over a 20–30-year period, absent other inputs, may significantly deplete soil nutrients, the impact of which on forest succession is not known but apparently reduces yields on subsequent cultivation. Finally, any estimates of forest biomass and carbon stocks must be understood in terms of past human impacts that have given rise to the structure of the current older growth forests in the region, especially the impact of the ancient Maya on species composition and of the logging industry on the presence of mahogany and Spanish cedar.

The ephemeral 'development' of the frontier region has shifted land-use dynamics from the rather simple expansion of *ejido*-adjudicated, swidden cultivation, modestly affected by the marketing of maize, to large-scale but failed mechanized rice projects, to a cacophony of land-use experiments. Among these, land speculation by way of pasture creation and a shift to commercial chili production have been especially important and difficult to address. The rationale for investing in pasture, other than payments that assisted this effort, is not well understood given that most pasture created by *ejidatarios* remains ungrazed owing to the paucity of livestock that they possess. And, pasture proves difficult to distinguish in the Landsat TM imagery from early stages of successional growth. Chili cultivation has profoundly affected the former *milpa* system in several ways: by mechanization of prime land that is currently devoted to chili production only, and by intensifying the swidden rotation through incorporation of several years of chili cultivation into the regular subsistence production cycle. Finally, farmers have moved from substantial intercropping in swiddens to parcel monocropping, be it chili, maize, or beans, rotating among different crops by year. Imagery analysis cannot yet distinguish among these different land uses, and the resulting intensification process must be observed at the regional scale in other ways, such as increased temporal resolution in the imagery from which crop–fallow cycles might be ascertained.

Beyond this complexity, the region is varied in the kinds of land uses that can be employed, determined in large part by the 'tenure' domain of the *ejido* (forest extension or agricultural), *ejido* location within the core or buffer zone of the Calakmul Biosphere Reserve, a few small private land holdings outside *ejidos*, and substantial federal lands in the center of the region. The general rules of land use by land-tenure type must and can be accounted for, but other, more fine-tuned land impacts have yet to be addressed. Of these, two appear to be especially important: parcel access by *ejido* household (usufruct) and *ejido* forest reserves. The location of these reserves is not known by individual *ejido*, although in principle creating this information is a matter of research time and support. More problematic, of course, is the demarcation of household parcels, given the enormous effort it would require. As illustrated in the modeling sections, however, much of the error in predicting deforestation within agricultural *ejidos* is probably linked to this issue (which pixel belongs to which household) as well as that of the forest reserve. Moving to this level of detail, which might be possible in the future, given federal demarcation of parcels in some *ejidos*, raises issues about data and household confidentiality.

Finally, land-use complexity has increased owing to the abundance of government and non-governmental programs applied in the region, especially in regard to the broader, externally developed goals for the region found in the creation of the Calakmul Biosphere Reserve and the 'green *municipio*' of Calakmul, and the various programs linked to the reserve, especially El Mundo Maya and the Mesoamerican Biological Corridor. These events have created a plethora of experiments to intensify production in extant cultivated and fallowed lands, potentially reducing agricultural expansion, as well as alternative 'forest-friendly' land uses. To simplify, the overall economy of the region is expanding beyond its forest extraction and agrarian roots, and it is too early to determine the implications of the various experiments on sustained land uses and on forest cover.

Work of this kind also pushes remote-sensing science to its frontiers. Repeated studies indicate that Landsat TM and other satellite imagery can reveal much more specificity about land cover and related land uses than the traditional categories of forest cover/open lands, and that information providing clues about land-manager decision-making, such as that gained by detailing crop–fallow cycles, can be gleaned from imagery analysis if targeted for detailed assessment. Phase one of the SYPR project perhaps overly pushed the production of a region-wide mosaic for change analysis, relative to fine-tuning the classification in order to increase the accuracy of the analysis. Phase two is taking steps in this regard, with the expected result that the errors in modeling results should be decreased.

Land-use and land-cover change science is, by definition, an integrated subject because it addresses a coupled human–environment system. Human intent and action in regard to land use—a partial product of the perceptions of the environment in the first place—are linked through ecological conditions and processes to environmental consequences—in this case, forest cover—and these consequences, in turn, are registered on adjustments in land use. Understanding the details of this coupled system is beyond that which can be gained by any single discipline or field of expertise and requires an integrative approach; and coupled systems maintain 'aggregate' characteristics and emergent properties that are exposed through synthesis. The complexity of land-change dynamics in what otherwise might be construed as a relatively simple and uniform case study—tropical deforestation driven by cultivation and tropical forest preservation driven by a biosphere reserve—heightens the need for synthesis and integration, as does the regional scale of analysis, which ultimately requires the use of remote sensing and GIS to tackle the questions posed.

It is, perhaps, a truism to note that the 'synthesis' developed to date by the SYPR project could not be accomplished, or at least not be accomplished as well as it has been, without an integrated approach, that which we label 'integrated land-change science'. The more critical issues are captured in the two following questions. What 'synthesis' understanding has been or can be gained by such efforts? And can fine-tuned, spatially explicit, integrated land-change models be developed that are useful to the myriad of research and application interests seeking to employ them?

The first question can be only partially answered at this point, but that answer is resoundingly positive. Each of the basic sciences employed—ecological, remote sensing, and social—generates detailed knowledge about different components of the problem. Integrated by design, this knowledge systematically builds an overall understanding of coupled human–environment system, in this case the *ejidatario*, cropped parcels, forest impact relationships and, in phase two of the project, the biophysical feedbacks on the *ejidatario* and forest system integrity. In this sense, the 'synthesis' has generated a range and connection of understanding that are difficult to gain by assimilating non-integrated research, and are surely beyond the expertise of the single researcher to produce, no matter how enlightened and holistic that research might be. The first question remains only partially answered, however, because the aggregate characteristics of land changes underway, such as the fragmentation of the landscape, and emergent properties of the coupled systems, such as the possible nutrient depletion in repeated swidden cycling and the tipping of soil conditions toward conditions favoring bracken fern invasion, are only now under systematic examination by the project.

The second question is also only partially addressed, recognizing that spatially explicit land modeling of the kind undertaken here remains in its infancy. The results to date provide powerful support that large improvements can be made over such simplifications as von Thünen-type 'friction of distance' to the market or to adjacent roads, or the environmental determinants of slope and elevation. By understanding the principal and intermediate factors influencing the land manager and by documenting the location of the lands to which that manager has access, some success has been demonstrated in explaining (and projecting in the IA model) 'where' land change has or will take place. For the most part, as the level of critical information increases by land manager type and area, so does the accuracy of spatially explicit models. Unfortunately, information is typically sparse in 'frontiers', and thus moving from specific data-rich and local cases to the data-poor and region-wide case involves a loss in the robustness of the models. What has yet to be determined, but is the subject of phase two research, are the marginal rewards of the models by added data detail. We suspect that these margins will be larger where the real-world outcomes are spatially fine-tuned, for example in the small, fragmented niches that serve special flora and fauna (questions of biodiversity), and smaller where those outcomes are less fine-tuned, such as estimating carbon stocks.

Phase one of the SYPR project has demonstrated the basic usefulness of synthesis science and integrated land-change understanding. This phase has only scratched the surface, however, in regard to the integrated questions it can inform and the kind of understanding it may glean. At its core, this kind of research has large start-up costs, and phase one of the project, reported in this volume, has been that beginning.

*The Editors*

# GLOSSARY AND ACRONYMS

***aguada*** a small surface cavity that usually retains rainwater throughout the year; usually a Maya-made reservoir

***akalche*** a solution feature, usually small, that retains sufficient water to constitute a wetland; geomorphically, a small *bajo* (see below)

**amsl** above mean sea level

**AVHRR** Advanced Very High Resolution Radiometer

***bajo*** literally, low; term for karst solution features (poljes) that form large sinks, infilled with clays, and serve as seasonal wetlands that sustain forest growth; wetlands with other vegetation or open water, even in poljes, have other, specific names

***bosque*** forest

**Bosque Modelo de Calakmul** (also **Bosque Modelo**) Calakmul Model Forest

***campesino*** peasant or any smallholder or landless individual from rural area

**CBR** Calakmul Biosphere Reserve

***chicle*** either the tree species *sapodilla* or *chicozapote* (*Manilkara zapota* L. Van Royen, formerly *Achras sapota* L.) or the resin from the species used for chewing gum

***chipotle*** smoke-dried jalapeño chili

**COINCE** Comisión Intersecretarial para Nuevos Centros de Población Ejidal—Inter-Secretarial Commission for New Population Centers

**CONASUPO** Compania Nacional de Subsistencias Populares—National Company for Popular Subsistence

***coyote*** in reference to 'scavenger', a term applied by farmers to purchase and transport middlemen in the chili market

**CRASX** Consejo Regional Agrosilvopecuario y de Servicios de Xpujil—Regional Council of Solidarity and Agriculture, Livestock, and Forestry Services of Xpujil (Campeche)

**CRIPX** Consejo Regional Indigena y Popular de Xpujil—Regional Indigenous and Popular Council of Xpujil

***doblado*** turning of the mature maize cob downward, while still on the stalk, to protect it from rain and birds

***ejido*** communal land set aside for rural impoverished people, primarily for agriculture activities; set up as a means of land redistribution, based on interpretations of indigenous land holdings

***ejidatario*** a member of an *ejido*

**FONAES** Fondo Nacional de Apoyo a Empresas Sociales—National Fund for Assistance to Social Enterprises

***garumbo*** invasive species; *Cecropia peltata* L.

**GATT** General Agreement on Tariffs and Trade

**GEF** Global Environmental Fund (United Nations)

**GIS** geographical information systems

**GTZ** Gesellschaft für Technische Zusammenarbeit—German Association of Technical Cooperation

***hacienda*** large land estate engaged in some form of agricultural production, often livestock

***helecho*** invasive species; bracken fern or *Pteridium aquilinum* L.

**INI** Instituto Nacional Indigenista—National Institute for the Indigenous

***laguna*** small, shallow body of water

**MBC** Mesoamerican Biological Corridor

**mechanized cultivation** plot initially disked before cultivation

***mediano/a*** mature or old growth forest on well-drained lands or uplands

***milpa*** literally, maize field; term replete in the literature as a mode of swidden cultivation involving maize, beans, and squash

***milpero*** one who engages *milpa* cultivation; a smallholder, swiddener

***montaña*** forest, locally forest over 15 years old

***municipio*** a political unit akin to county

**NAFTA** North American Free Trade Agreement

**NASA** National Aeronautics and Space Administration (US)

***nescafe*** *Mucuna pruiens* (L.) DC, or velvet bean

**NOAA** National Oceanic and Atmospheric Administration (US)

***norte*** cold front or mass of cool air pushing south across the Gulf of Mexico into the peninsula

**N$** new pesos, which has hovered at N$10 per $US1 throughout the late 1990s and currently

**NCPE** Nuevo Centros de Población Ejidal—New Ejidal Population Centers

***palancas*** political connections

**PCA** Principal Component Analysis

***pobladores*** rural landless who usually rent or borrow land or serve as agricultural labor

**PPF** Plan Piloto Forestal—Pilot Forestry Plan
**PRI** Partido de la Revolución Institucional—Institutional Revolutionary Party
**PROCAMPO** Programa de Apoyo Directo al Campo—Direct Rural Support Program
**PRODEFOR** Programa de Desarollo Forestal—National Forest Development Program
**PRONARE** Programa Nacional de Reforestación—National Reforestation Program
**ProNatura Peninsula de Yucatán** Pro-Nature Yucatán Peninsula
**PRONASOL** Programa Nacional de Solidaridad—The National Solidarity Program
**rendzina** basic upland soil in the region; a mollisol, composed largely of troporendolls
***sascab*** decomposed or decayed limestone found throughout the uplands of the region
**SAGARPA** Secretaría de Agricultura, Ganadería, Desarollo Rural, Pesca y Alimentación—Secretariat for Agriculture, Livestock, Rural Development, Fishing and Nutrition
**SARH** Secretaria de Agricultura y Recursos Hidraulicos—Secretary of Agriculture and Hydraulic Resources
**SCT** Secretaría de Comunicaciones y Transportes—Secretariat for Communications and Transportation
**SEDESOL** Secretaría de Desarollo Social—The Secretariat for Social Development
***selva*** forest
**SEMARNAT** Secretaría del Medio Ambiente y Recursos Naturales—Environment and Natural Resources Secretariat
**SEP** Secretaría de Educación Pública—Secratariat for Public Education
**smallholder** originally, farmer with very little land (e.g. several hectares or less); because these farmers tend to be fiscally restricted and semi-subsistence, the term is also associated with an economic or decision-making behavior consistent with these conditions
***solar*** (pl. *solares*) garden, usually a house garden with trees
**swidden** slash-and-burn or shifting cultivation
**SYPR** Southern Yucatán Peninsular Region Project; the acronym is used in the text in reference to this project only, not to the region
**tajon** invasive species; *Viguiera dentata* [Cav.] Spreng
**TM** Landsat Thematic Mapper

***tular*** sedge, reeds, and grass vegetation complex found on edges of, or near, permanent wetlands

**UEPMC** Unión de Ejidos Pueblo Maya de Campeche—Union of Maya Peoples' Ejidos of Campeche

# INDEX

Notes
1 **Emboldened** page numbers indicate **chapters** and references.
2 Acronyms with very long names can be found in the Glossary.